TROUBLED WATERS

Dieter Telemans

RIGHT Ethiopia, Somali region. A boy walks in the Wayb river at sunrise. In April 2006, the water levels in the river had receded drastically. Successive seasons of failed rains had contributed to a humanitarian emergency across the Horn of Africa, putting millions of lives at risk.

Book design
Eric de Mildt

Translations
Guido Telemans, Ligna

Published by
BAI, Schoten

Printed by
Blonde Printing nv, Wommelgem

Distributed by
Exhibitions International, Leuven

ISBN 9085860245
D/2007/5751/01

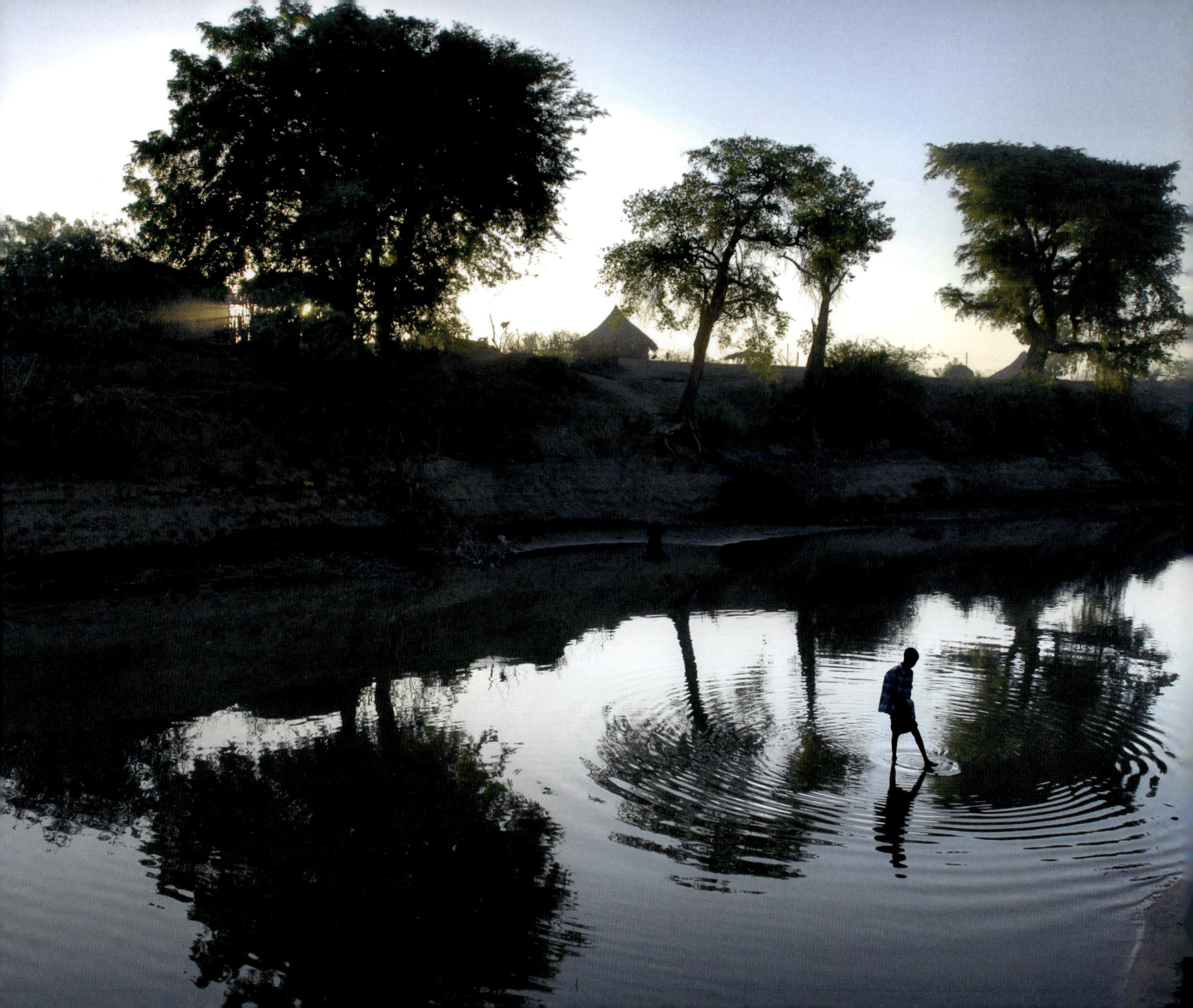

Contents

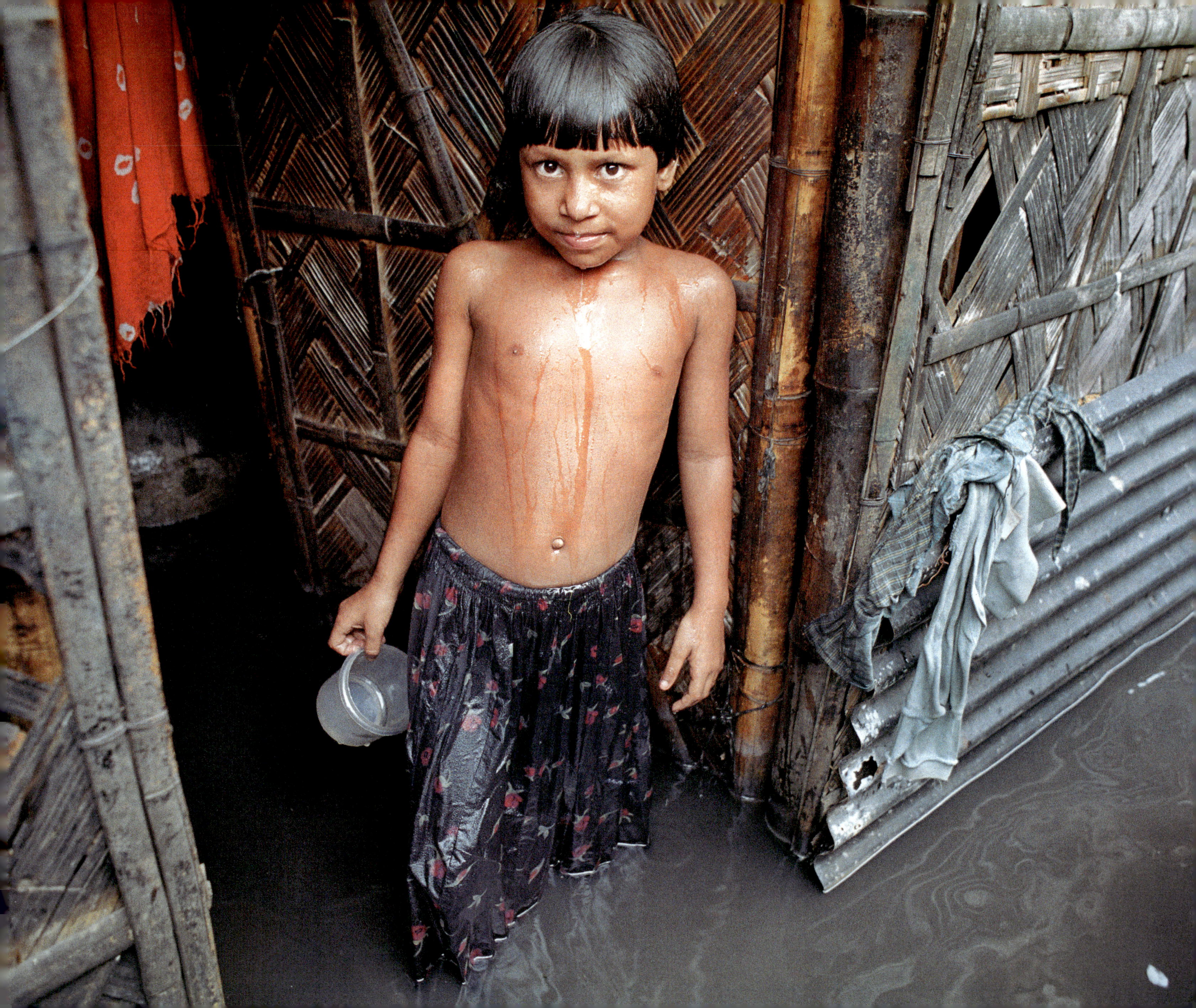

Dieter Telemans Preface

"Aral's kaput, Aral's gone," Sergei moaned while I was photographing him on the Swan, the ship whose captain he had been for eight long years. Today, the Swan is one of the many rusting shipwrecks of a once proud fishing flotilla of 300 ships. The Aral Sea used to be the world's fourth largest inland sea, and with its disappearance, the region has lost its economic lifeline.

The ecological catastrophe of the Aral Sea shows how essential water is for human subsistence and how carefully it should be treated. Sergei's story was my inspiration to further focus on world water issues.

I have photographed farmers in India watching their water resources dry up incessantly and hopelessly. Thousands of them have already committed suicide by drinking some of the pesticides that were destined for their crops. In 2006, most herdsmen in the Horn of Africa lost their cattle to drought. The older generations believe that the centuries-old nomadic tradition is on its way out. In some regions, floods destroy people's houses and spread diseases, such as in Bangladesh, where I found a sea of fresh water, but not a drop of drinking water.

Some leaders still believe they can control Mother Nature. Gigantic water projects are under way in Egypt and Turkey, amongst others, with the goal of growing thirsty export crops in desert areas. This may benefit the local populations, but downstream, the painful consequences will soon be felt.

I have tried to classify these photo reports according to major water-related themes. In reality, however, drought-stricken areas may become flooded areas tomorrow. The same regions in Ethiopia and Kenya, where I saw people begging for water, were inundated only a few months later. And wherever drinking water is a problem, people's health is at stake and life becomes a struggle for survival. Climate change, population growth and especially the "we'll never run out of it" mentality are the biggest threats to fair and sustainable access to water for everybody.

Agriculture uses more than 70% of the available water worldwide. Israel is a leader in efficient drip-irrigation techniques and desert agriculture. These technologies may reduce water consumption significantly everywhere. In Israel itself, however, the water grab goes on undiminished: 80% of the water from the Palestinian West Bank goes to the state of Israel.

Experts and scientists have been warning us for years: the major conflicts of the 21st century will be about water, not oil. Only sustainable and fair management of world water resources can prevent this doom scenario. Gandhi once said: "There is enough water for everybody's need but there is not enough water for everybody's greed".

PREVIOUS PAGE Bangladesh, Dhaka. A girl washes herself in the doorway of her flooded hut. She lives in one of Dhaka's many slums. The poor are often excluded from the city's safer sites, and have no option but to live in areas that are prone to flooding. The mixture of flood waters, sewage and rubbish causes widespread health problems.

Riccardo Petrella The right to life

Over 1 billion people do not have access to drinking water, the fundamental source of life. This is unacceptable. There is now a huge risk that in the year 2020, when the world population reaches approximately 8 billion human beings, there will be more than 3 billion people without access to drinking water. We can and must prevent the unacceptable from becoming possible.

As the fundamental and irreplaceable "source of life" for the ecosystem, water is a vital and common good, which belongs to all the inhabitants of the Earth. None of them, individually or as a group, can be allowed the right to make it private property. Water is the heritage of mankind. Individual and collective health depends upon it. Water is not like any other resource: it is not an exchangeable or marketable commodity.

While the sharing of water has often been a major source of social inequality in the past, today's civilisations recognise that access to water is a fundamental, inalienable, individual and collective right. The right to water is part of the basic ethics of a "good" society and a "good" economy. It is up to society as a whole to guarantee the right of access without discrimination based on race, sex, religion, income or social class.

Fresh water resources are unequally distributed on the Earth. The same is true for wealth. But this does not mean that there must also be inequality of access to water. People with an abundance of water and wealth cannot make use of water as they please, by selling it or buying it to derive maximum profit or pleasure. In many regions of the world, water remains a source of inequality between men and women, the latter bearing the burden of having to provide their families with the necessary water. All these inequalities must be tackled.

Creating the necessary conditions to ensure the most effective and sustainable access to water should be everybody's concern. It is also an inter-generational issue, and it is up to the present generations to use, value, protect and conserve water resources so that future generations can enjoy the same freedom of action and choice that we wish for ourselves today.

Users have a key role to play through their choices and practices in ensuring environmental, economic and societal sustainability. "Partnerships for water" is the inspiring principle behind all the plans, such as "the river agreements", that have enabled the efficient resolution of conflicts between riverside communities sharing the same hydrographic basin.

Society must collectively assume all the costs of water in view of supplying and guaranteeing access to water in the quantities and qualities considered as being the indispensable minimum. Beyond the vital minimum, progressive pricing must be applied: the more water one uses, the higher the price per water unit. Finally, all abusive and excessive use must be considered illegal.

Riccardo Petrella is President of the Group of Lisbon, Secretary of the Committee for the World Water Contract
and Professor of Human Ecology at the Academy of Architecture, University of Italian Switzerland.

> nearly a third of the world's land surface is at risk of extreme drought by 2100

> desertification threatens one-fifth of the global population

> Spain, Greece and Italy will suffer ever-more frequent regional water shortages due to climate change and rising demand

> it is very likely that high temperatures, heat waves and heavy precipitation will continue to become more frequent

> by the middle of this century, at worst 7 billion people in 60 countries will lack water

> Lake Chad has shrunk by 95% since the mid-1960s

> dust from Asia invades North America, and dust from Africa reaches Europe via the trade winds

> the Sahara is advancing southward at a rate of 5 to 10 kilometres per year

1 DESERTIFICATION

> temperatures over the Mediterranean Sea could rise by over 4°C by 2100

> thirsty times are likely to become the norm in the Horn of Africa

> desertification is devouring more than 20,000 square miles of land worldwide every year

> two-thirds of the African continent can already be classified as desert or dry land

> Hans Blix says that global warming is a greater threat than terrorism

> desertification and drought have caused more human suffering than any other environmental problem

> the UN warns that fast climate change assessment is vital for safeguarding the world's food supply

> the economic impact of desertification, more than $40 billion per year, is horrendous

> annual precipitation is projected to decline by 10 to 40% for much of Africa and south-eastern Spain by 2100

> desertification could create millions of refugees

> the effects of desertification especially threaten the 22 million people living on Africa's desert borders

> at least 33 countries in Africa are seriously affected by the problem

> many locations in the Sahara region experience an average annual rainfall below 25mm

> desertification is being blamed for forced migration

> experts estimate millions will be forced to move from newly desertified areas

> the UN says desertification has led to worsening poverty in Africa

> much desertification is attributed to overcultivation, deforestation and overgrazing

> climate change extends the areas prone to desertification

> the WWF says the world lost about 30% of its natural wealth between 1970 and 1995

> droughts caused 42% of all deaths due to natural disasters

> by 2025, two-thirds of the world's people are likely to be living in areas of acute water stress

> the glaciers on Mount Kilimanjaro have shrunk dramatically

> the highest shade temperature recorded in the world is 58°C in a locality in Libya

> desertification affects 74% of North America's dry land

> increasing human populations and poverty contribute to desertification as poor people are forced to overuse their environment

> human activities are frequently the proximate cause of desertification

> dust particles are causing health problems and have been shown to boost death rates

> the spread of the Gobi desert is forcing the populations of three provinces in China to abandon their homes

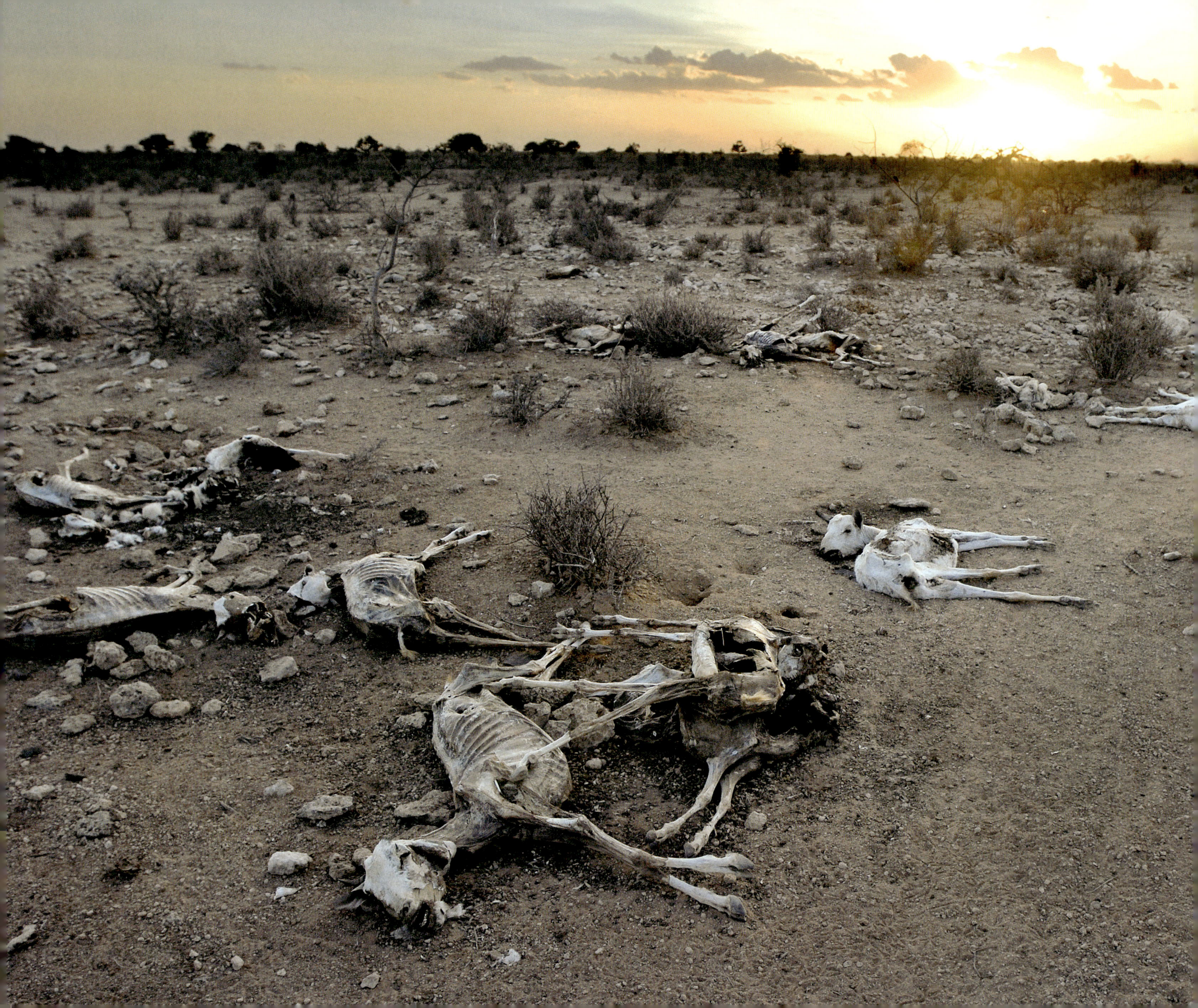

Kenya & Ethiopia
Drought in the Horn of Africa

Early in 2006, East Africa was suffering the worst drought in over a decade. An estimated 11 million people in Djibouti, Ethiopia, Kenya, Somalia and Tanzania were on the brink of starvation. Years of low rainfall were exacerbated by the almost complete failure of the rainy season of 2005.

Millions of nomadic herdsmen lost an estimated 70% of their cattle stock. The Kenyan government declared it the worst natural disaster to hit the country since the drought of 1971.

The drought cycle has shrunk from once every eight years to once every three years. Climate change, together with overpopulation resulting in overgrazing and the cutting down of trees for fuel and charcoal, has transformed the land. Former pasture areas have become desert land. Nomads now find themselves without pastures to return to.

Community leaders wonder whether their tradition of pastoralism is sustainable. But to ask communities such as the Masai and the Somali to herd fewer cattle, seems impractical. 90% of these people's livelihood depends on livestock animals. This has been the tradition for many generations. To break with this tradition, alternative income-generating activities have to be developed.

Many adolescents are eager for change and believe that education is a better investment than livestock. But these drought-stricken regions lack basic infrastructure, and good education and jobs are faraway dreams.

Politicians in the capitals discuss these issues, but without providing long-term development projects. As long as these regions remain isolated, young people cannot escape their ancestors' harsh existence, and seasonal droughts will continue to take a deadly toll.

RIGHT Kenya, North Eastern Province. Sacking, cardboard and sheeting cover the roofs of huts at the El Hache Camp, which shelters 380 households of herdsmen who have lost their livestock due to the drought. Because they used to be nomads, these people are not recognized as refugees or IDPs (internally displaced persons), and thus fail to fit with international donor statistics. They are surviving with hardly any food aid. The numbers of these drop-outs will only increase with the ongoing desertification.

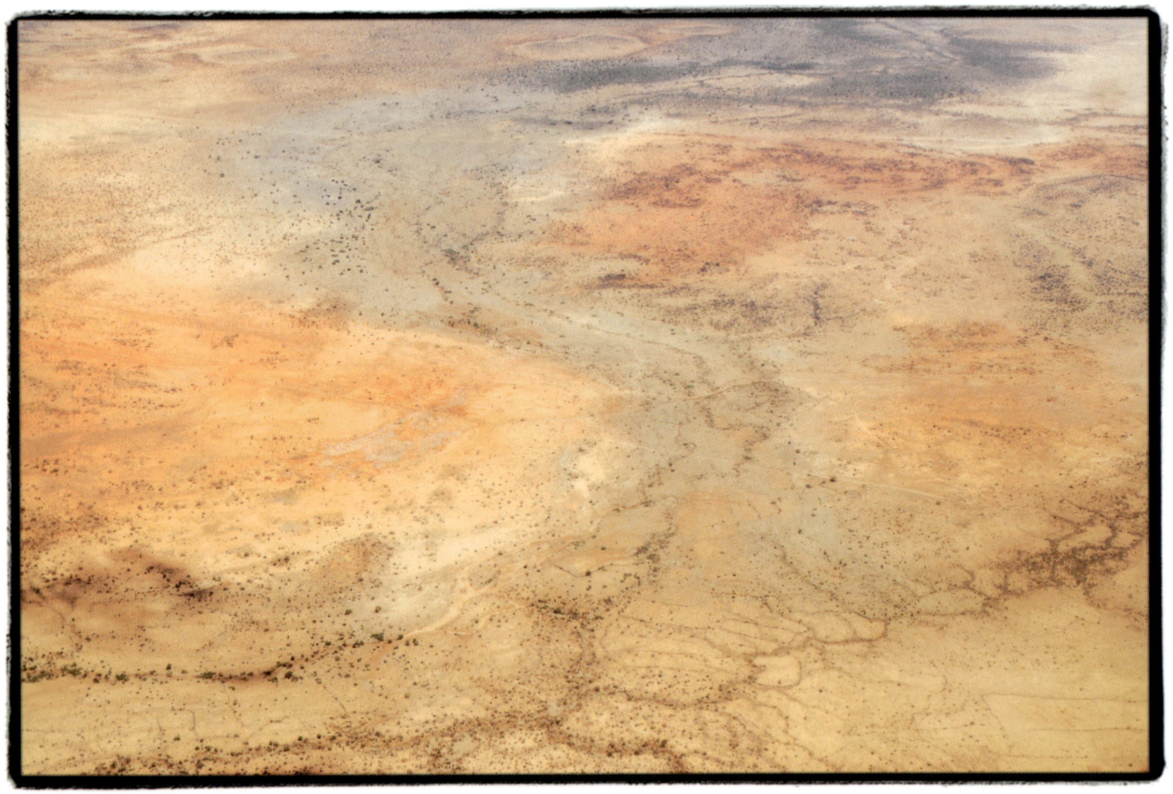

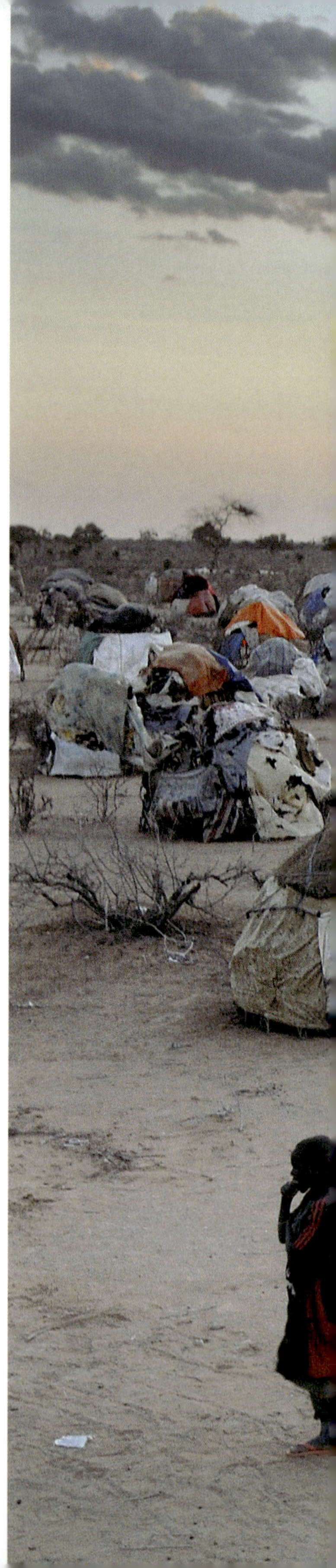

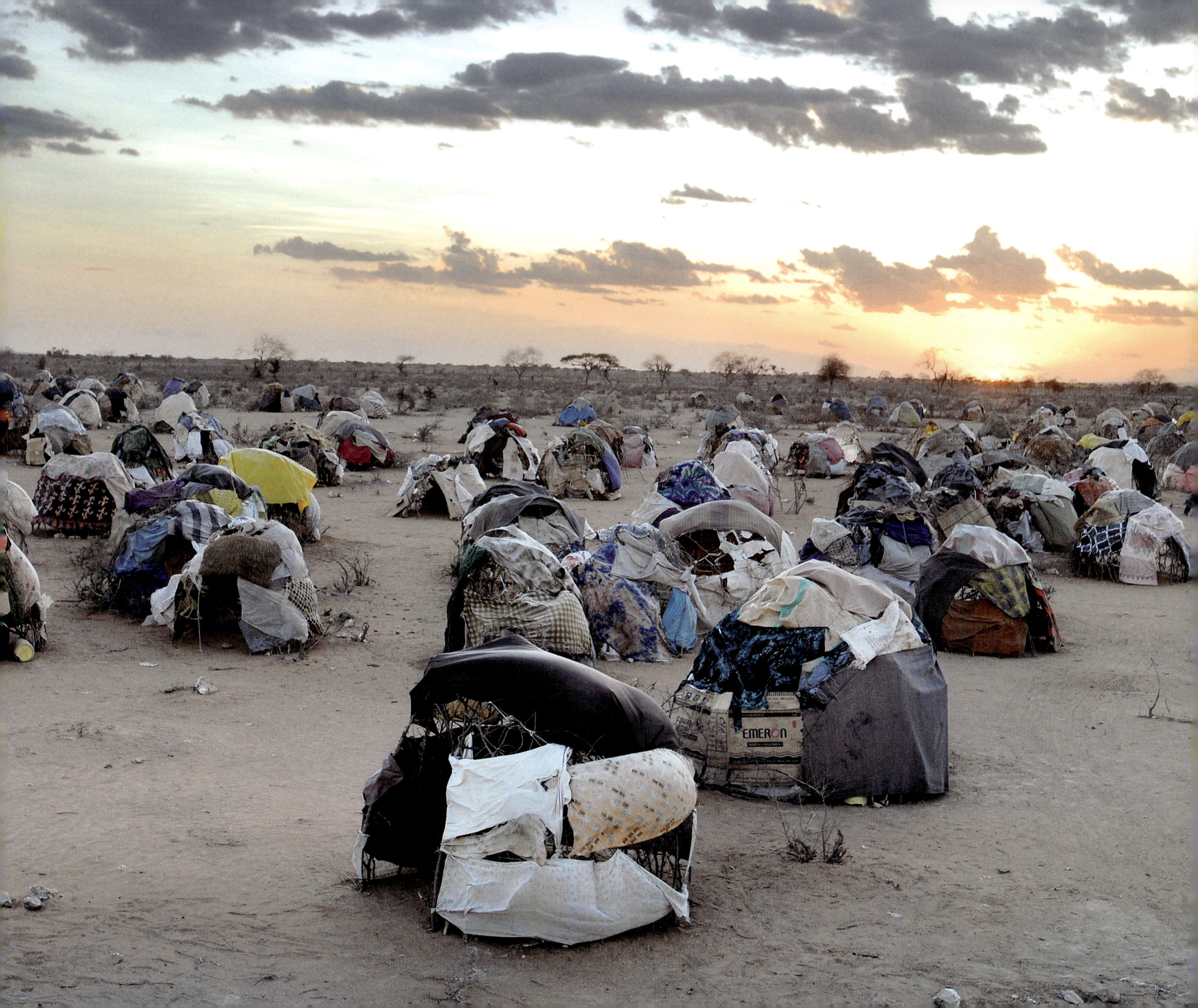
EMERON

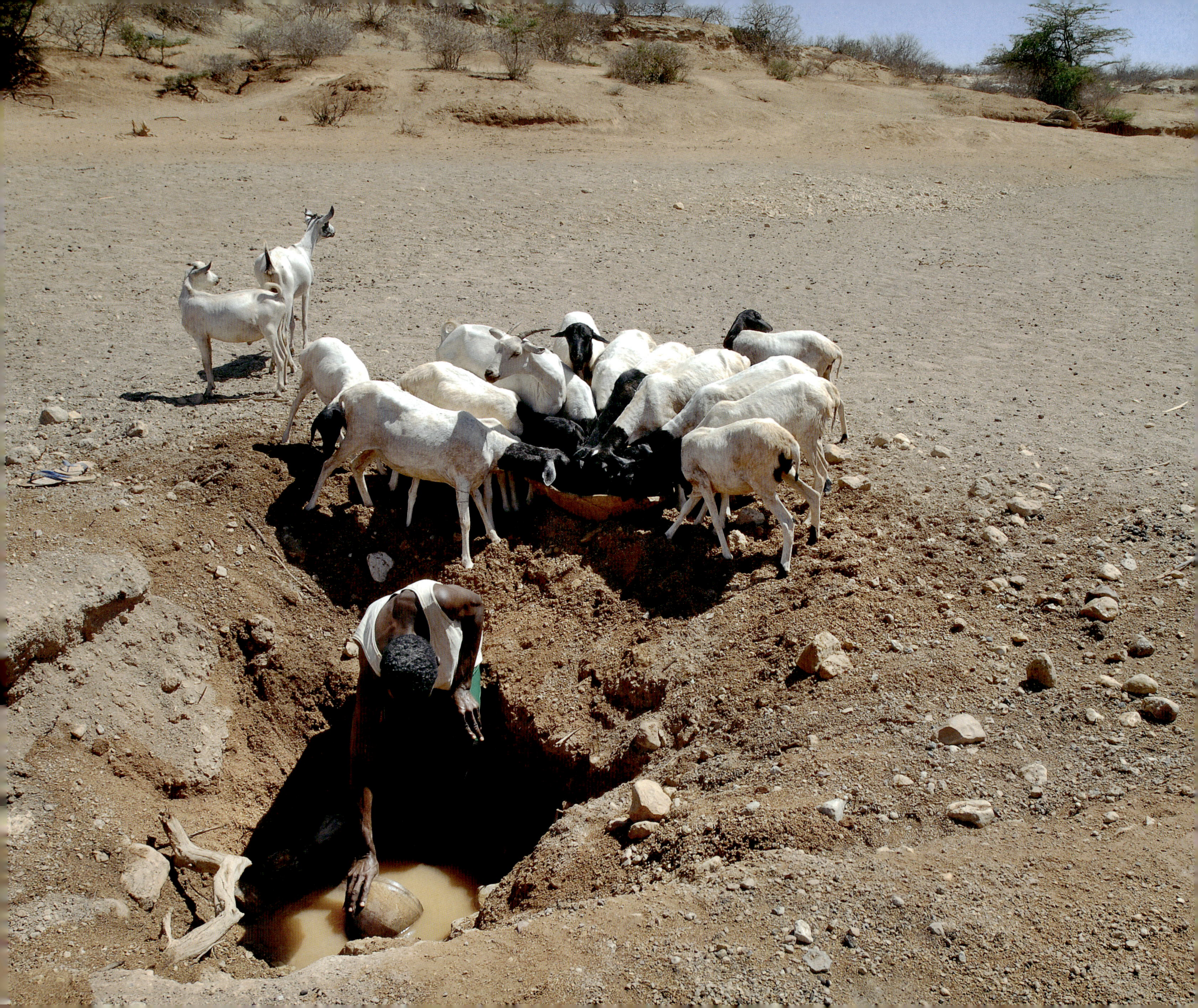

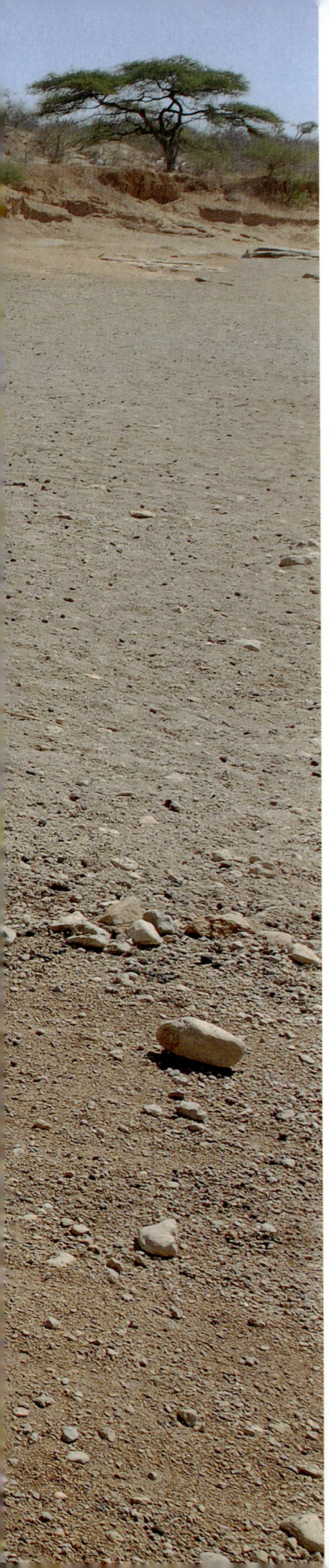

LEFT Ethiopia, Somali region. A goat herdsman fills a vessel with water from a hole in a dry river bed. For many people, this is the only available water source. By sharing these water sources with their livestock, water-born diseases are easily spread amongst the population.

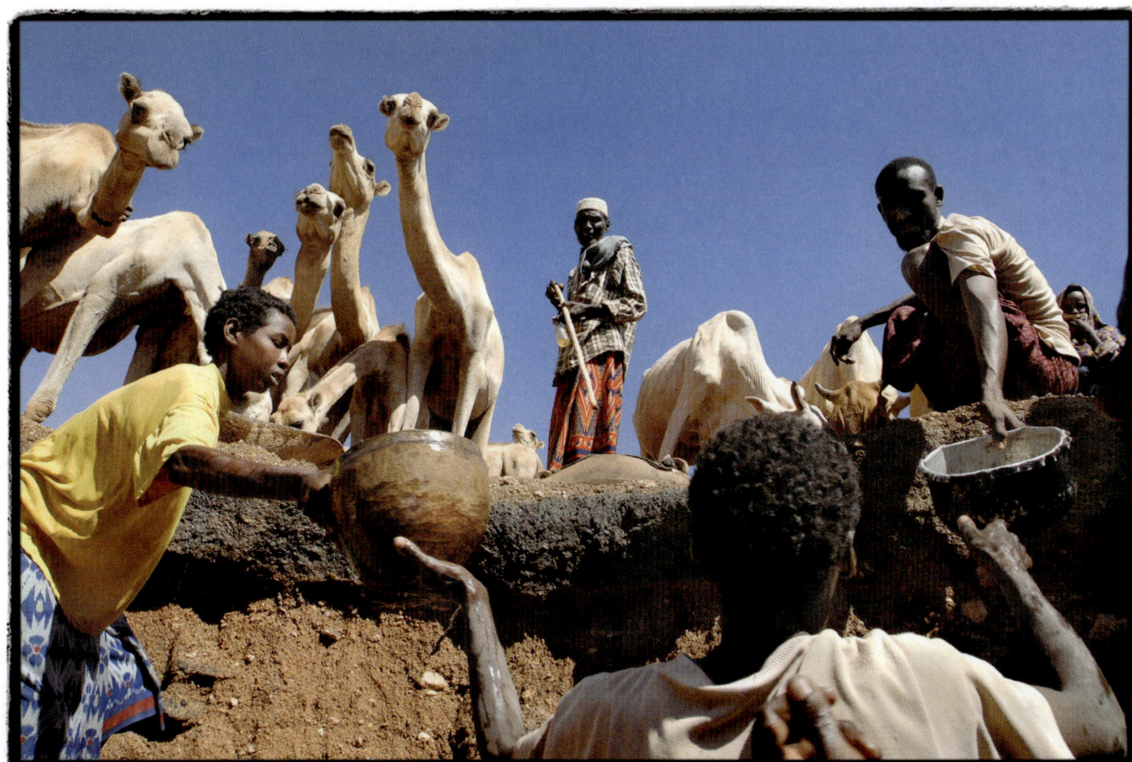

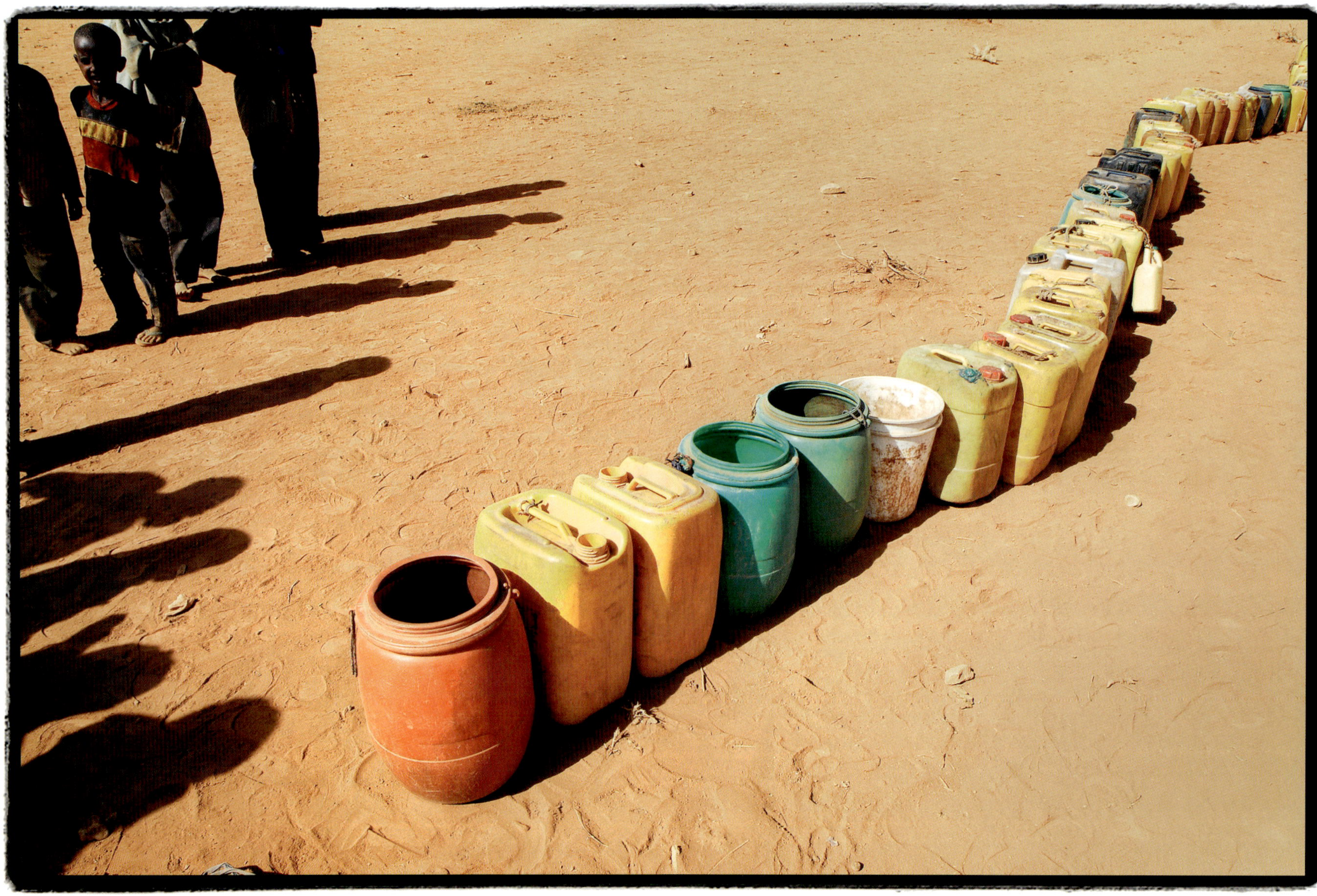

LEFT Ethiopia, Somali region, Daraye. Plastic containers are lined up in anticipation of the arrival of a Red Cross water truck. Most of the wells in the area have dried up, or the water has become too salty to drink.

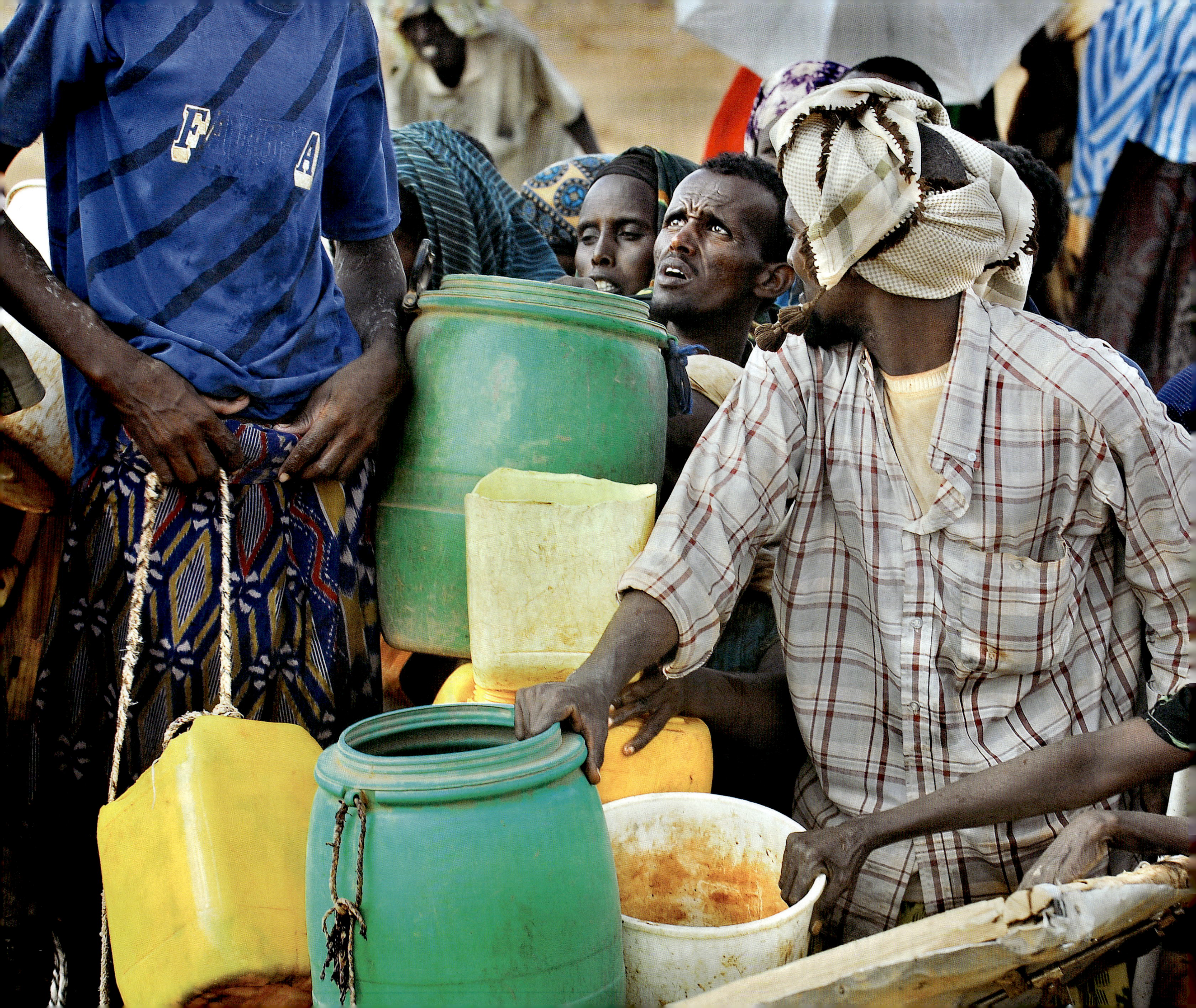

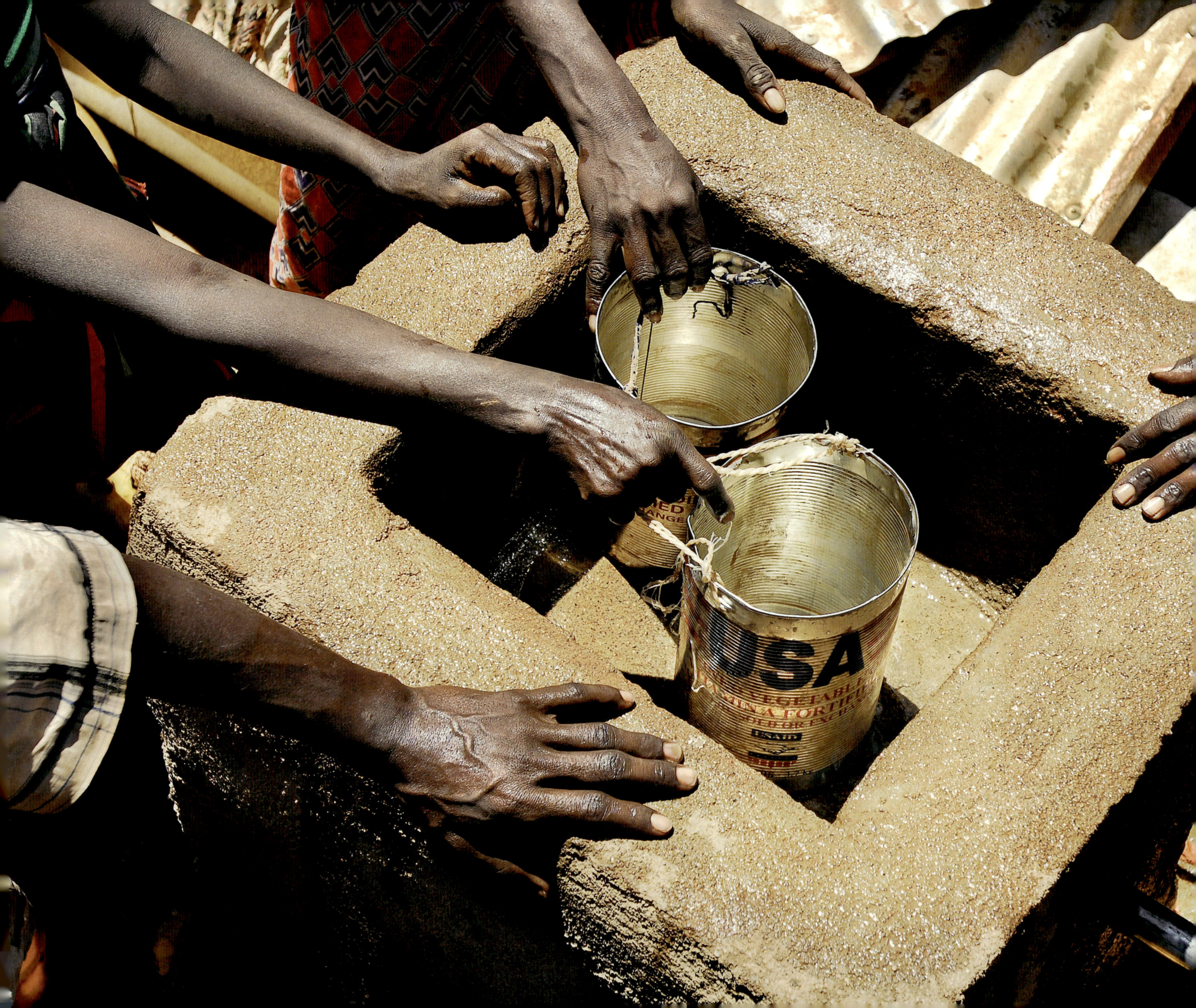
USA

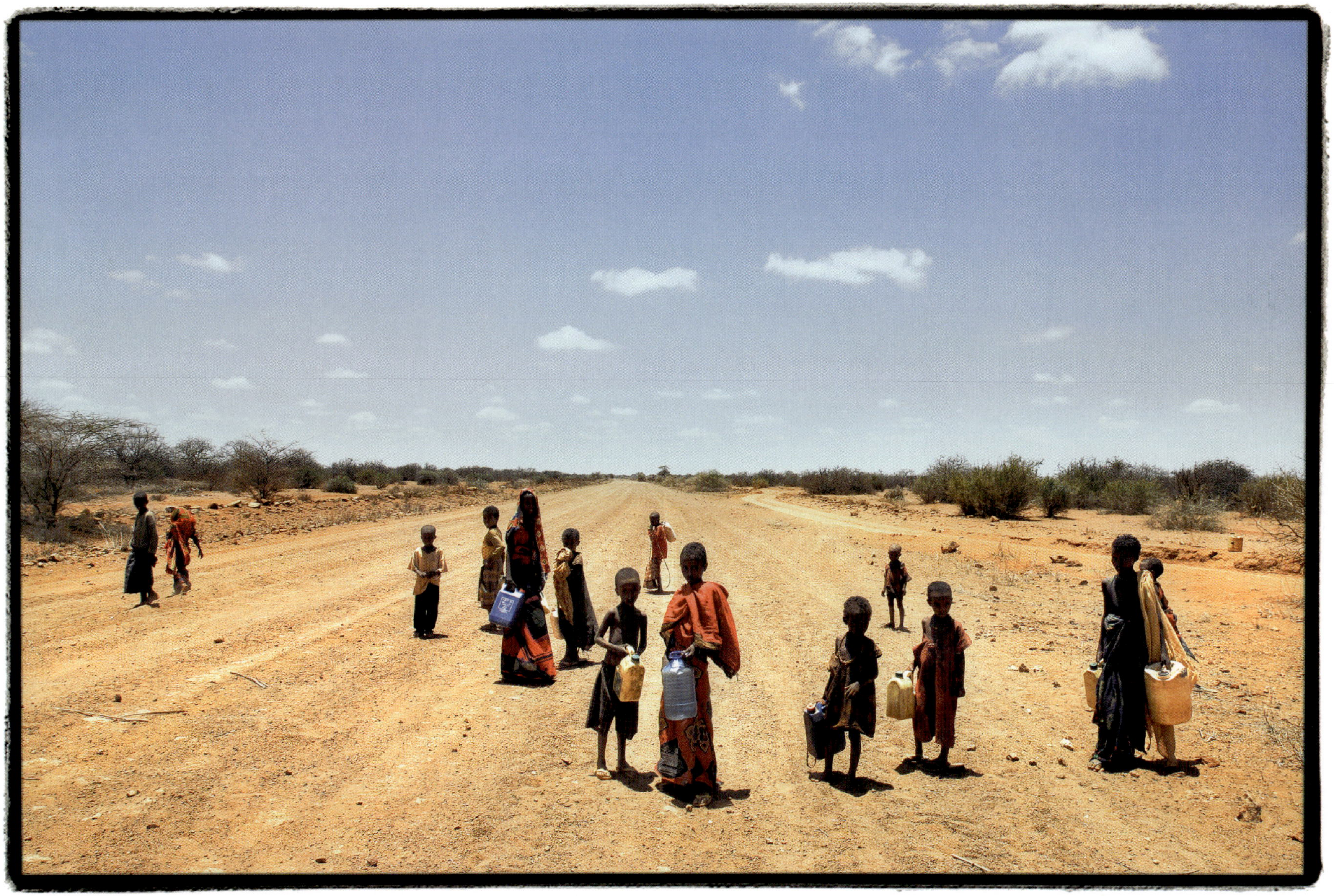

RIGHT Kenya. Women and children wait with containers at the roadside between Wajir and El Wak in the hope that a passing truck will be able to offer them drinking water.

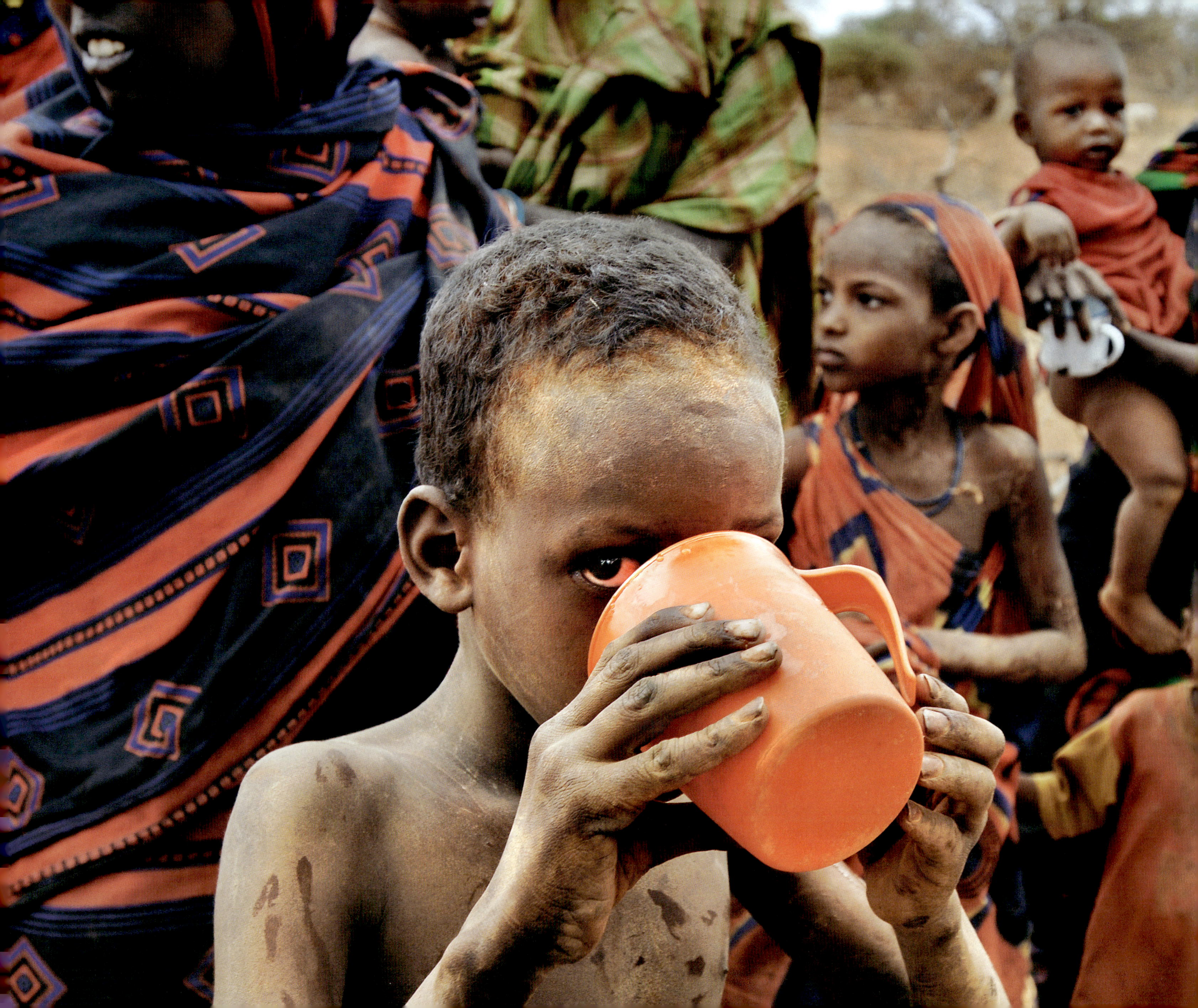

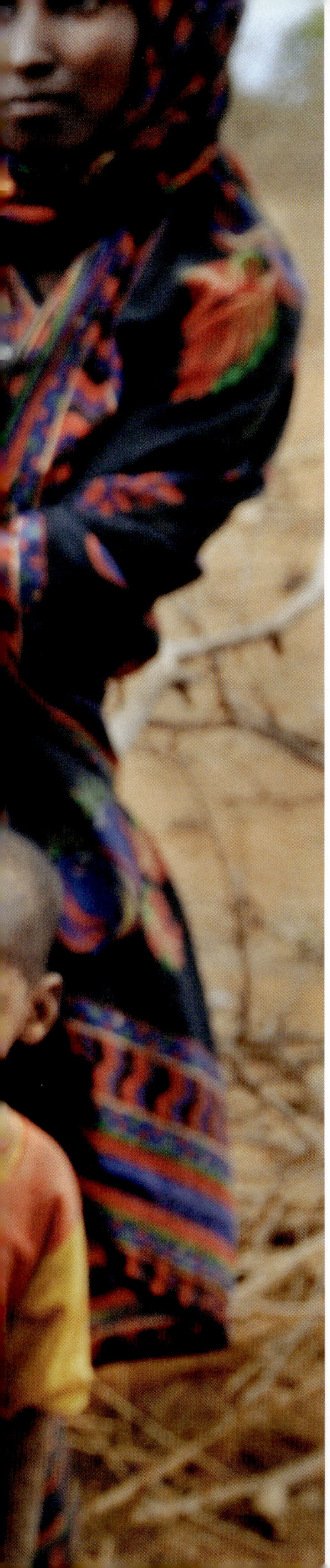

LEFT Ethiopia, Somali region. A boy receives a cup of drinking water after being vaccinated against measles during an immunisation campaign. In a population weakened by the effects of drought, the outbreak of measles is a threat.

RIGHT Ethiopia, Somali region. Four-year-old Sabtiyd Abdi Khalif, a severely malnourished girl, weighs 7.3 kilograms. The letter 'R' for RED is written on her hand to indicate that she is severely malnourished. She has also been suffering from tuberculosis during the past three years, but cannot be treated because the closest TB clinic is 50 kilometres away.

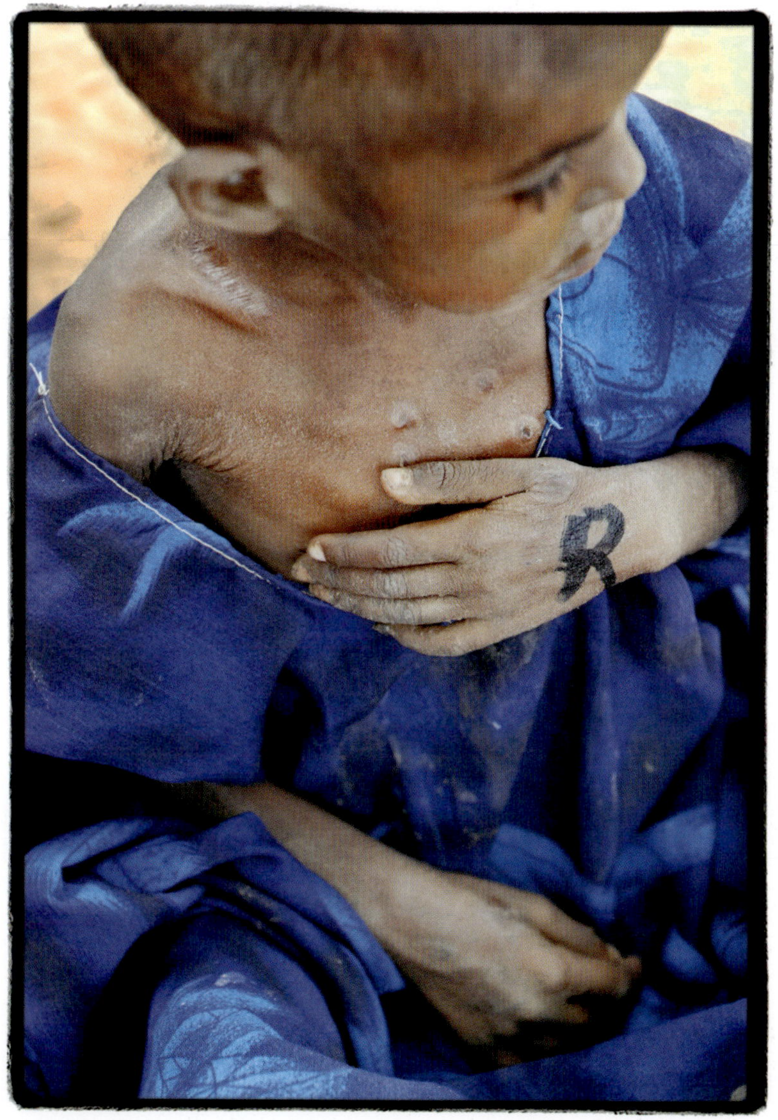

LEFT Kenya, North Eastern Province. Khero Abdi (20), with her malnourished son Abdullahi Mohamud (1) at the therapeutic feeding centre run by Médecins Sans Frontières (MSF). They are Somali refugees who came to the region during the drought in search of healthcare which was not available in their country of origin.

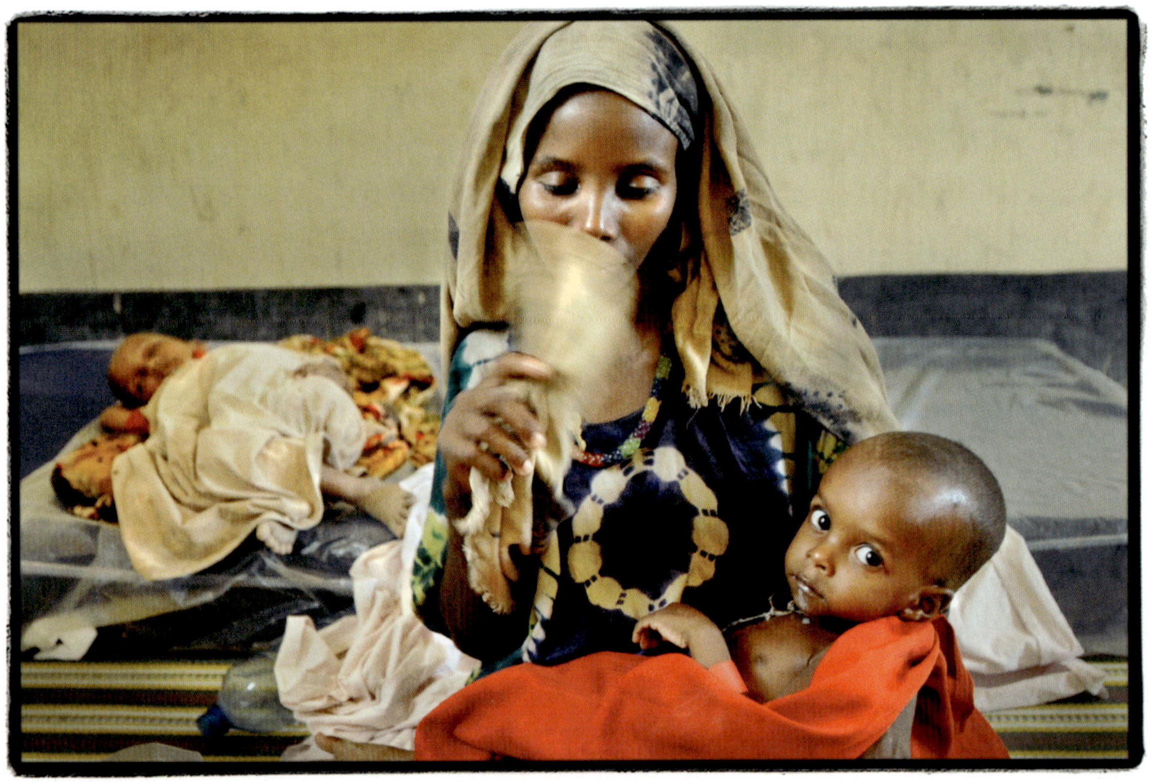

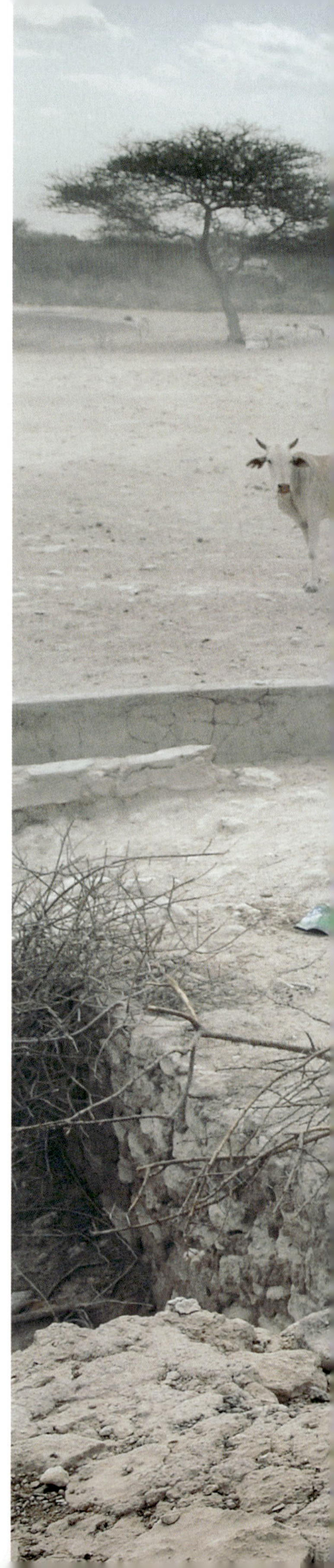

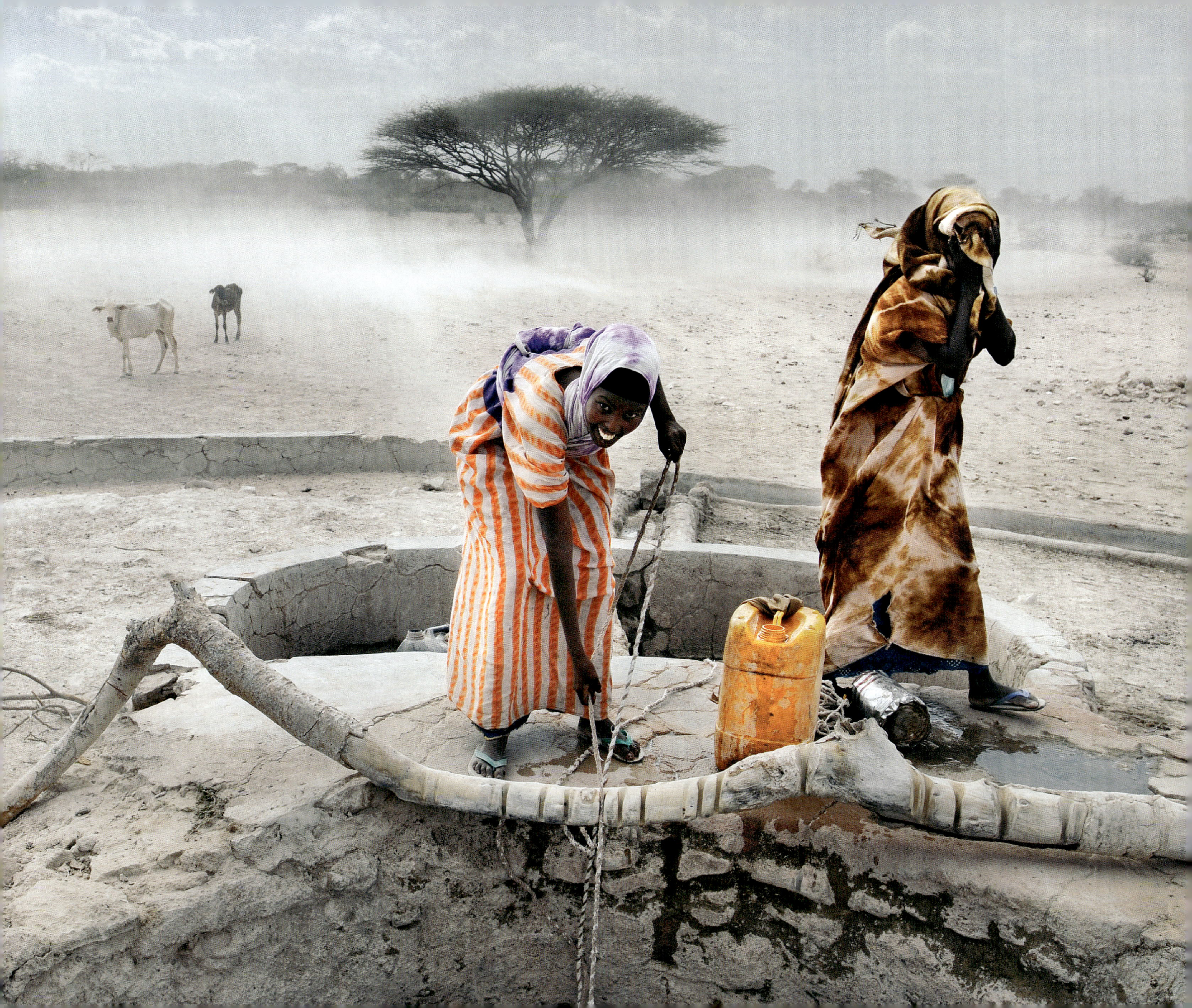

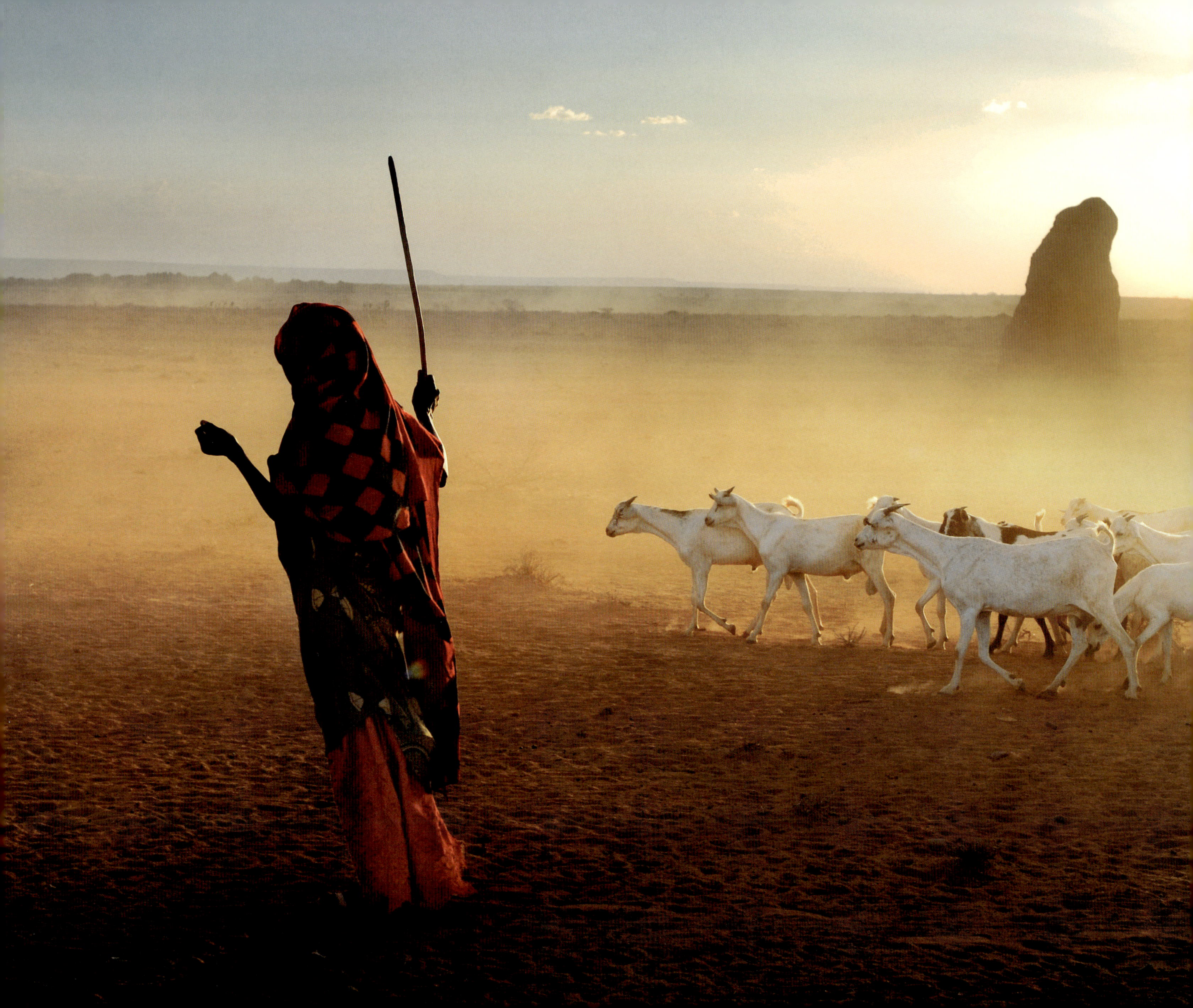

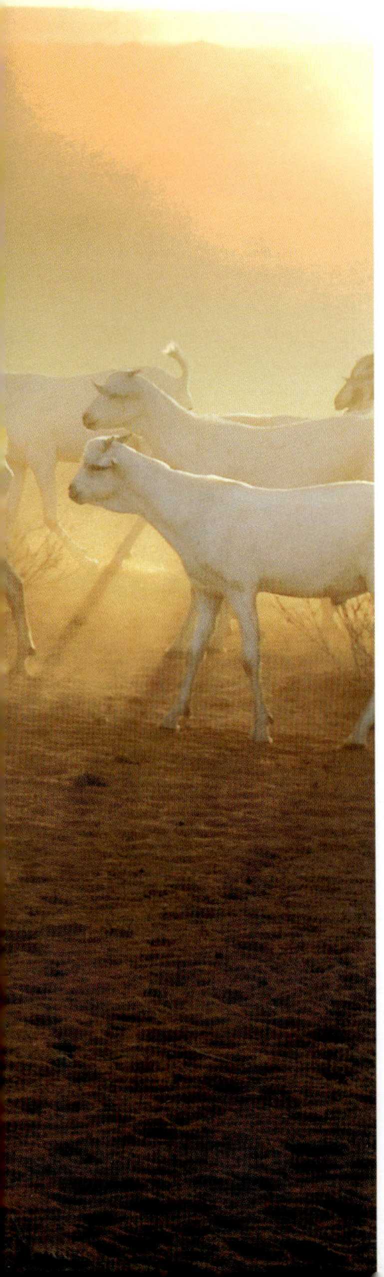

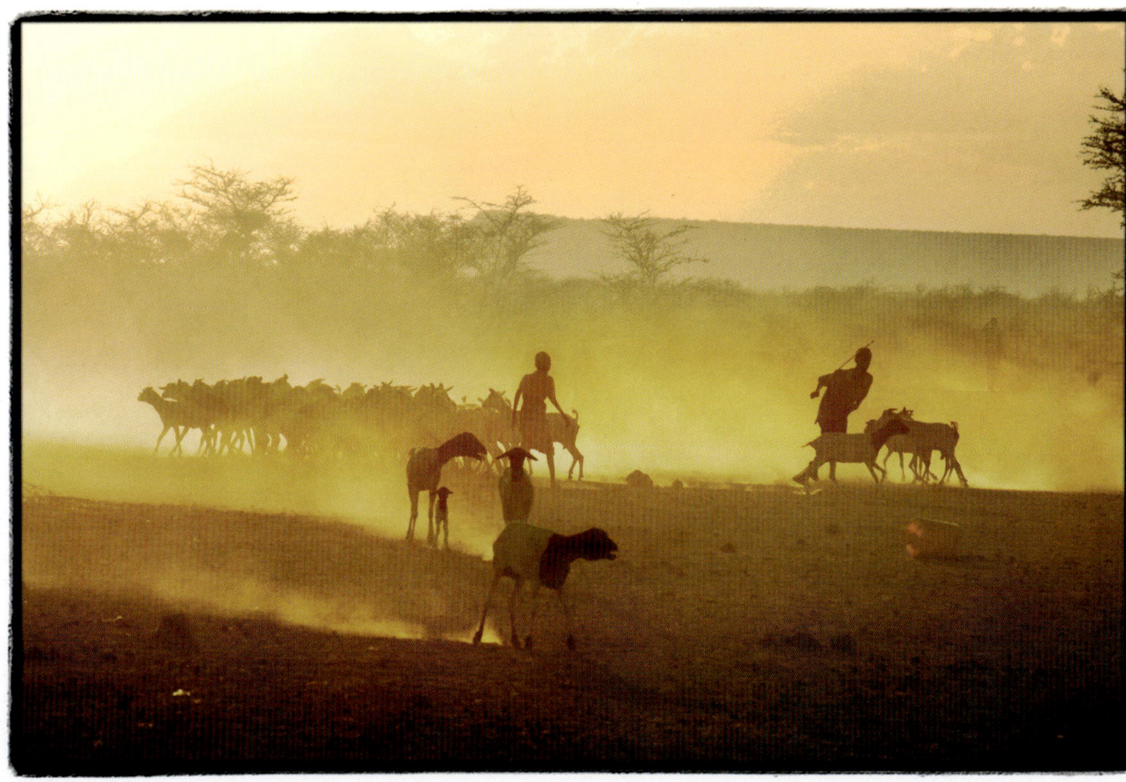

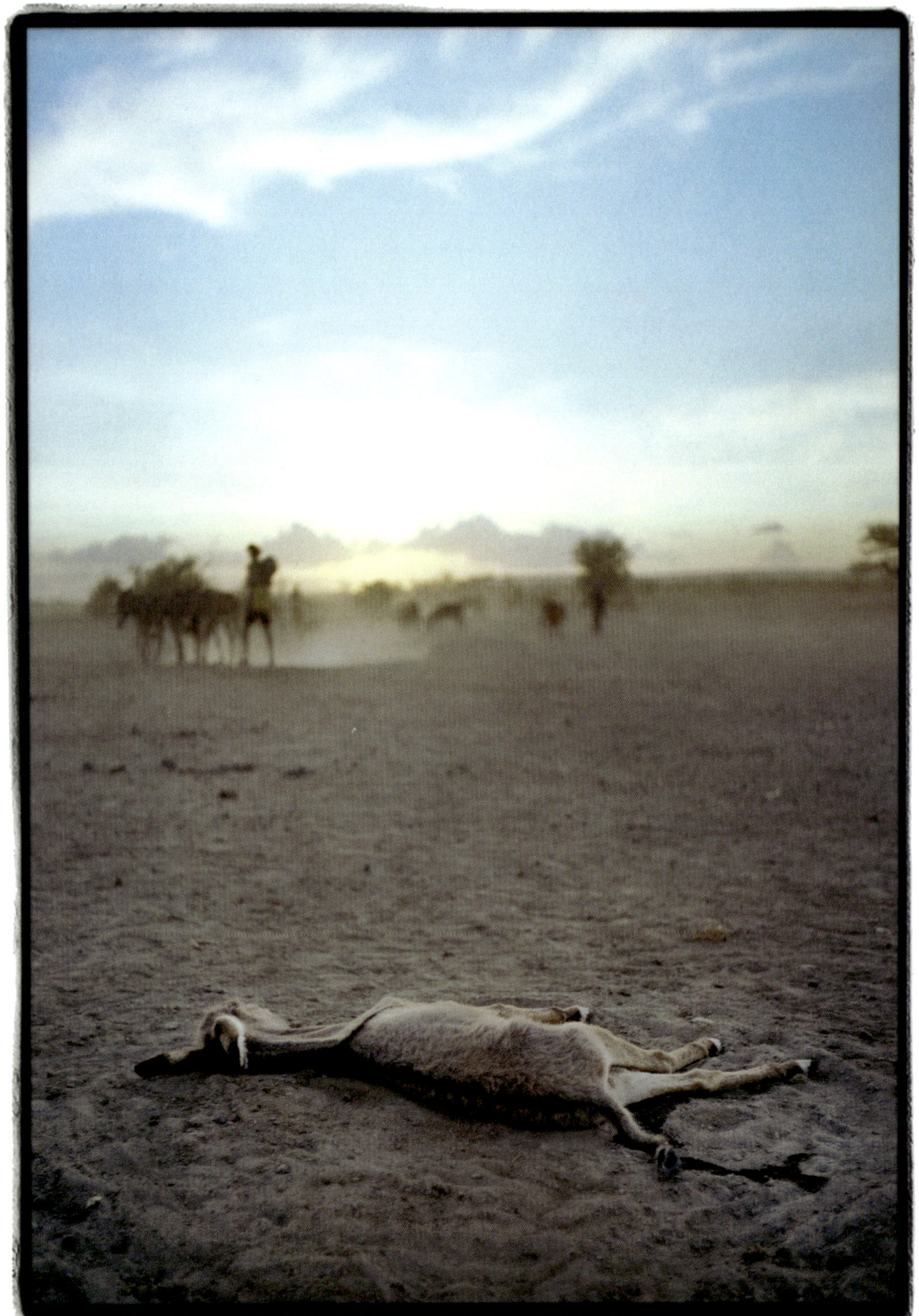

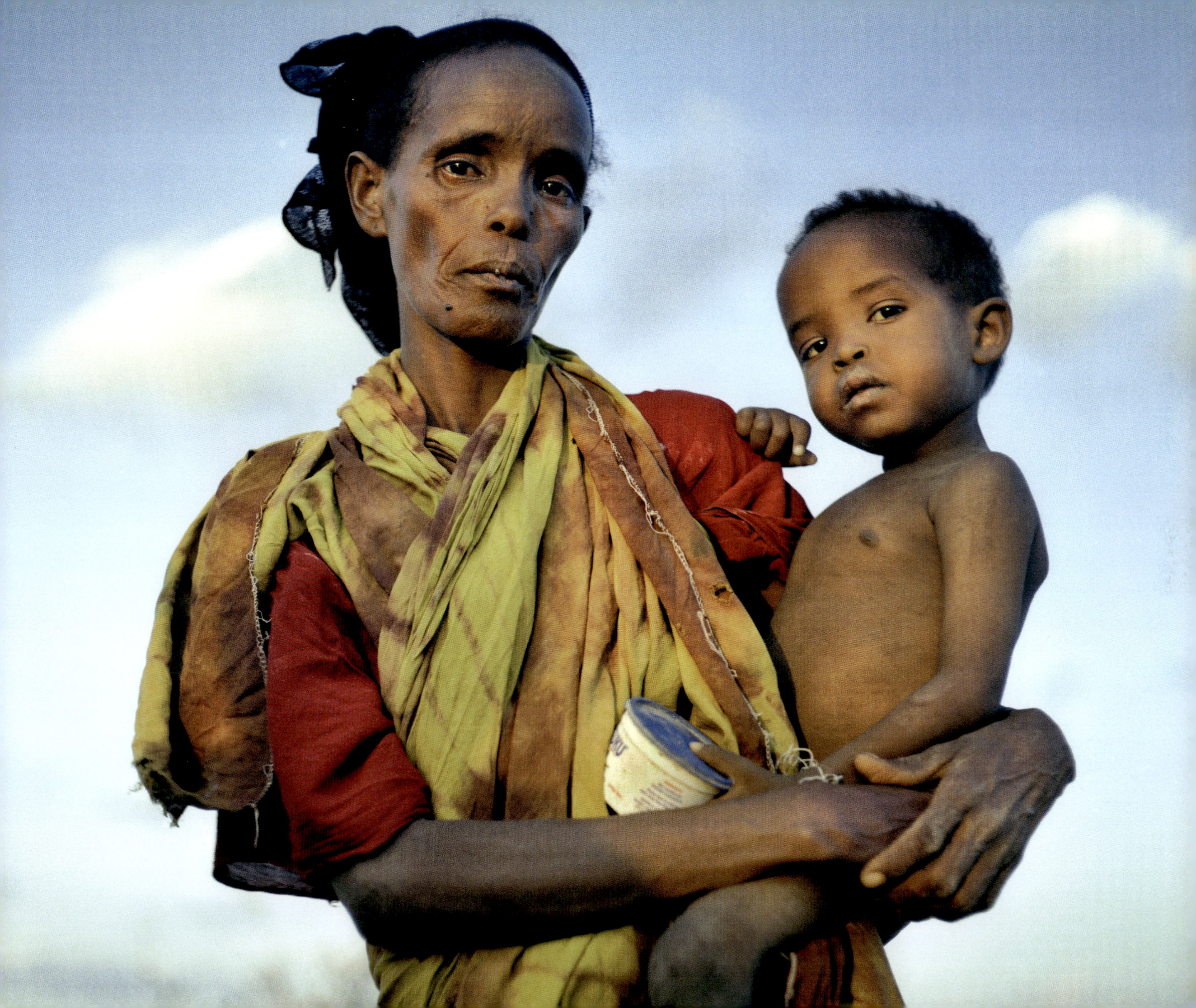

> there are already an estimated 25 million environmental refugees - more than half the number of political refugees

> floods, storms and rising tides are starting to inundate low-lying countries such as Bangladesh

> rising sea levels could leave 150 million people permanently displaced in less than 50 years

> flooding is a major problem in all informal settlements in Nairobi

> major floods that used to only happen every 100 years, may now start to occur every 10 or 20 years

> average Arctic temperatures increased at almost twice the global average rate in the past 100 years

> in some projections, Arctic late-summer sea ice will disappear almost entirely by the latter part of the 21st century

> developed countries are very much at risk because so much infrastructure is at sea level

2 FLOODS

> even if harmful emissions were cut now, many parts of the world would face a greater risk of natural disasters

> in Bangladesh, 17 million people live less than one metre above sea level

> the flood season may become longer and there will be flooding in places where it has never occurred before

> of the 15 largest cities in the world, 13 are on coastal plains and will be threatened as sea levels rise

> the island of Tuvalu in the Pacific is already experiencing severe flooding - nearly 3,000 Tuvaluans have left so far

> mountainous glaciers are retreating in many parts of the world

> around 500 cubic miles of ice have been lost from Alaska in the last 50 years

> since 1850, half the volume of Europe's Alpine glaciers has disappeared

> in the last 100 years, sea levels have risen by 10-20cm globally, and future rises are predicted to be much greater

> in 2002, the Larsen B, an ice shelf weighing 500 million tonnes, collapsed in Antarctica

> global warming causes seawater to expand, contributing to sea level rise

> between 1991 and 2001, the number of people affected by natural disasters rose from 147 million per year to 211 million per year

> if the entire Western Antarctica ice sheet melts, the water produced will raise sea levels by 5-6 metres

> 5 million people in England and Wales alone are now at risk from yearly flooding

> around much of the Mediterranean basin, sea levels could rise by close to 1 m by 2100

> Venice will be one of the worst affected areas if sea levels continue to rise

> rising sea levels and the penetration of salt water will leave some water supplies unusable

> the poor are the most affected because they live in marginal areas that are afflicted by floods and scarce water supplies

> there will be millions more poor people living in high-risk areas of the world by 2100

> recorded economic losses from natural catastrophes have grown from US$30 billion in 1990 to US$70 billion in 1999

> the number of floods has more than doubled since 1996

> as more and more people live on marginal land, there is increasingly greater risk from flooding

> unseen and uncounted, millions are already on the move in search of greater water security

> global warming is bringing chronic flooding to cities

> due to climate change some areas will increasingly suffer from destructive floods, while others will hardly see any rain at all

> in Ganges-Brahmaputra delta, up to a dozen islands, which are home to 70,000 people, are immediately threatened by the rising seas

> rising sea levels have caused the people of islands near New Guinea to start evacuating their homes

> New York could be subject to flooding by major storms

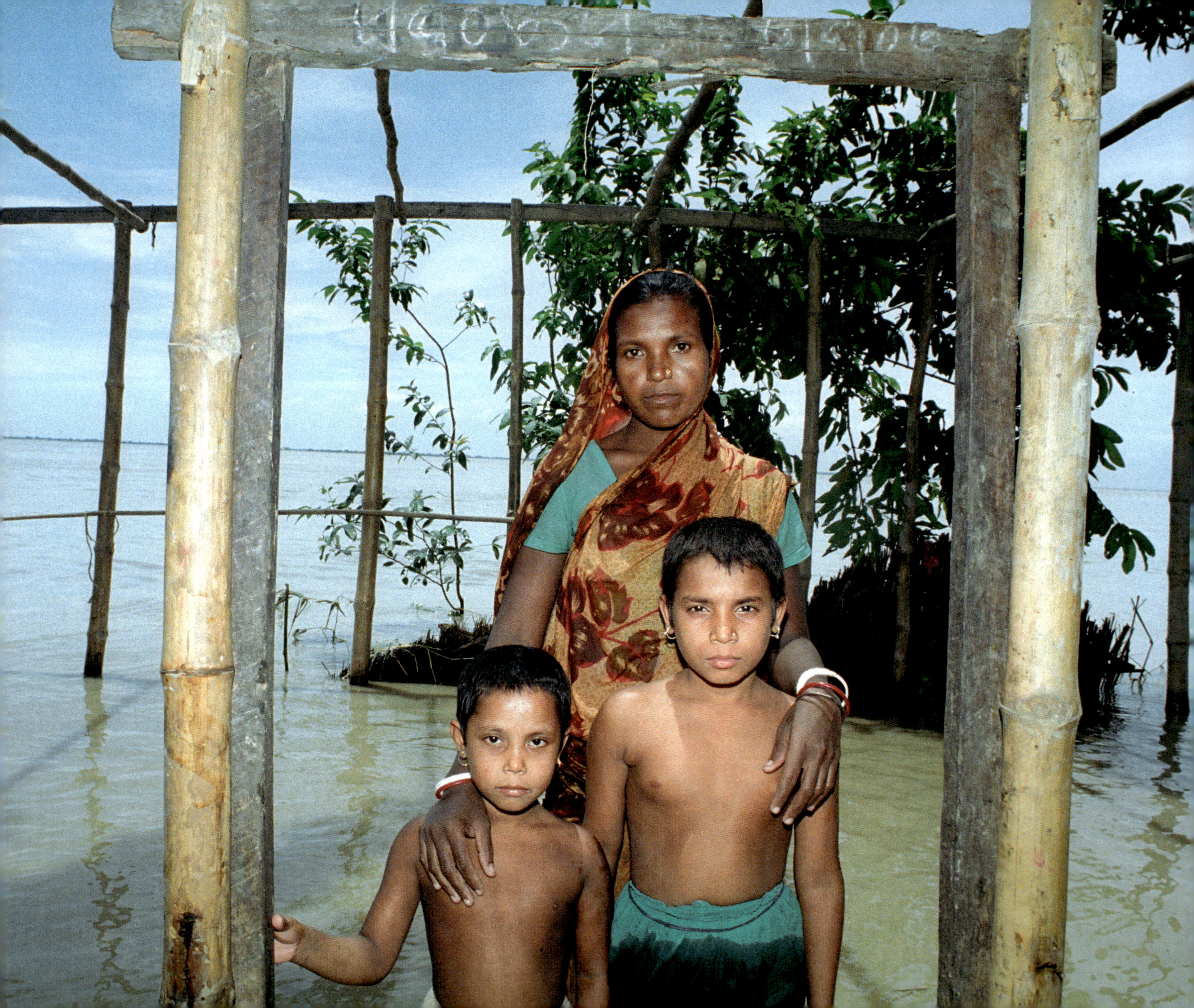

Bangladesh
Flood Victims

Floods are natural recurring phenomena in Bangladesh. The country is situated in the delta of the Ganges (Padma), the Jamuna (Brahmaputra) and the Meghna. These rivers drain excess water from monsoon rains and the melting of Himalayan snows into the Bay of Bengal. The flooding deposits fertile silt that replenishes the soil. This makes the river system both the country's principal resource and its greatest hazard.

Bangladeshis have learned to live with the floods, except when they cause massive devastation. In recent times, heavy floods took place in a 10-year cycle, with the last major floods in 1998 and 1988. But in 2004, they arrived three years early. By the end of the monsoon season, nearly half of the country was flooded. About 30 million people were forced to leave their houses for safe havens in higher areas. 10 million Bangladeshis were left homeless and about 700 people died. The UN World Food Program assessed the damage to farms, factories and infrastructure at US$ 7 billion.

It was not only the countryside which was flooded. 40% of the capital, Dhaka, a megacity with a population of 13 million, was under water as well. The floodwater mixed with the city's sewage system and left black stinking mud in the streets. Water-borne diseases spread fast and the city was hit by an epidemic of cholera. Many children died from diarrhoea and respiratory infections. The crisis was not over with the receding floodwater. People returned to their ravaged homes but remained without income or assets. They faced a massive food crisis which was only overcome by the distribution of food aid.

Some experts say irrigation for farming is to blame, because it causes rivers to silt up. This reduces their capacity to hold flood water or drain it into the sea. Other experts believe global warming is speeding up the melting of Himalayan snows and increases the monsoon rainfall. They predict Bangladesh might in the future also have to deal with large-scale flooding of its coastal areas.

If no action is taken on carbon emissions, global temperatures are expected to rise between two and three degrees celsius over the next 50 years. The Bangladeshi coast, where the terrain is generally at sea level, will then face rising sea levels and vast coastal flooding.

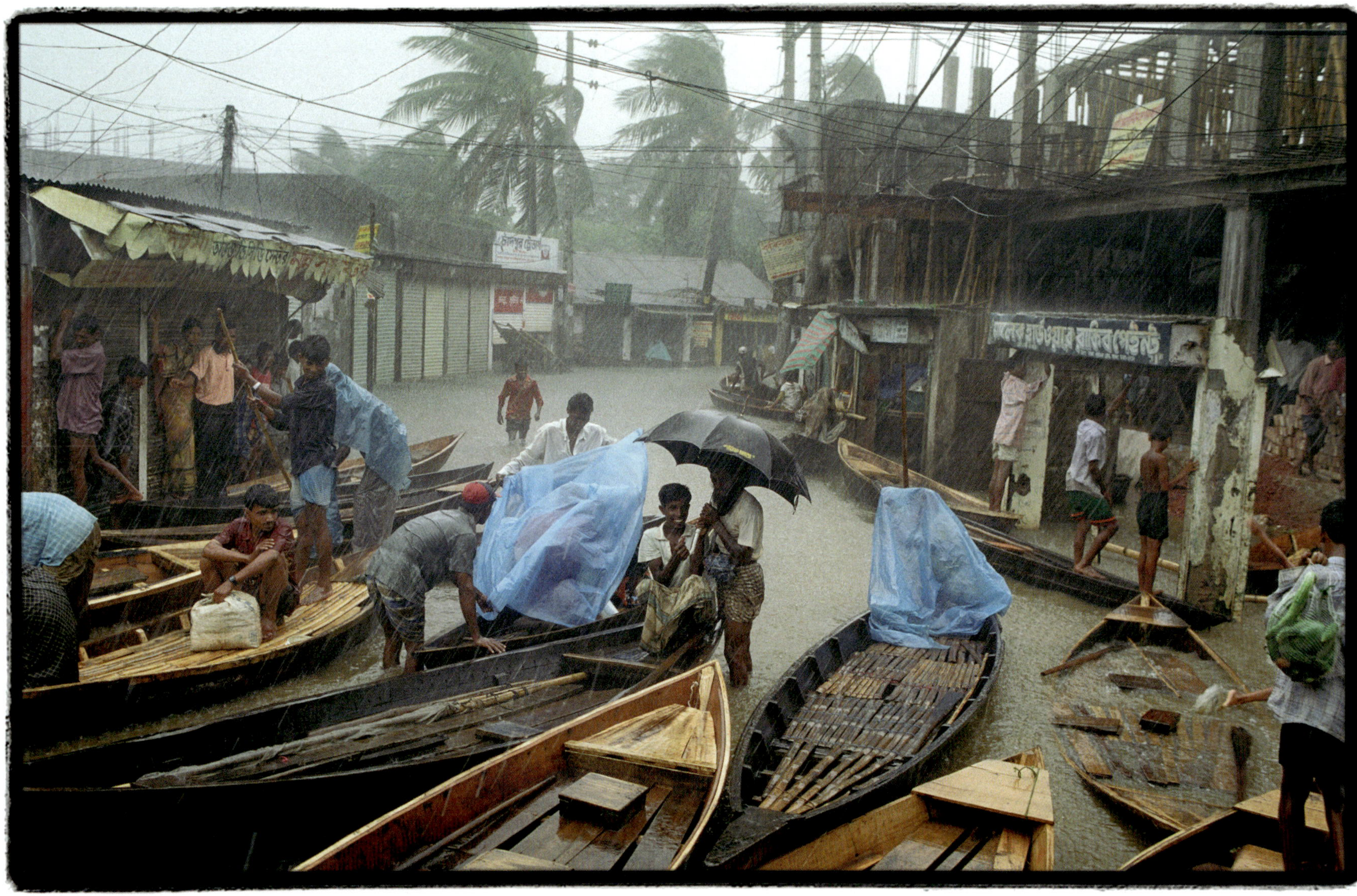

Dhaka. The impact of this flood on the population is increased by the fact that Bangladesh, apart from some micro-states, is the world's most densely populated country.

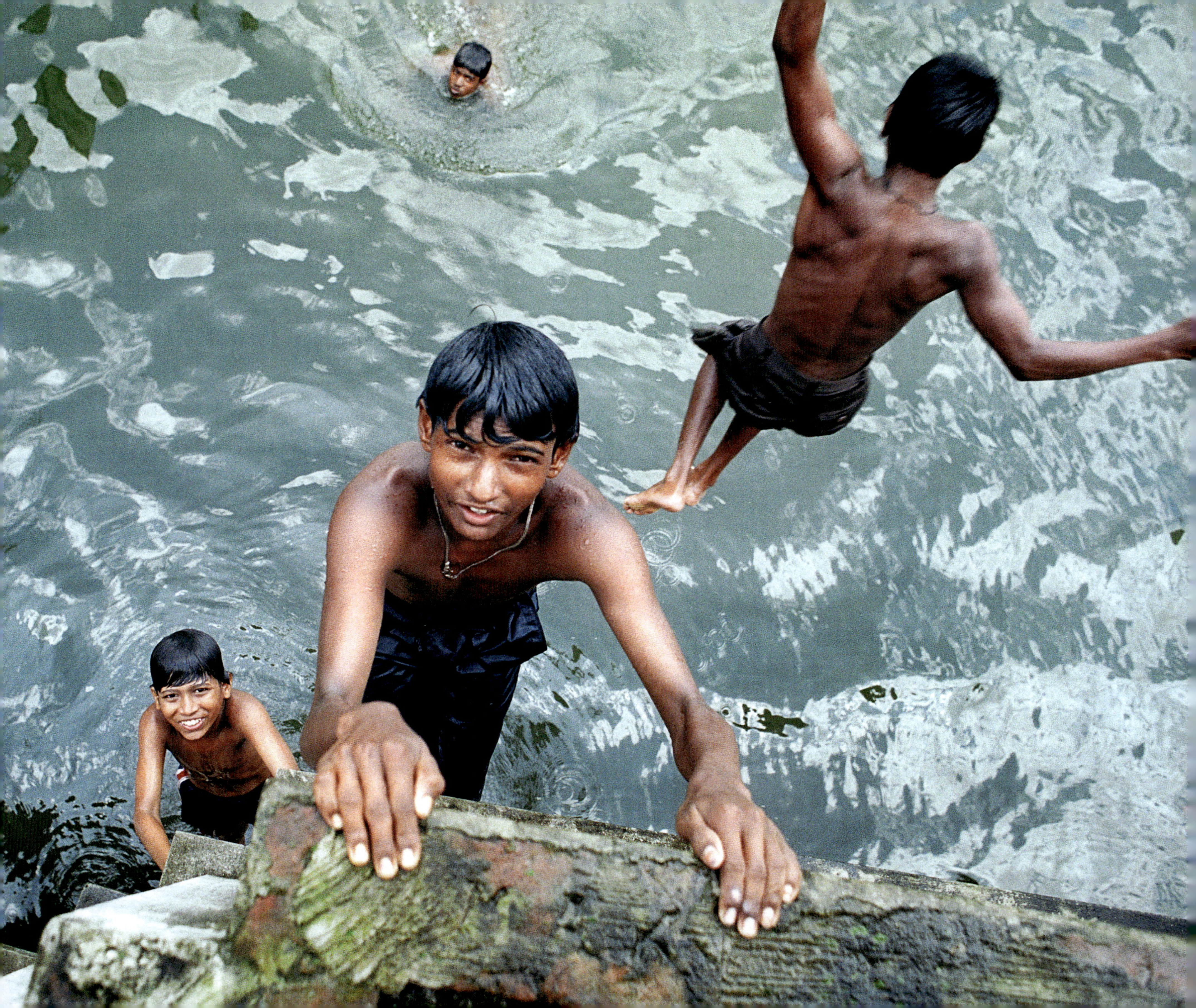

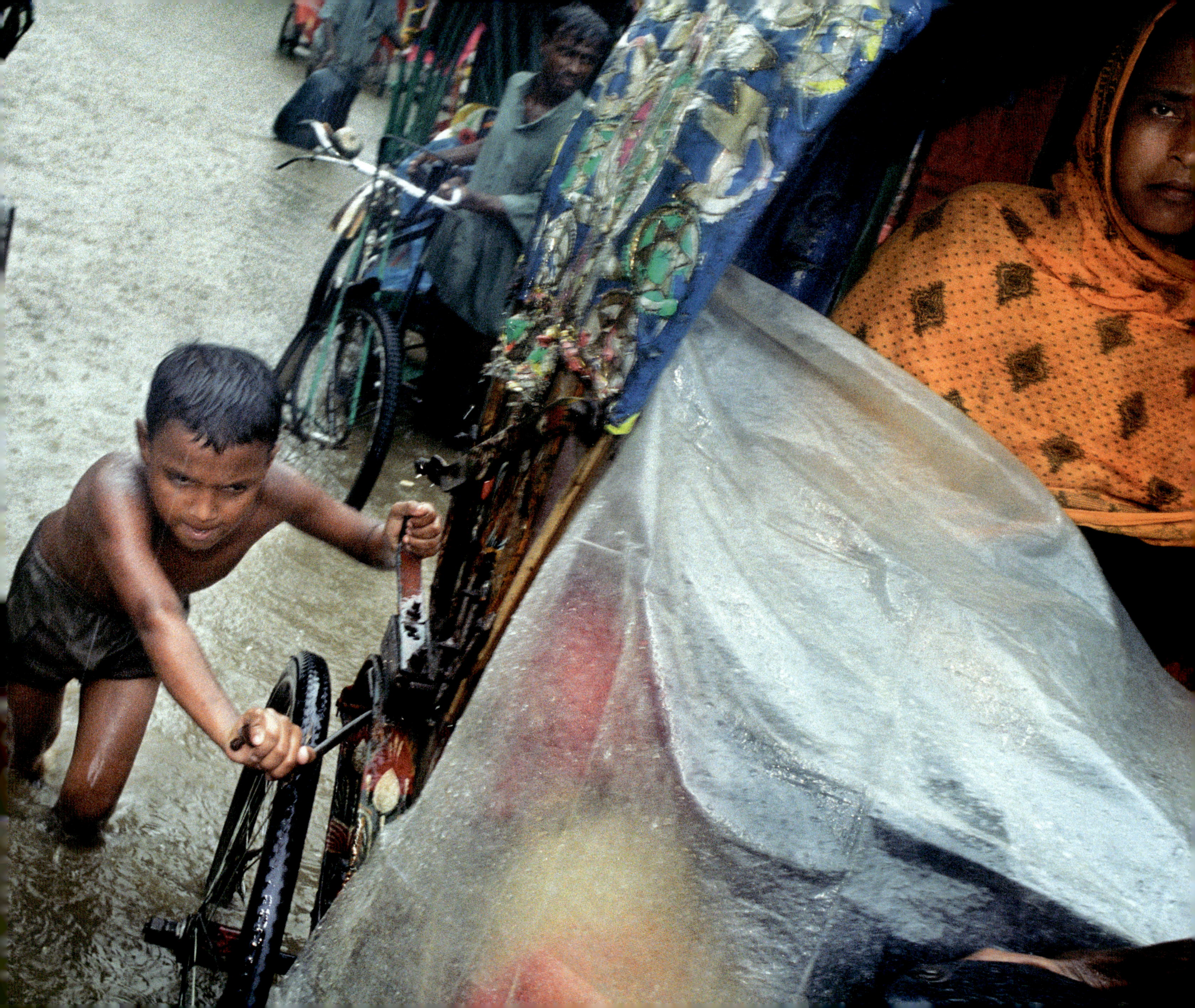

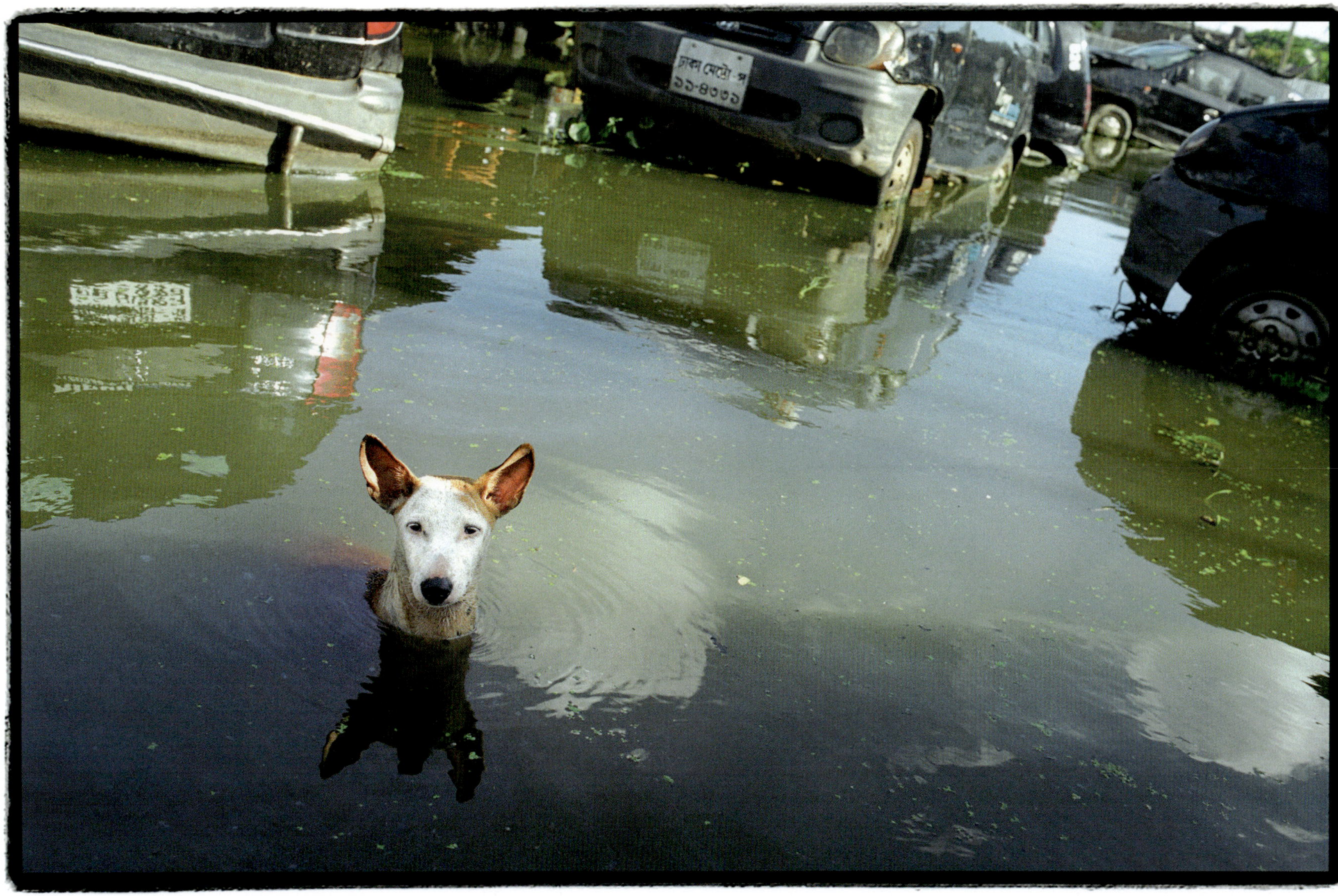

Dhaka. The flooded areas make for an unhealthy environment where people face problems such as lack of food and clean water, gastro-enteric disease, pneumonia, snake-bite and drowning.

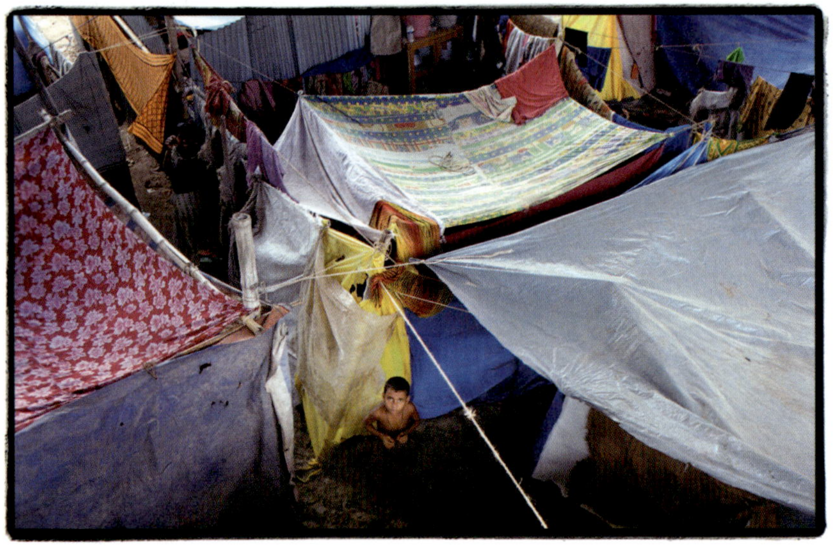
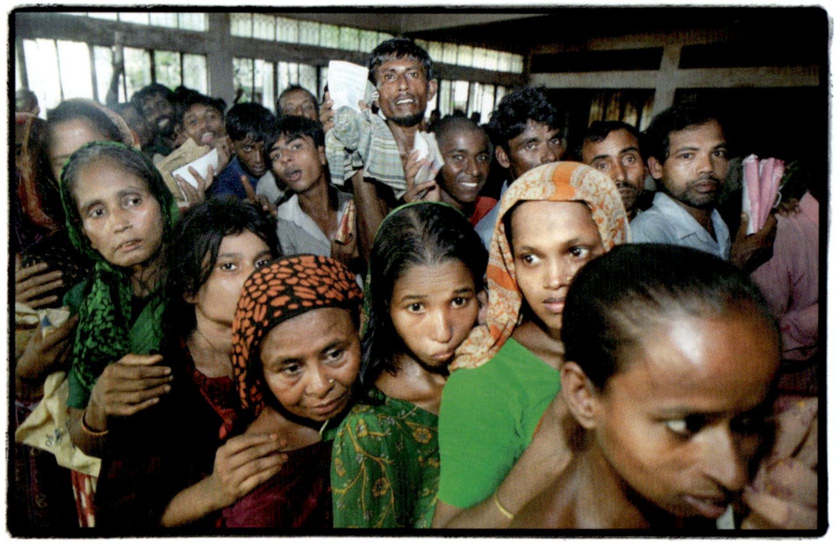

Dhaka. Millions of people left their flooded houses and found shelter in schools, under bridges or under plastic sheets along the roads. They depend on the goodwill of local NGOs, organisations and religious groups for the provision of food and drinking water. Many complain that the government is not providing any assistance.

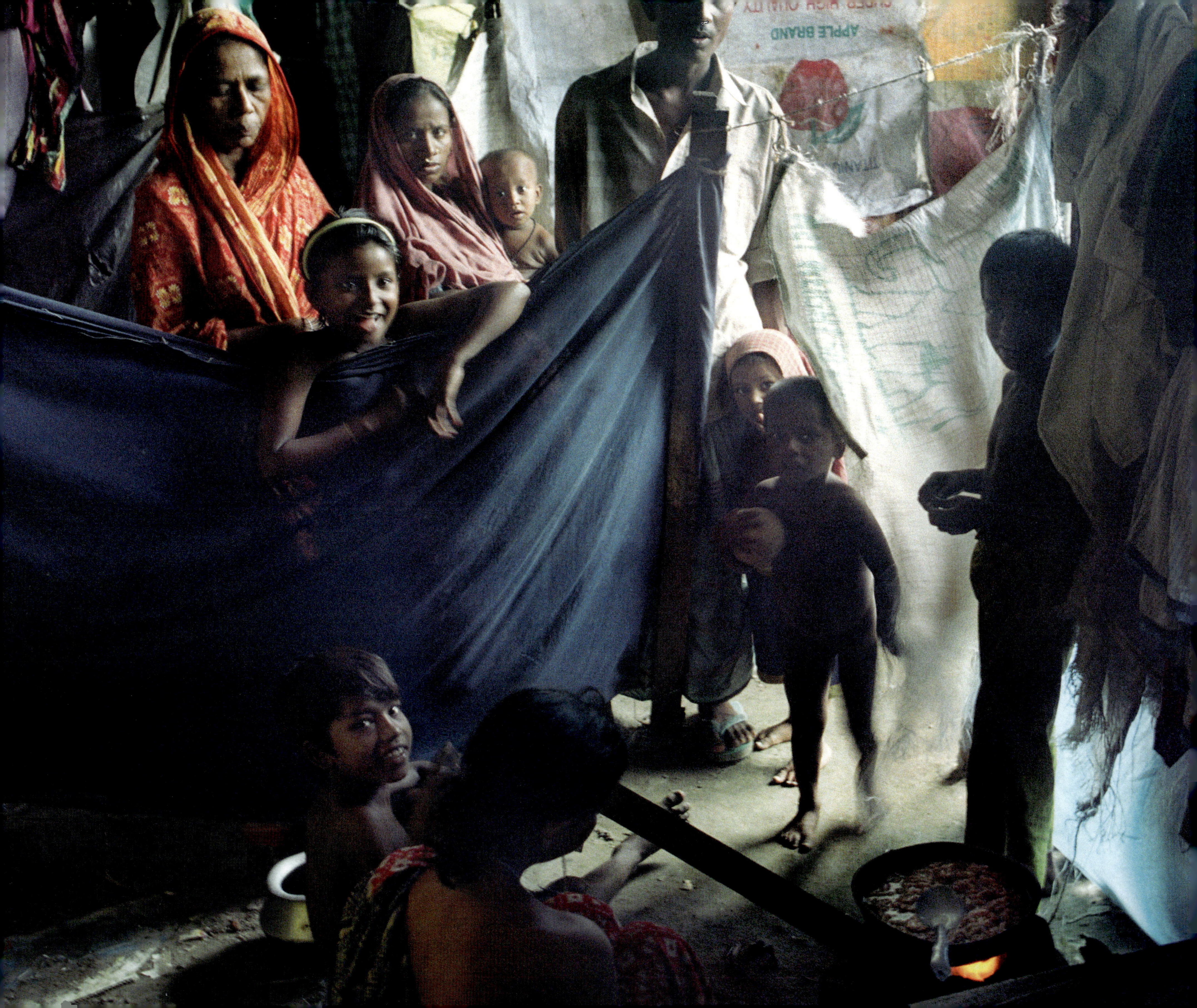
APPLE BRAND

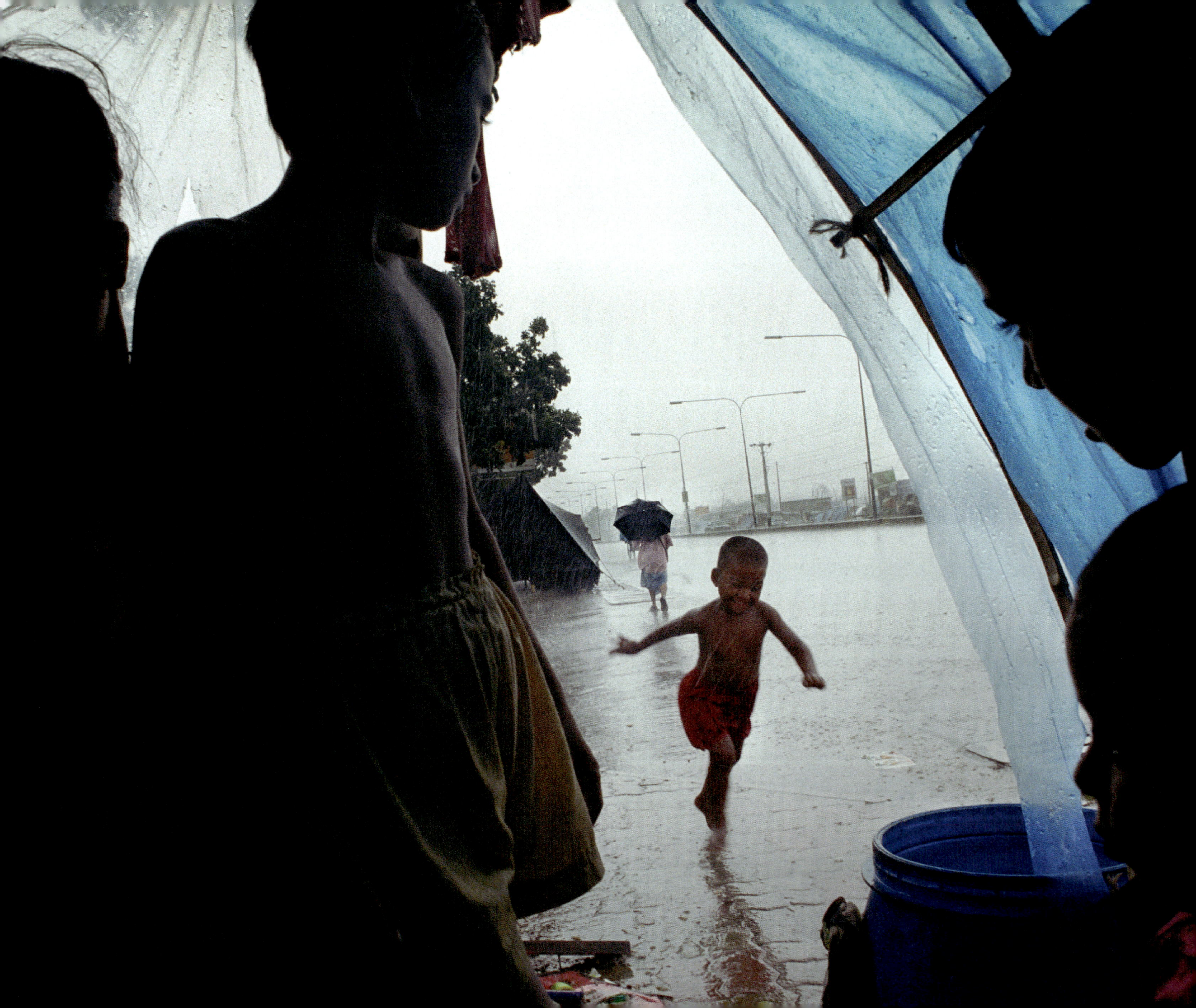

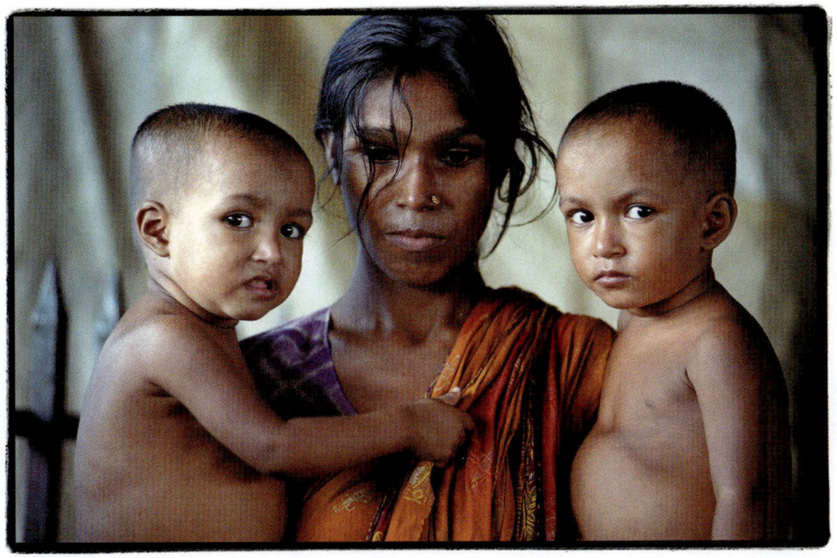

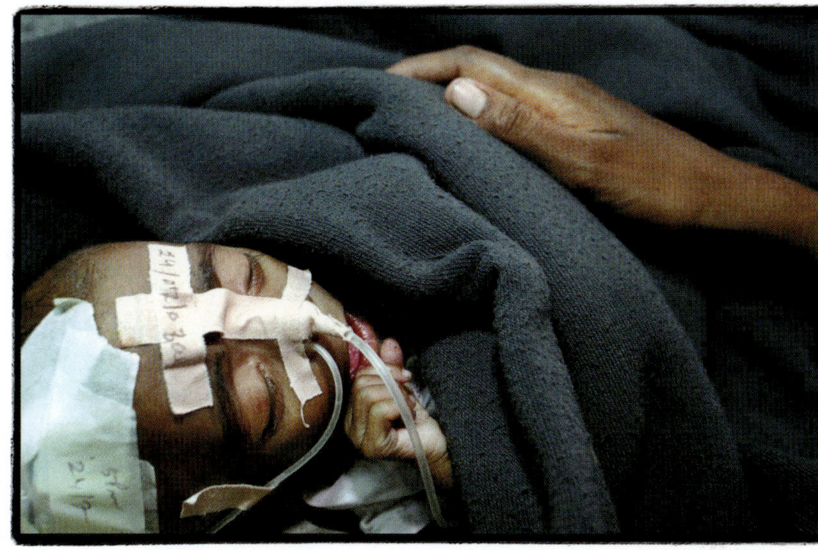

RIGHT Dhaka. A baby is treated in the Bangladeshi Centre for Diarrhoea Disease Research, known as the cholera hospital. During the flooding, the centre received more than 300 new patients every day. In some areas of the capital, the floodwaters, mixed with sewage, have contaminated the city's piped water, resulting in the spread of water-related diseases.

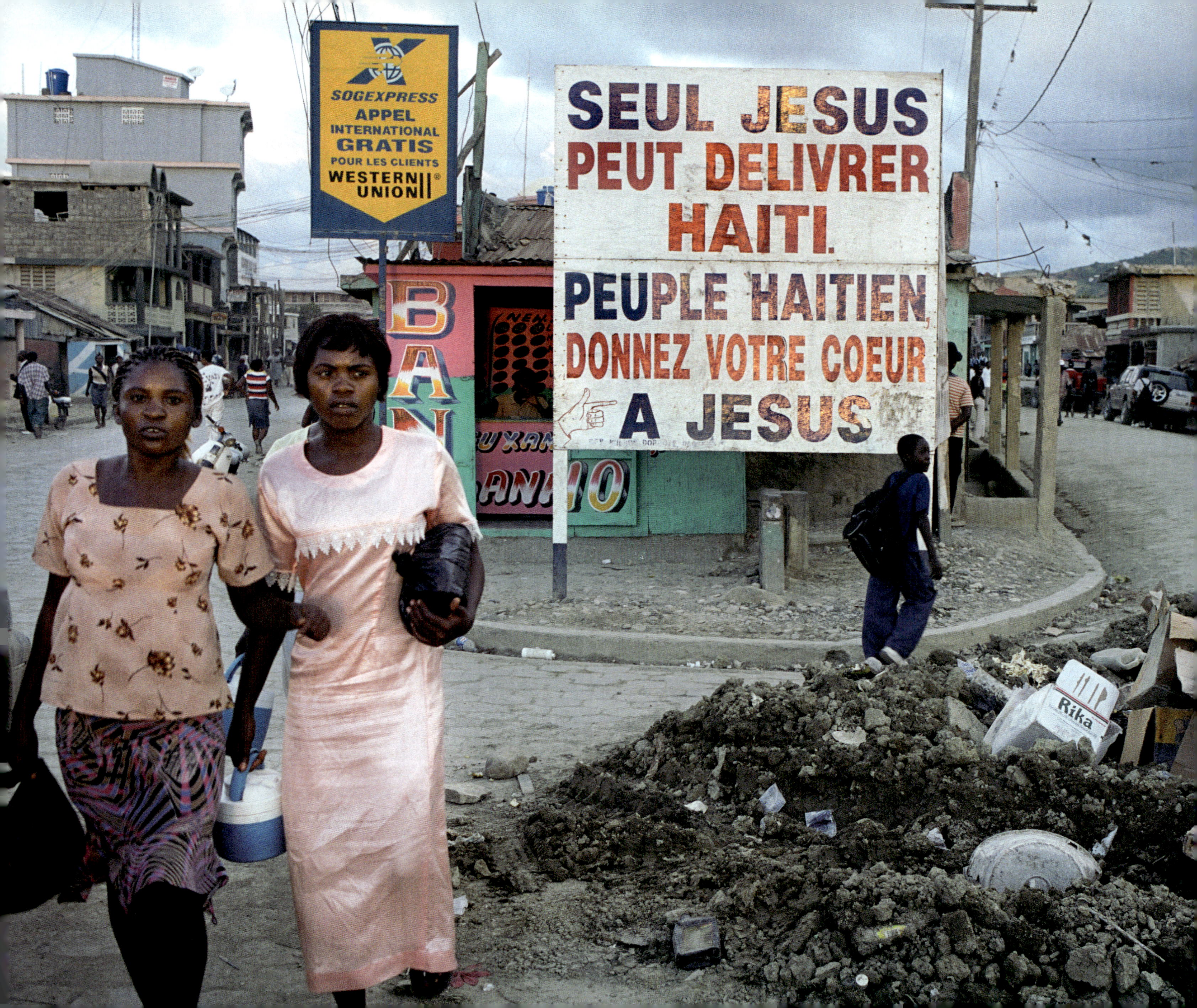
SOGEXPRESS
APPEL
INTERNATIONAL
GRATIS
POUR LES CLIENTS
WESTERN UNION
SEUL JESUS
PEUT DELIVRER
HAITI.
PEUPLE HAITIEN,
DONNEZ VOTRE COEUR
A JESUS
Rika

Haiti
Flood Ravages

Haiti is the poorest country in the Western Hemisphere. It has economic and social indicators that are comparable with those of many Sub-Saharan countries: 80% of the rural population is living below the poverty line, more than half of the population is illiterate and 50% is under-nourished.

The health system is poorly organised and insufficient. A substantial part of the population does not have access to basic healthcare. One in every 20 Haitians has HIV/Aids and life expectancy at birth is below 50. Only half of the population has access to clean water and even fewer to sanitation facilities.

Haiti used to be covered with tropical rainforest, but it has now become one of the most deforested countries in the world. The French colonisers were the first to cut thousands of acres of forest for sugar production and to ship fine furniture wood to Europe. Today, the poor population takes whatever it can from the remaining forest. The wood is being converted into charcoal for cooking and the forest goes up in smoke.

Deforestation has made Haiti particularly vulnerable to storms. When there are no forests to absorb the heavy rain and no tree roots to bind the soil together, the water runs off the mountains in huge volumes and may cause mudslides. Rivers burst out of their banks and create floodplains. The consequences for villages in the paths of the mudslides or in the floodplains are dramatic.

In September 2004, the coastal town of Gonaives in northwestern Haiti was flooded after tropical storm Jeanne lashed the area with 30 hours of rain. The city was covered with waist-deep mud. Thousands of people were left stranded on rooftops. 700 people died and hundreds were reported missing. For weeks, the city was left without electricity or medical aid. The risk of epidemics was great. While trucks with relief aid were struggling to reach the affected areas, an estimated 175,000 people were left without food or clean water.

This disaster happened four months after other heavy rains had caused another 500 deaths in the south of Haiti. With each flood, the poor infrastructure that was available is destroyed and there is even less land to use, resulting in yet more poverty. This vicious circle could worsen as global warming is predicted to cause even more tropical hurricanes.

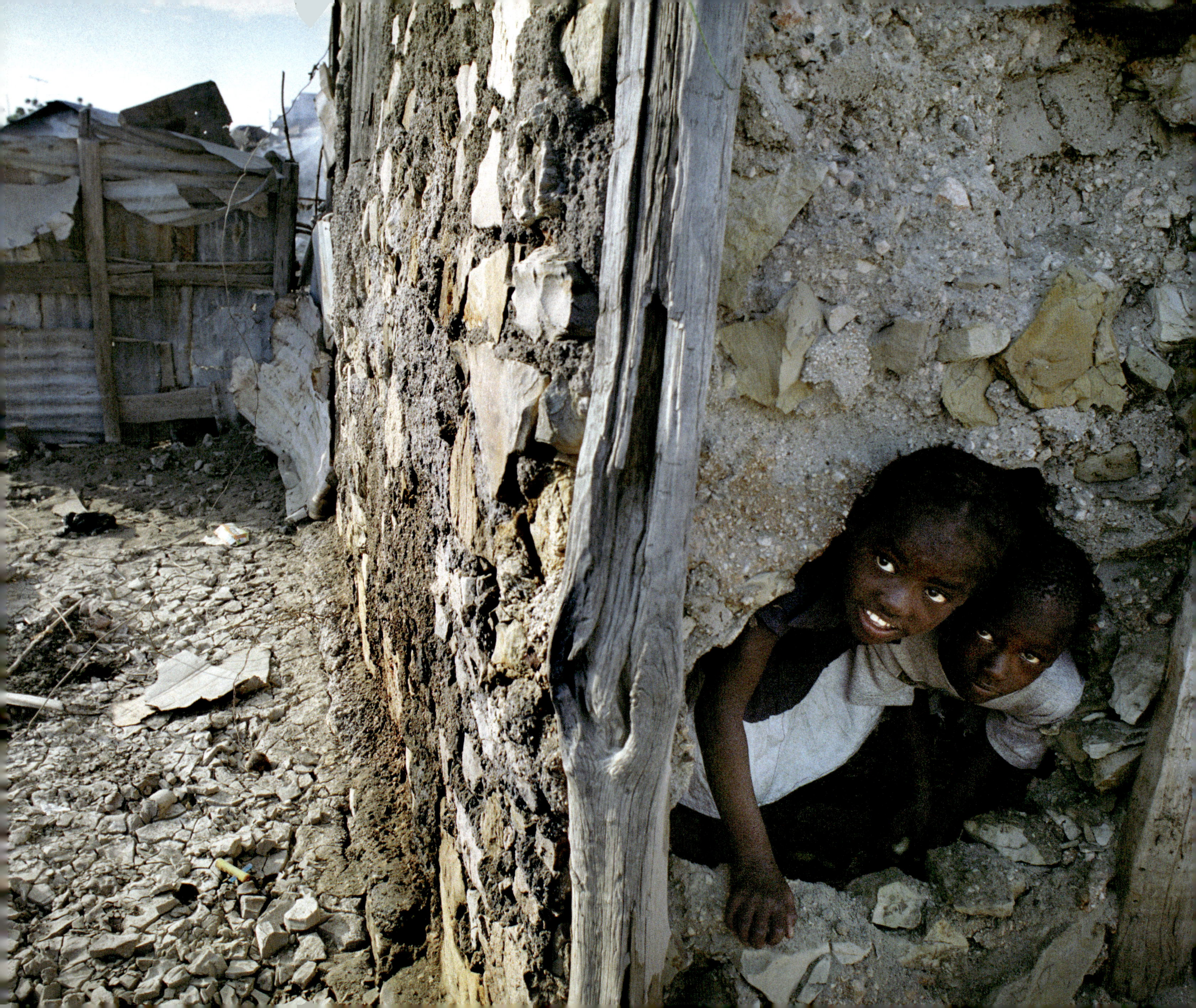

RIGHT Gonaives. Madame Moe outside her home where she lives with her 6 children. She used to sell rice in a market, but lost her stock of 10 sacks in the flood. All her household goods were also destroyed.

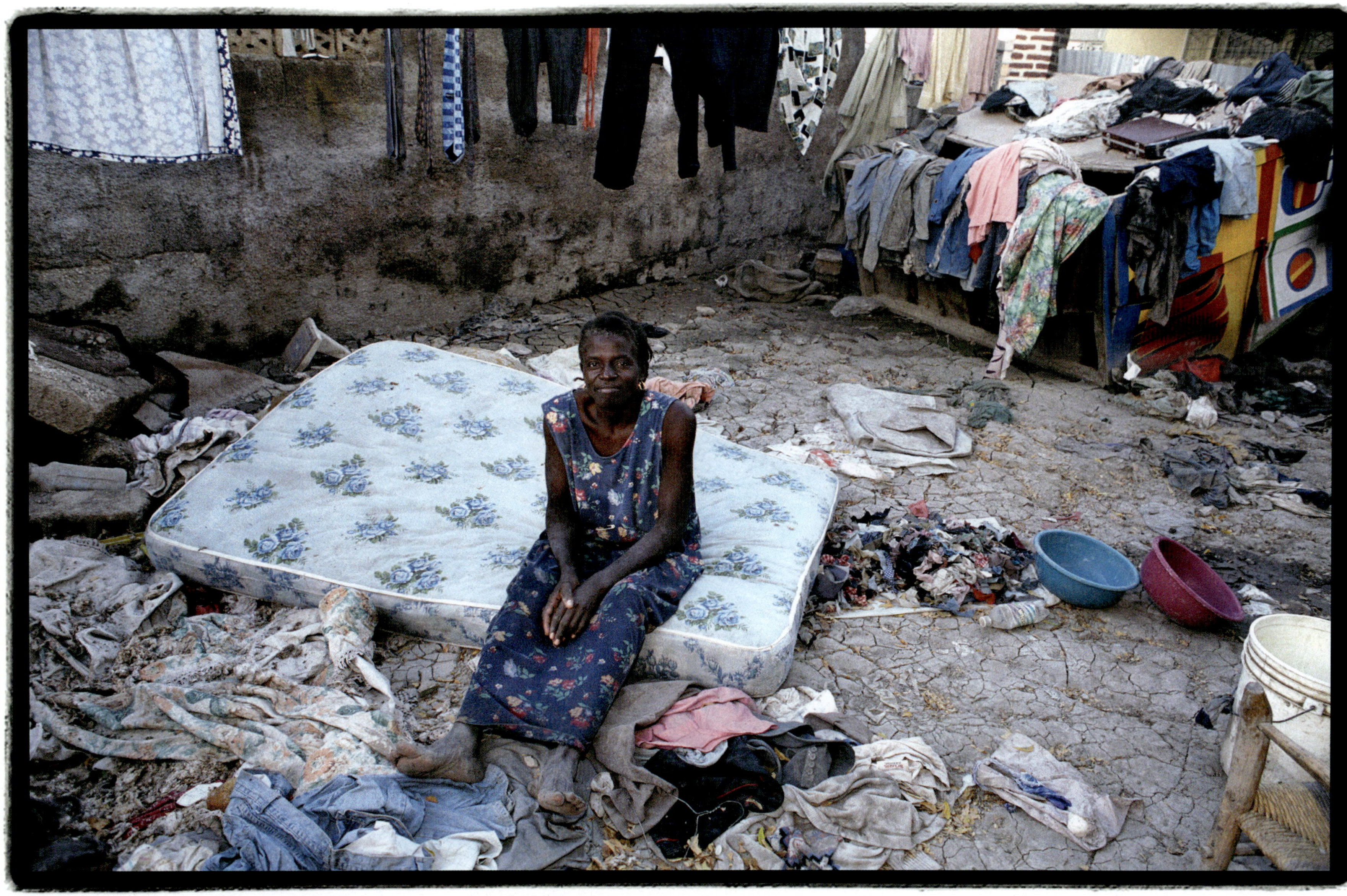

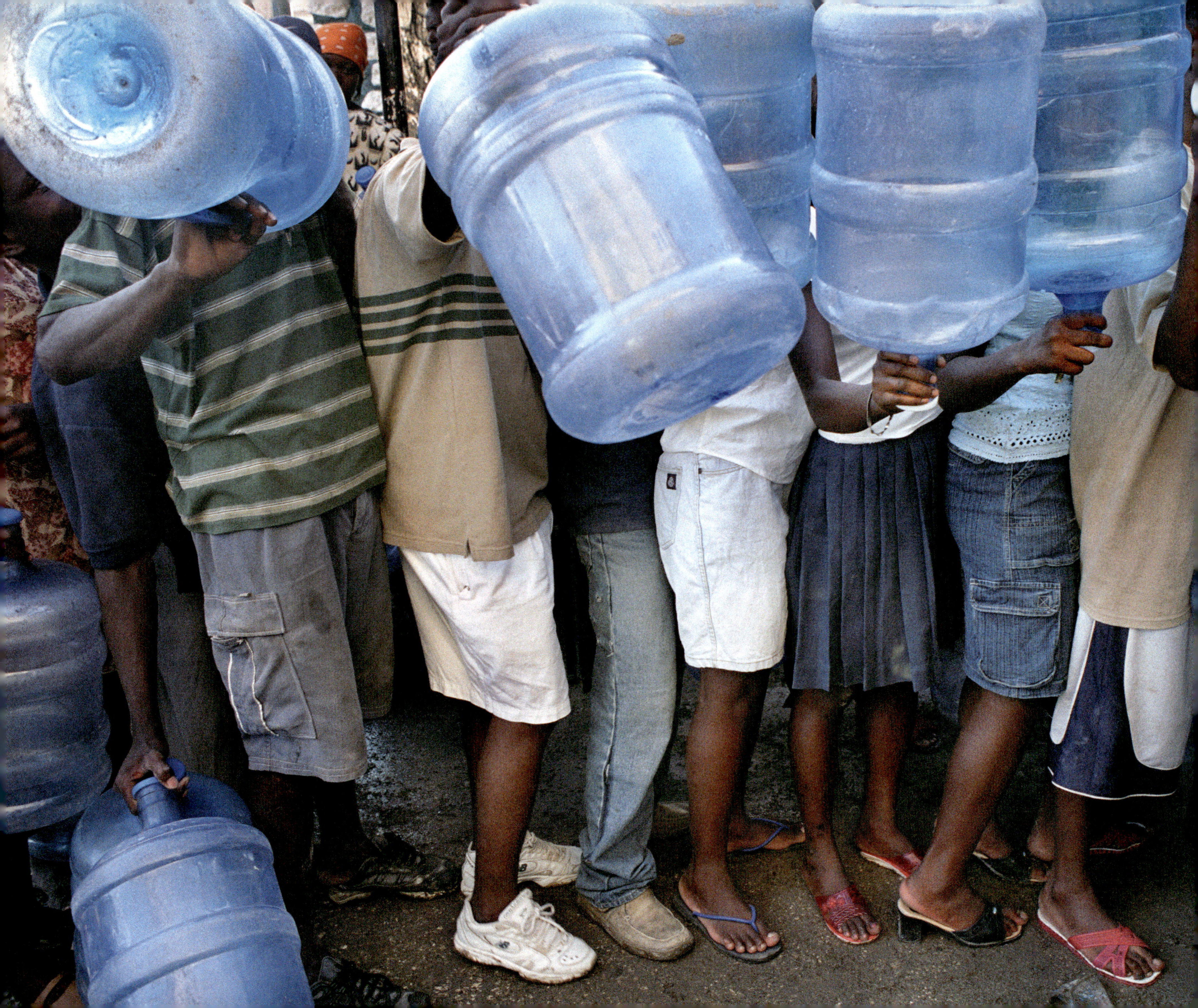

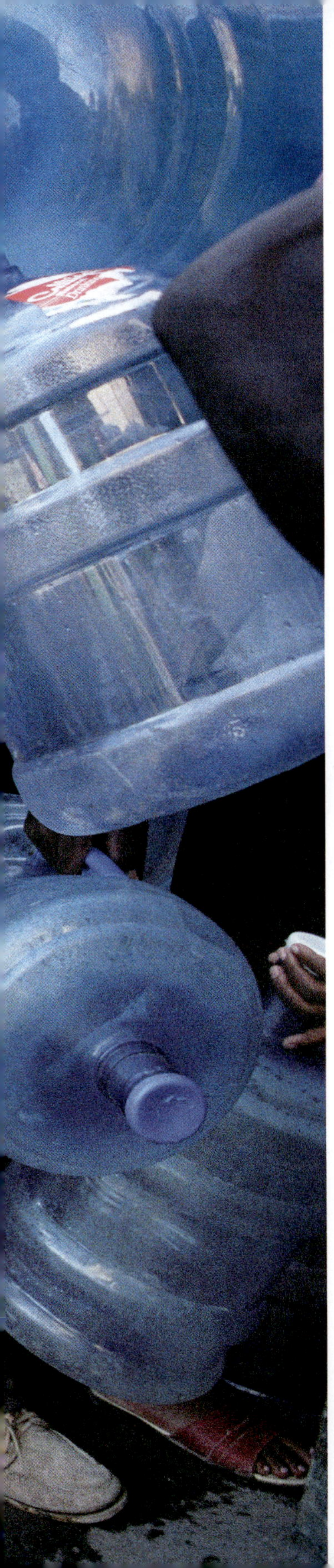

Gonaives. Women waiting in queues for up to 8 hours to receive food and relief supplies donated by the United Nations World Food Programme (UN WFP).

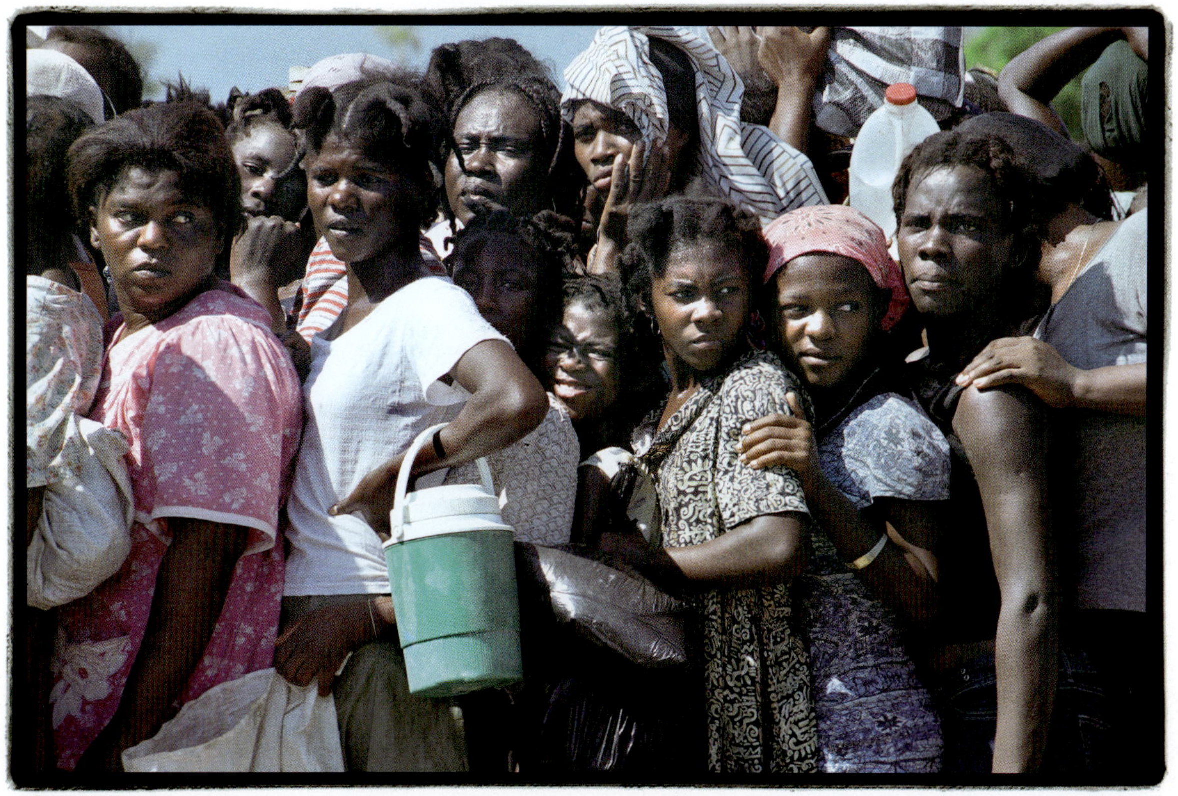

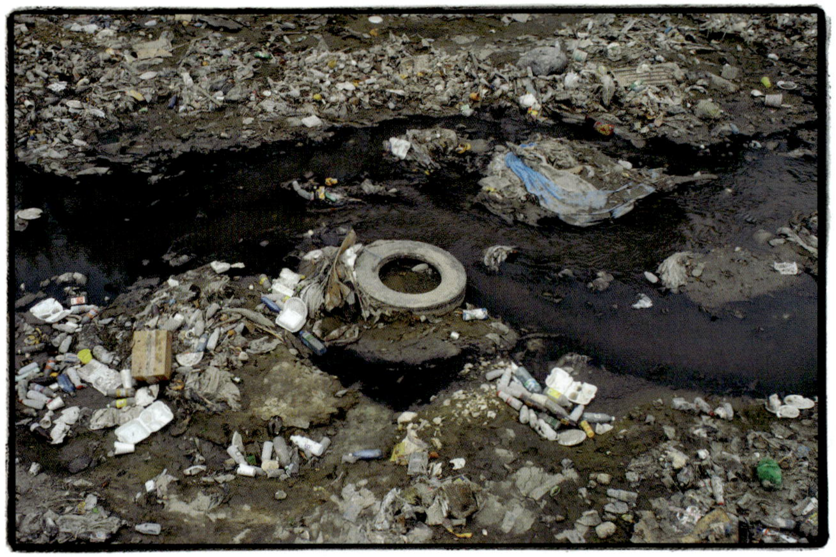

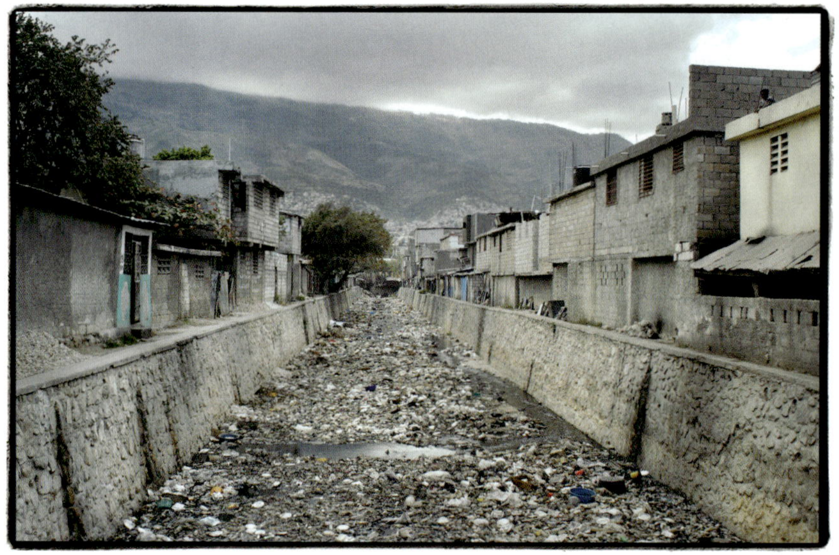

Port-au-Prince and Port-de-Paix. Badly polluted rivers, filled with dumped rubbish. Most sewage produced in Haiti is discharged into aquatic systems without treatment. If these aquatic systems are near or upstream from human habitations, sewage can pose a serious public health risk.

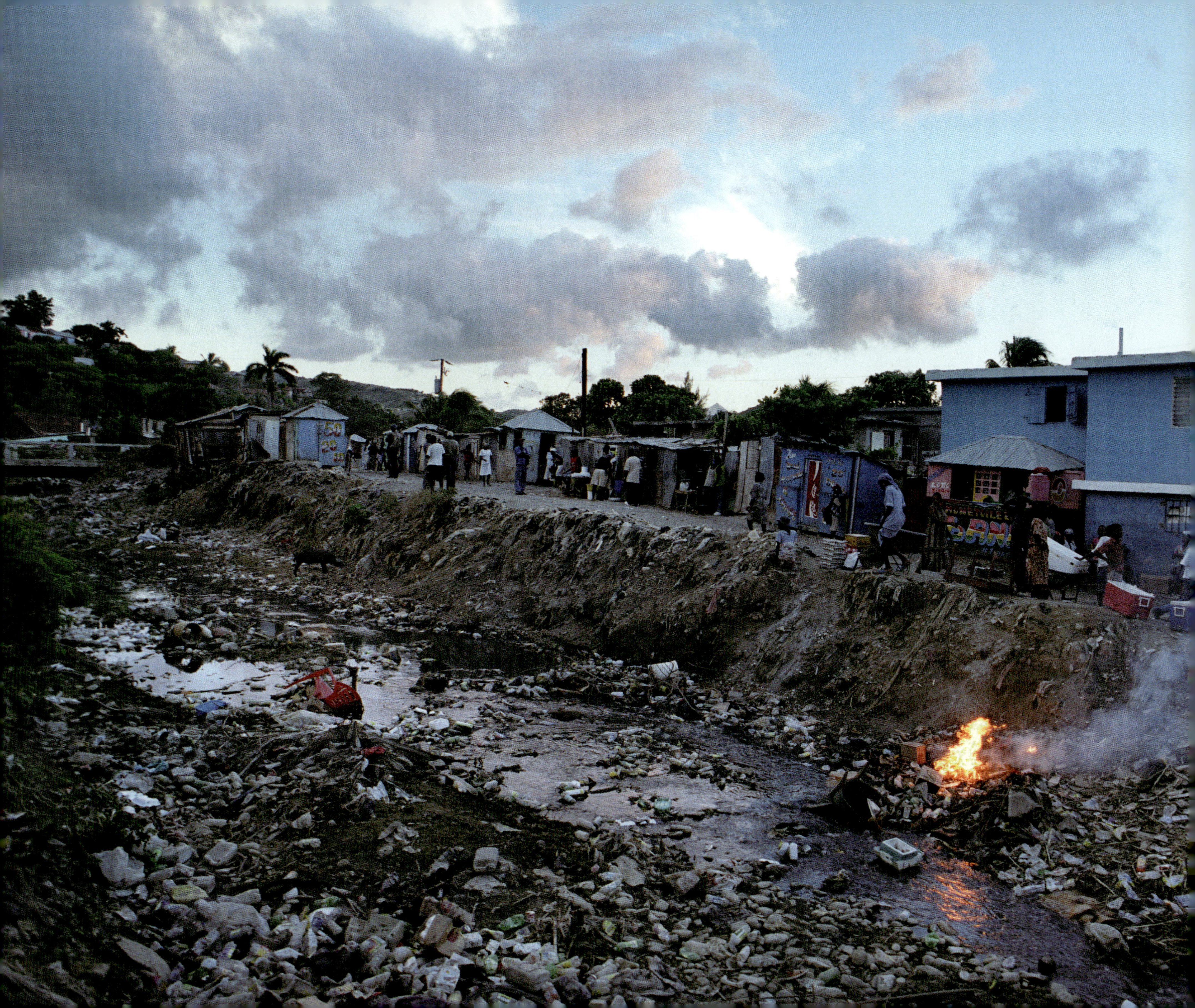

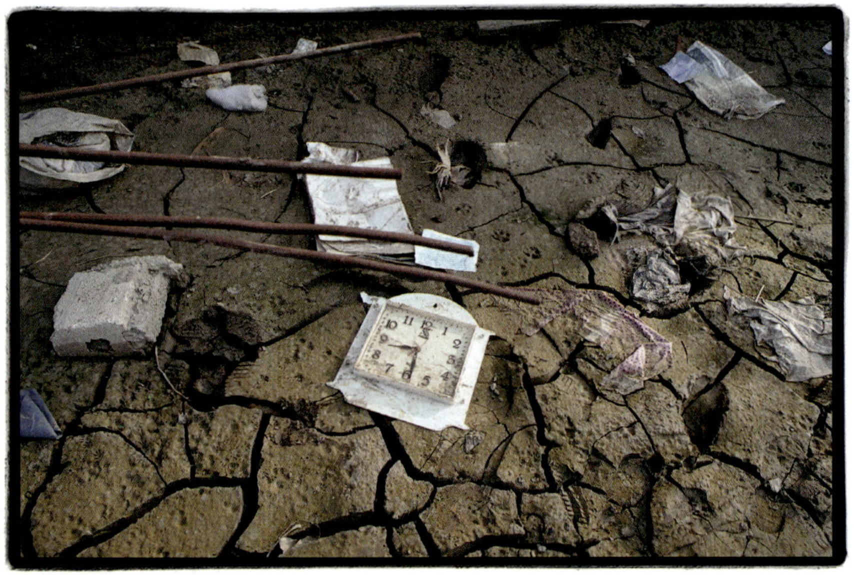

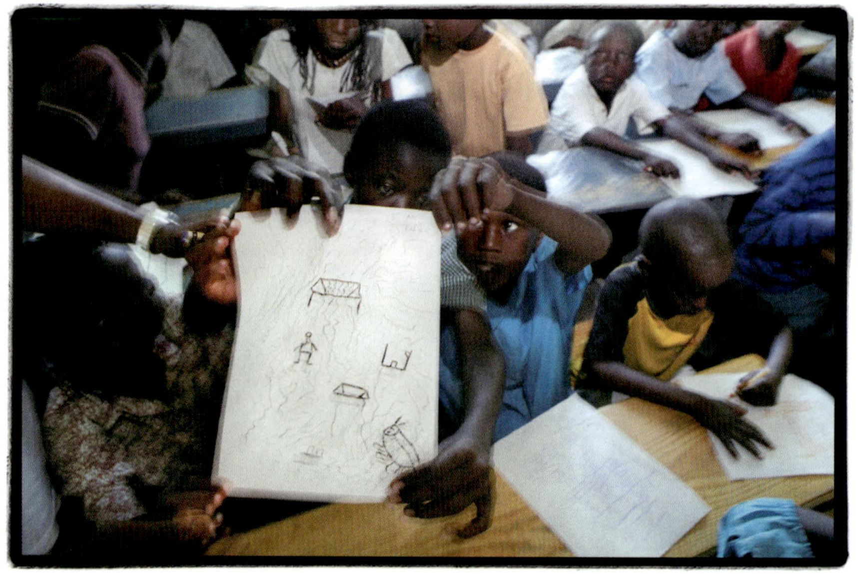

MIDDLE Gonaives. Children during a session with psychologists from Médecins Sans Frontières (MSF): they draw destroyed houses as they relive the trauma of the flood.

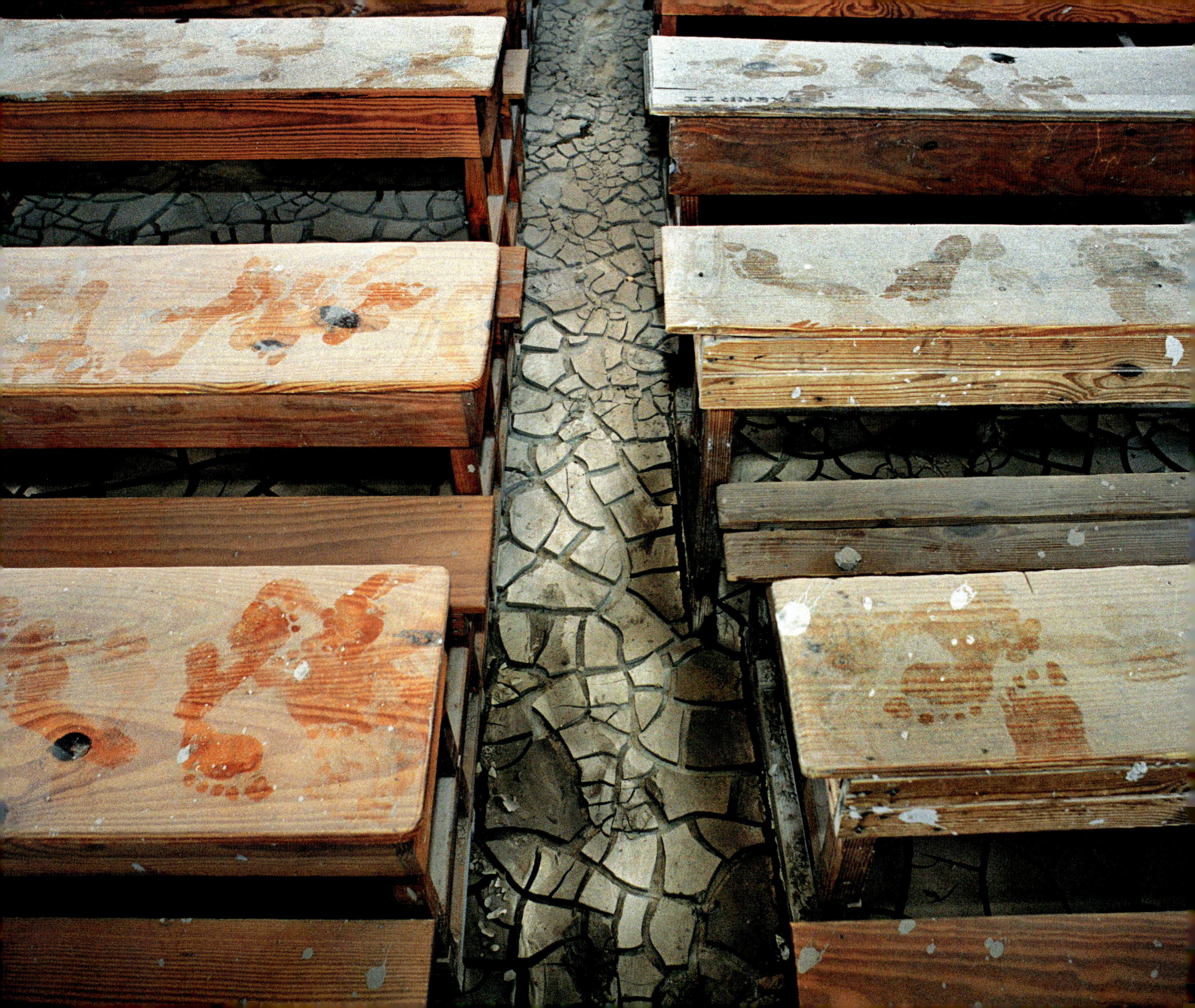

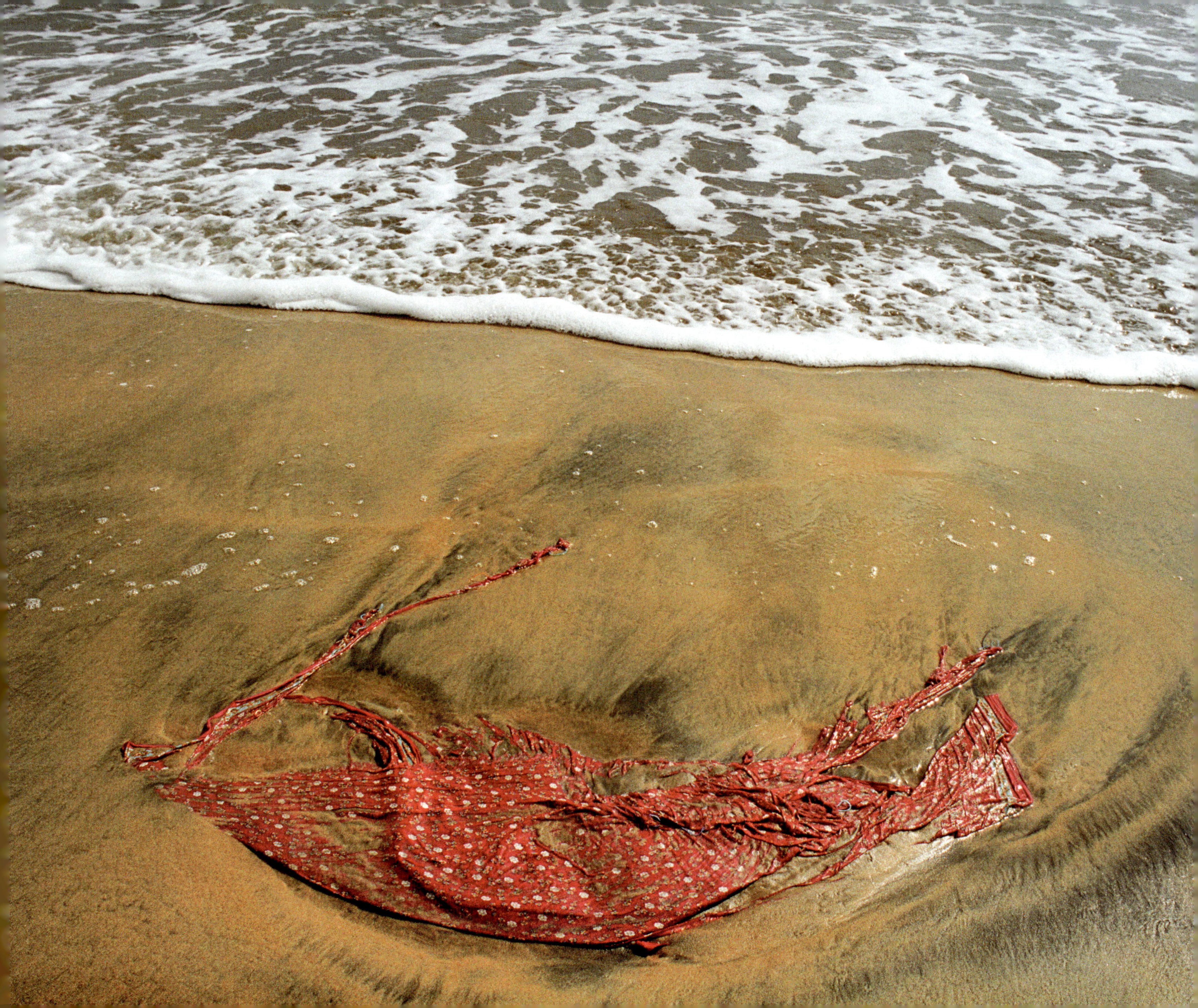

India
Tsunami

On December 26, 2004, the world witnessed the terrible power of the Indian Ocean tsunami. An underwater earthquake, measuring 9.0 on the Richter scale, moved the seabed between India and Southeast Asia and forced billions of tons of seawater upwards. This resulted in a series of giant waves that caused devastation when they struck dry land.

Places as far apart as Sri Lanka, Thailand and Somalia were hit by the disaster. More than 220,000 people were killed and millions were left homeless. Worst affected was the Indonesian province of Aceh, where more than two-thirds of the deaths occurred. The waves reached Aceh with heights of 20m and destroyed the city of Banda Aceh in just 15 minutes.

The westbound series of waves killed more than 31,000 people in Sri Lanka and 10,000 people in India. The wall of water stripped away fishing boats, crops, trees and houses. Thousands of people lost their jobs in the fishing and tourism sectors. Today, tens of thousands are still living in tents or shelters.

According to the World Conservation Union (IUCN), the tsunami revealed the consequences of severe degradation of coastal ecosystems over the last decades: where mangrove forests had been lost, the killer waves did their worst. In the southern Sri Lankan village of Kapuhenwala, which was surrounded by 200 hectares of dense mangroves and scrub forest, the tsunami killed only two people. In contrast, nearly 6,000 people died in the nearby village of Wanduruppa, where mangrove forests had largely been wiped out.

Almost 40% of the world's mangroves are concentrated in Asia. However, a quarter of this natural protective barrier has been destroyed since the 1980s due to development and the rapid growth of shrimp and fish farms. It is, therefore, these largely unprotected coastlines that were badly hit by the tsunami.

Initial efforts by aid organisations to replant mangroves faltered because many lacked the technical expertise. But in 2005, a coalition of environmental groups teamed up with governments across the Indian Ocean basin to launch programmes aimed at rebuilding coastal ecosystems and reviving local economies. The IUCN, for example, launched "Mangroves for the Future", a US$45 million programme that aims to build natural barriers of mangroves in twelve countries in Asia and Africa.

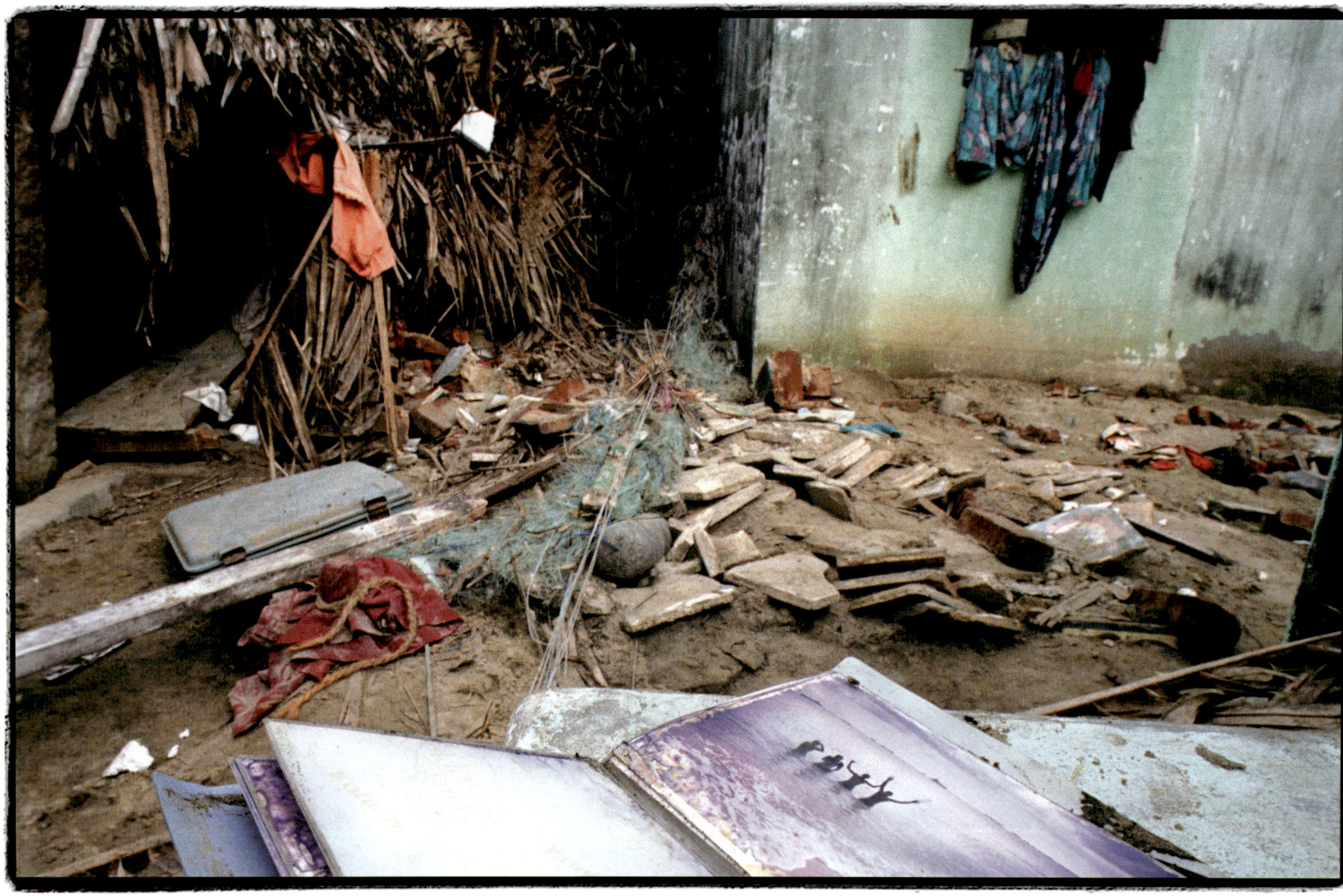

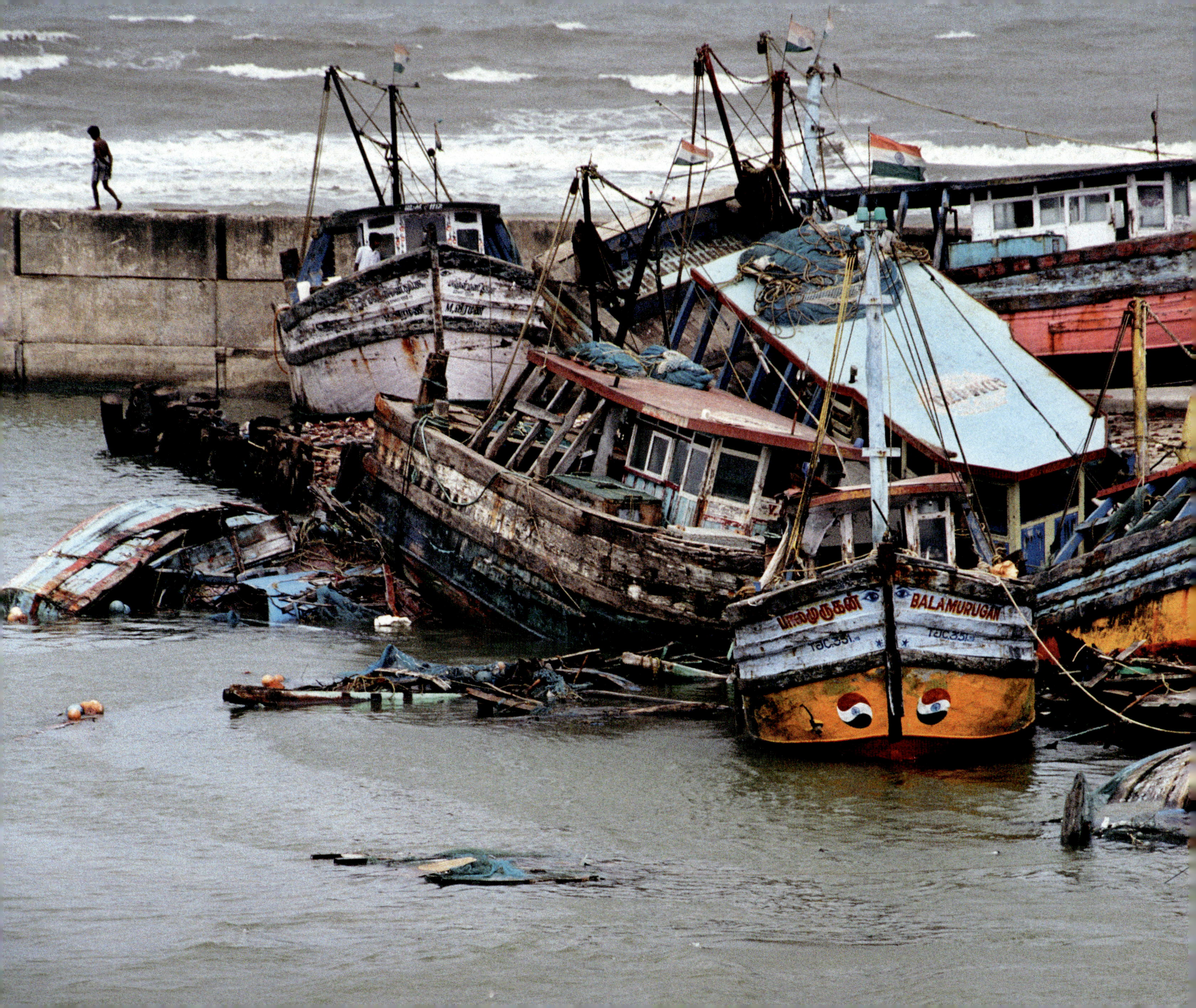
BALAMURUGAN

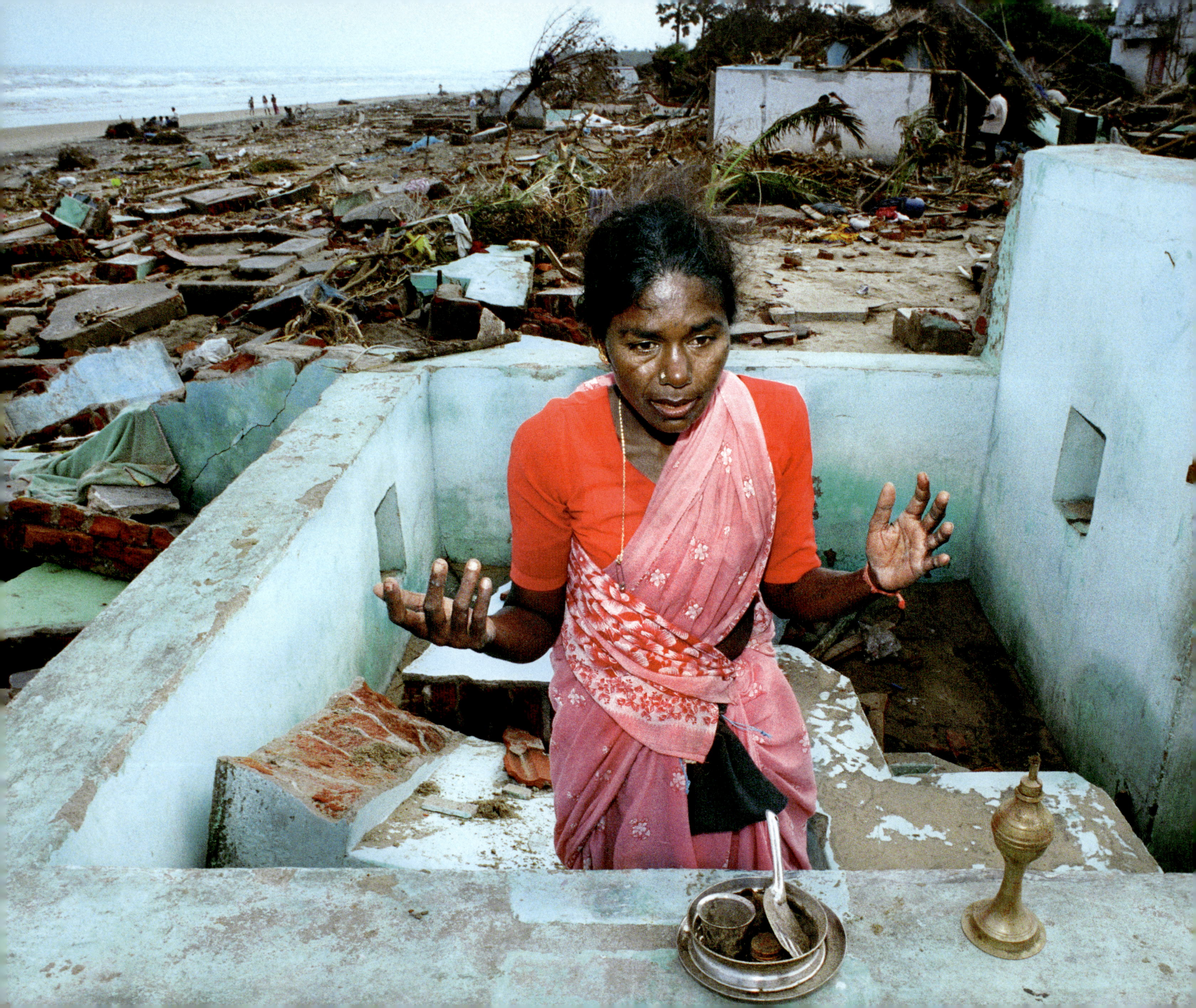

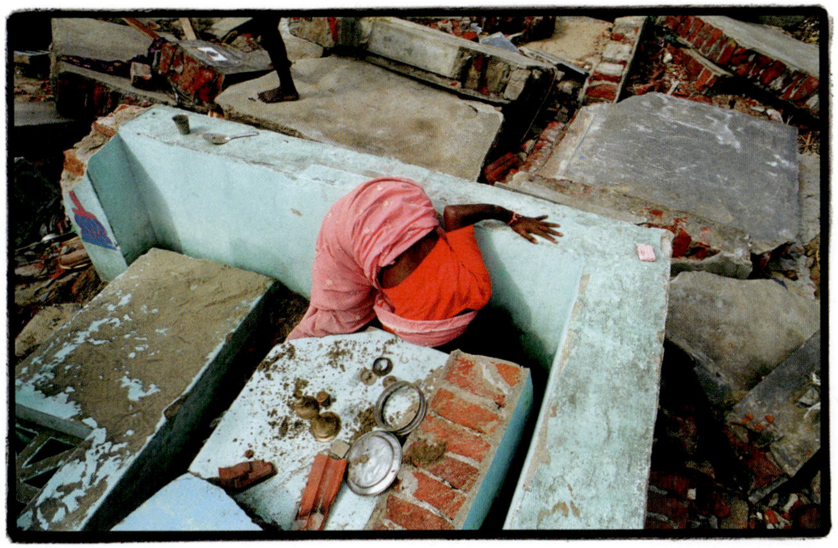

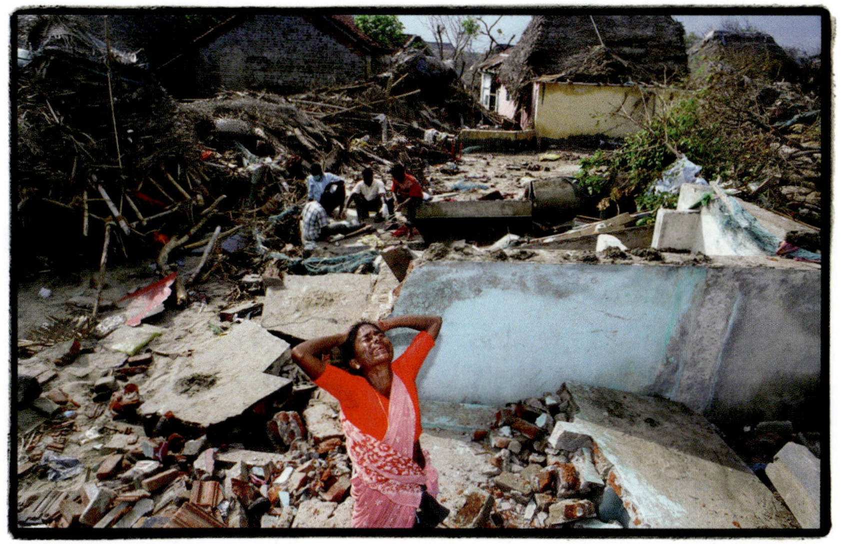

Tamil Nadu, Nagapattinam. A woman recovers what few possessions remain in the ruins of her home following the tsunami.

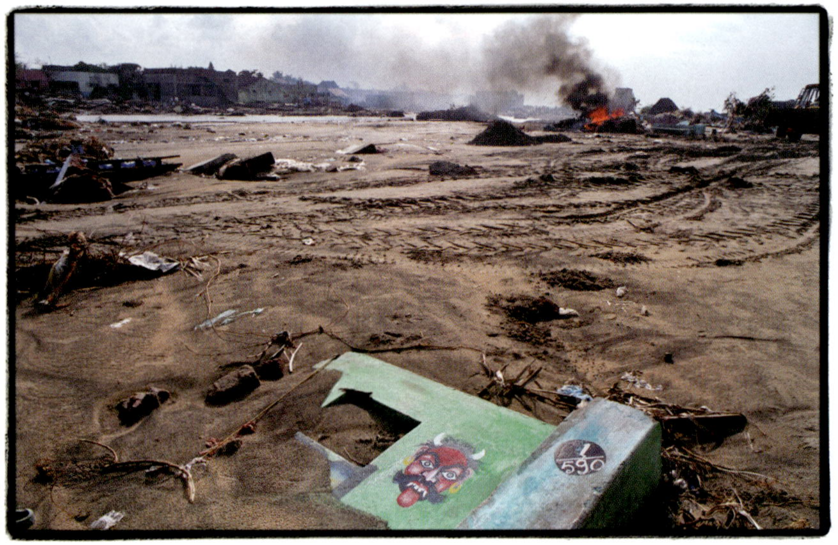

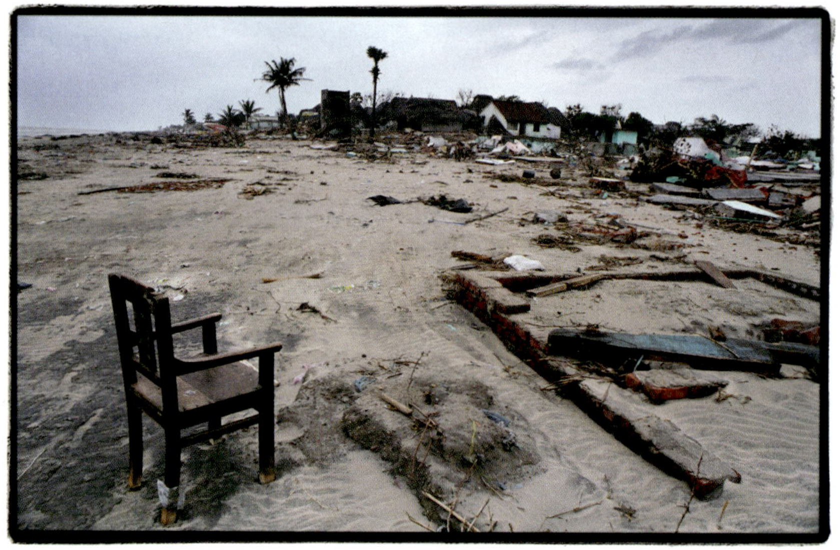

Tamil Nadu, Nagapattinam. In this district alone, 73 coastal villages were completely destroyed, causing massive loss of lives and property.

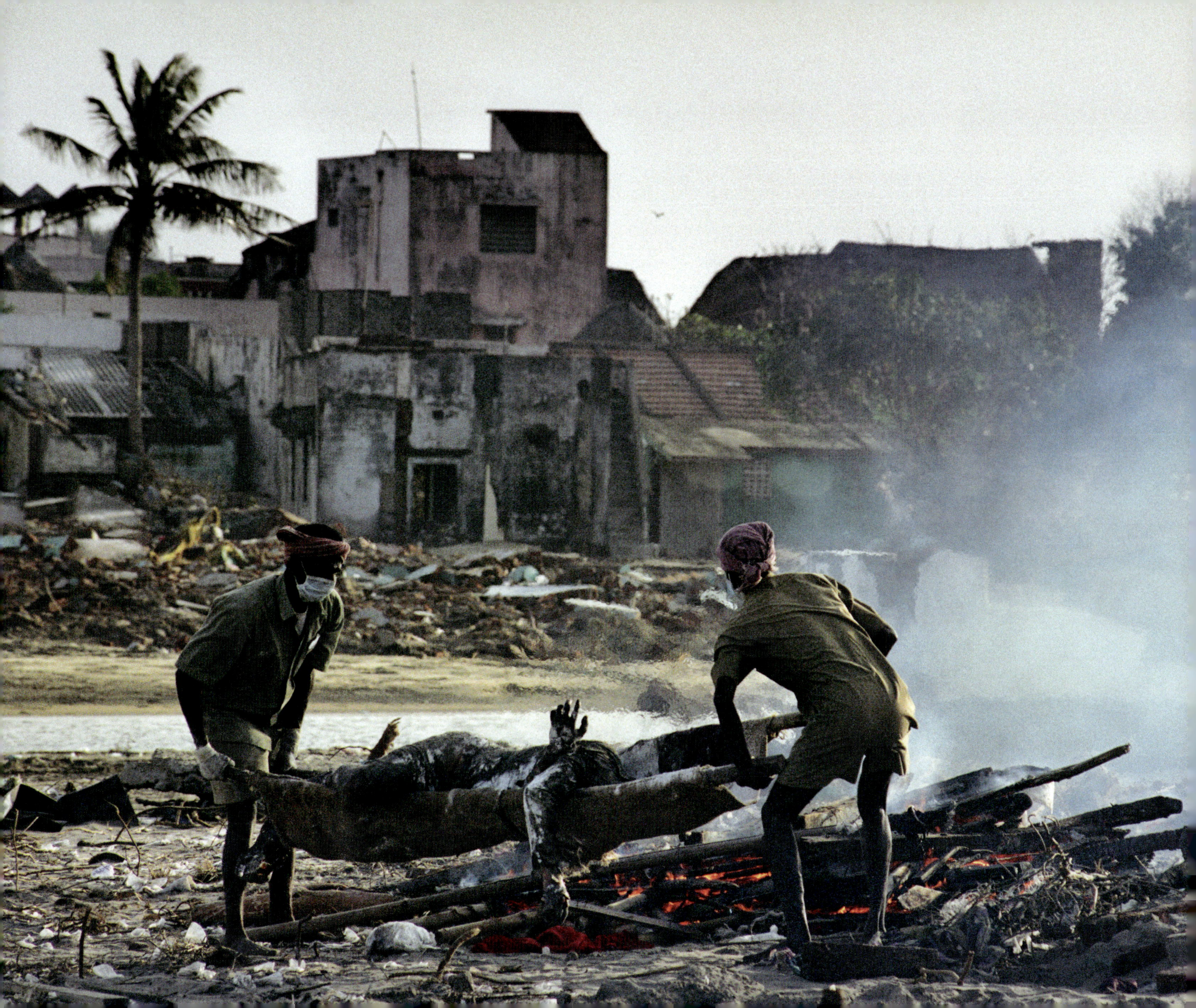

> Millennium Development Goal: Halve, by 2015, the proportion of people without sustainable access to safe drinking water

> invariably, those who lack adequate and affordable water supplies are the poorest in society

> transport of water is one of the most arduous and time-consuming tasks of rural women in Africa

> some women are forced to spend four to six hours a day on water collection

> women and girls suffer most from water shortage

> distances of ten miles or more to fetch water are not unusual

> the urban-rural disparities are most obvious in sub-Saharan Africa, where less than half of the rural population has access to safe water

> in Africa, 90% of the work of gathering water and wood is done by women

3 LACK OF OPPORTUNITIES

> a September 2006 report from Unicef concludes that currently available resources will not suffice to reach MDG 7 on a global scale

> the water task prevents women from being decision-makers

> young girls carry up to 20 litres of water (20 kg)

> in sub-Saharan Africa, neither the sanitation nor the drinking water Millennium Development Goal targets will be met by 2015

> in Africa, only 24% of households have access to piped water

> millions of the world's poorest people survive on less than a fifth of the water they need, according to the UN

> 80% of people living in rural areas do not have access to essential sanitation facilities

> the provision of safe drinking water and proper sanitation has the biggest impact upon economic development

> clean water is getting to more and more people, but the number of people benefiting is outstripped by the growth in human numbers

> the real tragedy of the water crisis is the effect it has on the everyday lives of poor people

> the poverty of a large percentage of the world's population is both a symptom and a cause of the water crisis

> providing the poor with improved access to better water can make a big contribution to poverty eradication

> the Millennium Development Goals cannot be achieved without adequate access to fundamental resources such as water and energy

> pressure on the inland water system is increasing with population growth

> factors such as illness and daily water transport keep girls in particular out of school and uneducated, perpetuating the cycle of poverty

> in India, 69% of people in non-irrigated areas are poor, and in irrigated areas this figure falls to 26%

> 777 million people in developing countries are undernourished, and the target of halving this will not be met before 2030

> to meet water supply and sanitation Millennium Development Goal targets, 342,000 people need to receive provisions every day until 2015

> the Jordanians survive on a per capita water availability that is far below the minimum specified as absolute water scarcity

> a child born in the developed world consumes 30 to 50 times as much water as one in the developing world

> during dry periods, women have to walk longer distances for even smaller quantities of water of lower quality

> Kenya's poverty reduction programme is committed to providing water and sanitation to the poor within a 'reasonable distance': i.e. 2 km

> 33% of all children in Mali are underweight

> Asia shows the highest number of people without either water supply or sanitation, yet proportionally this group is even larger in Africa

> in sub-Saharan Africa an estimated 40 billion hours are spent collecting water each year

> US$100 billion is spent every year on bottled water: only a fraction of that amount is needed to provide clean drinking water worldwide

> more water is lost through leaking taps in the rich countries, than is needed by the one billion poorest on this planet

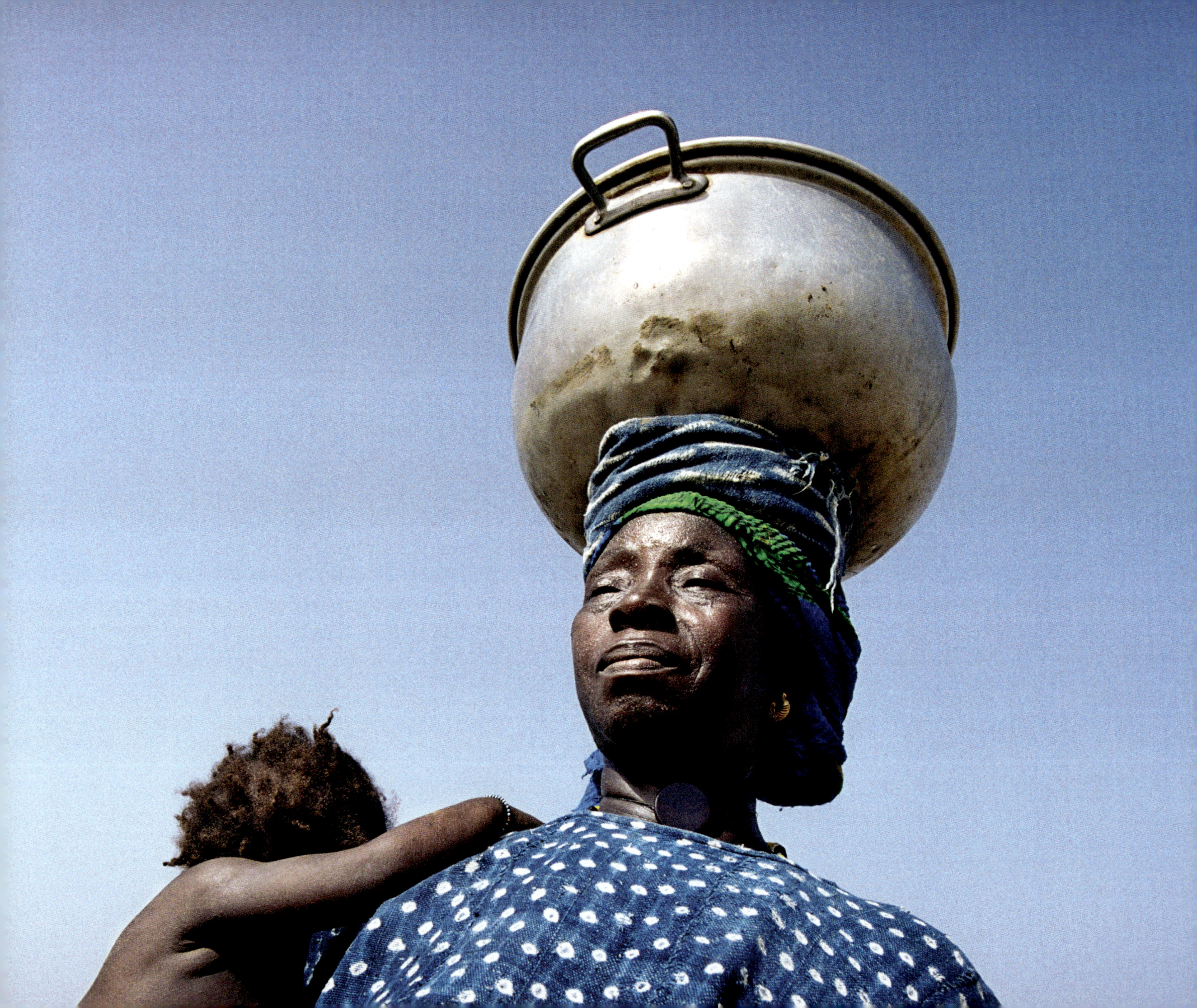

Mali
Water transport

A very large part of Mali –more than twice the size of France– is situated in the Sahara desert. It is expanding slowly but surely with desperate implications for a country where 80% of the workforce is employed in farming or fishing. The Niger River, which is Mali's lifeline, is drying up rapidly in the ancient city of Timbuktu.

Women and girls in particular suffer most from the water shortage. According to the tradition, they are in charge of collecting the family's water. Every morning they get up at the crack of dawn to walk to the nearest water hole. Distances of ten miles or more to fetch the water are not unusual. In the dry season, this distance can even double.

In the Dogon valley the water level can drop 80 meters. The situation then becomes catastrophic. Most villages have only one well that is deep enough, and sometimes several villages share the same well. In addition to the six hours or so the women need for the water transport, they also have to queue for the water, which may add another five hours to the journey. To avoid the queue, many women get up in the middle of the night to get to the well in time.

The amount of energy women spend on the water task is huge. Their water containers usually hold about 20 litres of water (20 kg). Carrying such heavy weights day after day has severe health implications. Pulling the water out of the well is not an easy task either, as indicate the deep scars left by the ropes in the wooden beams.

This gruelling task and journey prevent women and girls in developing countries from having an education or from being decision-makers in their communities.

Providing clean and accessible water for women everywhere not only ends needless drudgery and indignity, but improves their health and that of their families. Women's time is freed up for education, childcare, agriculture or other income-generating work, or simply for relaxation.

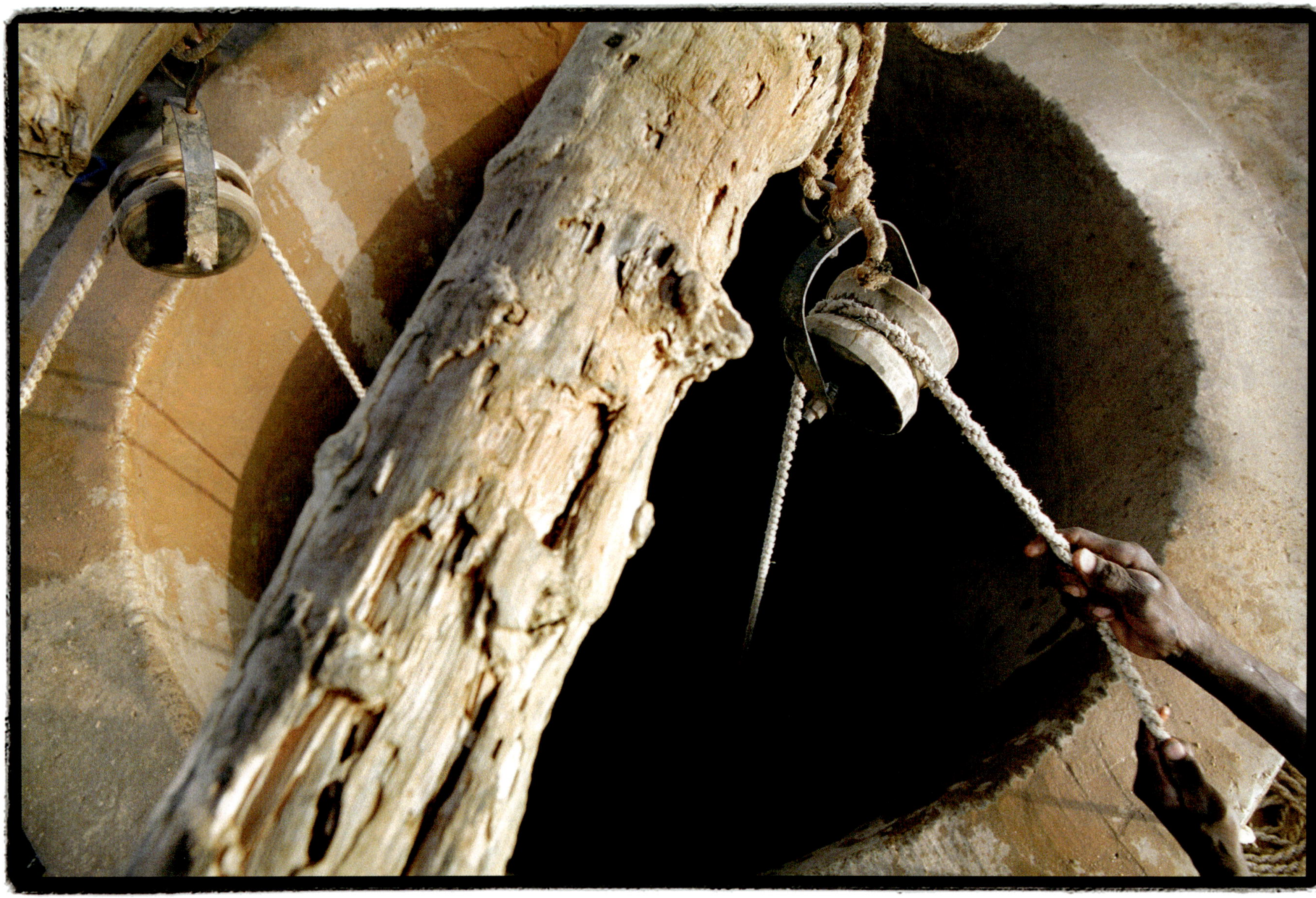

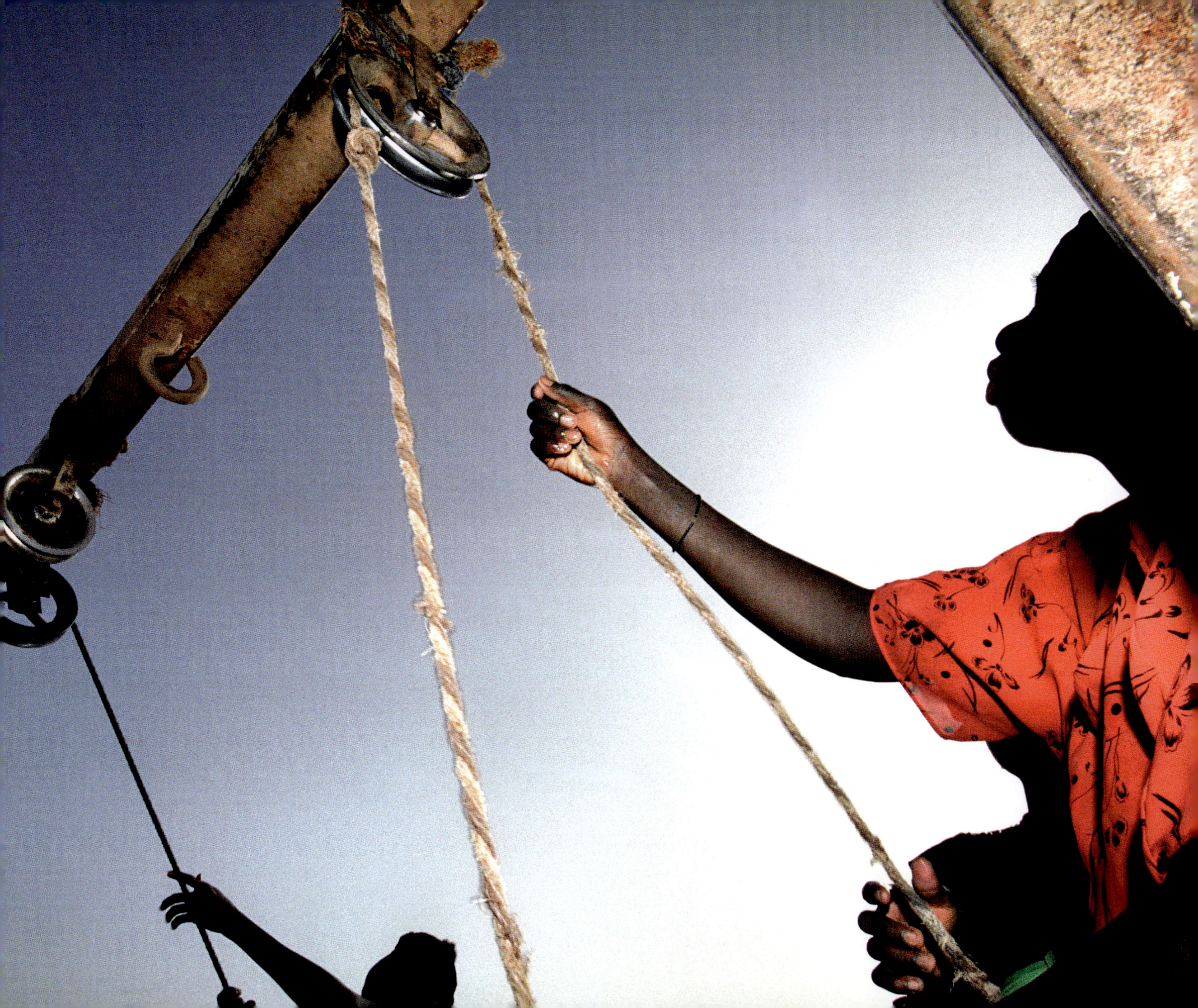

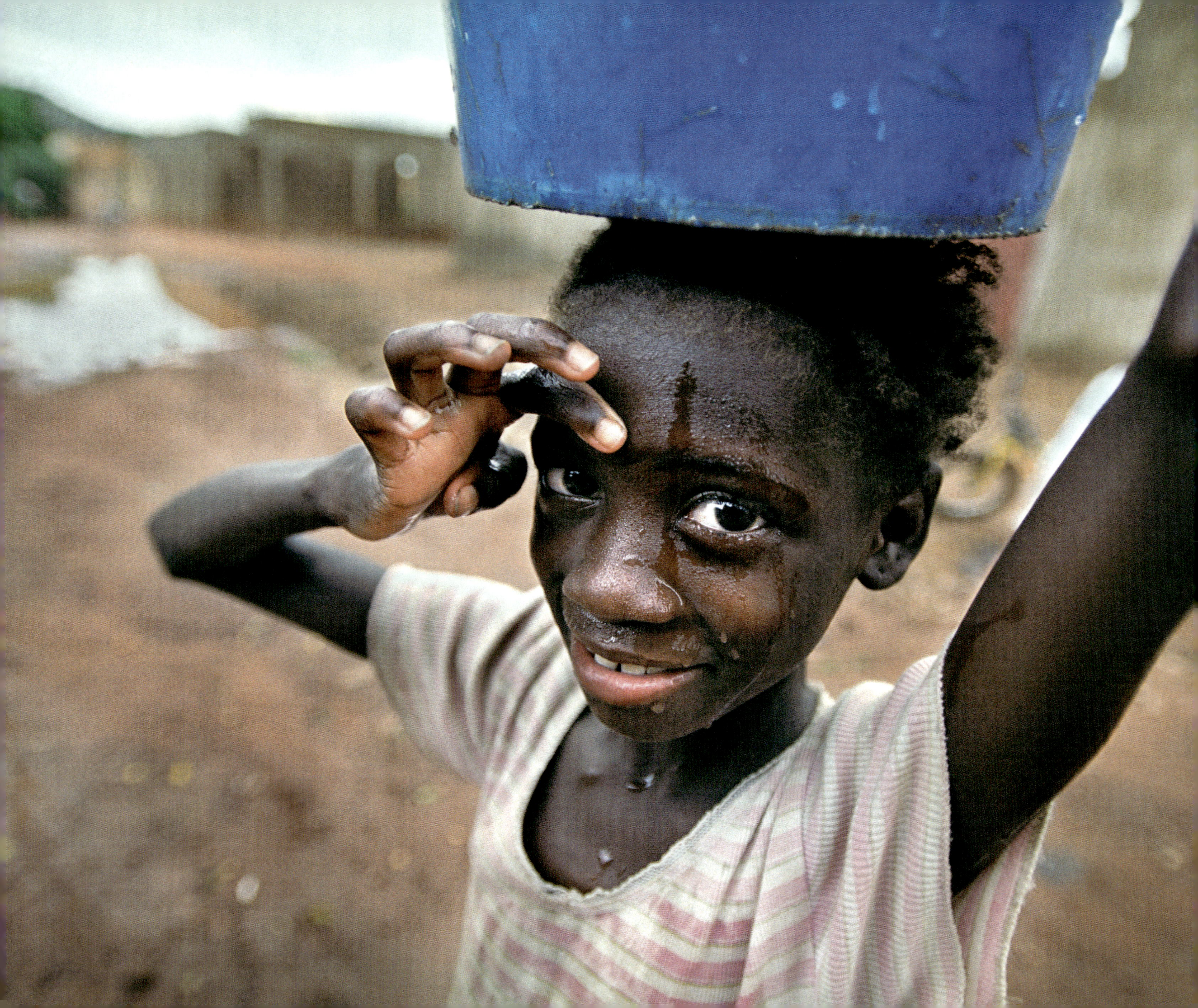

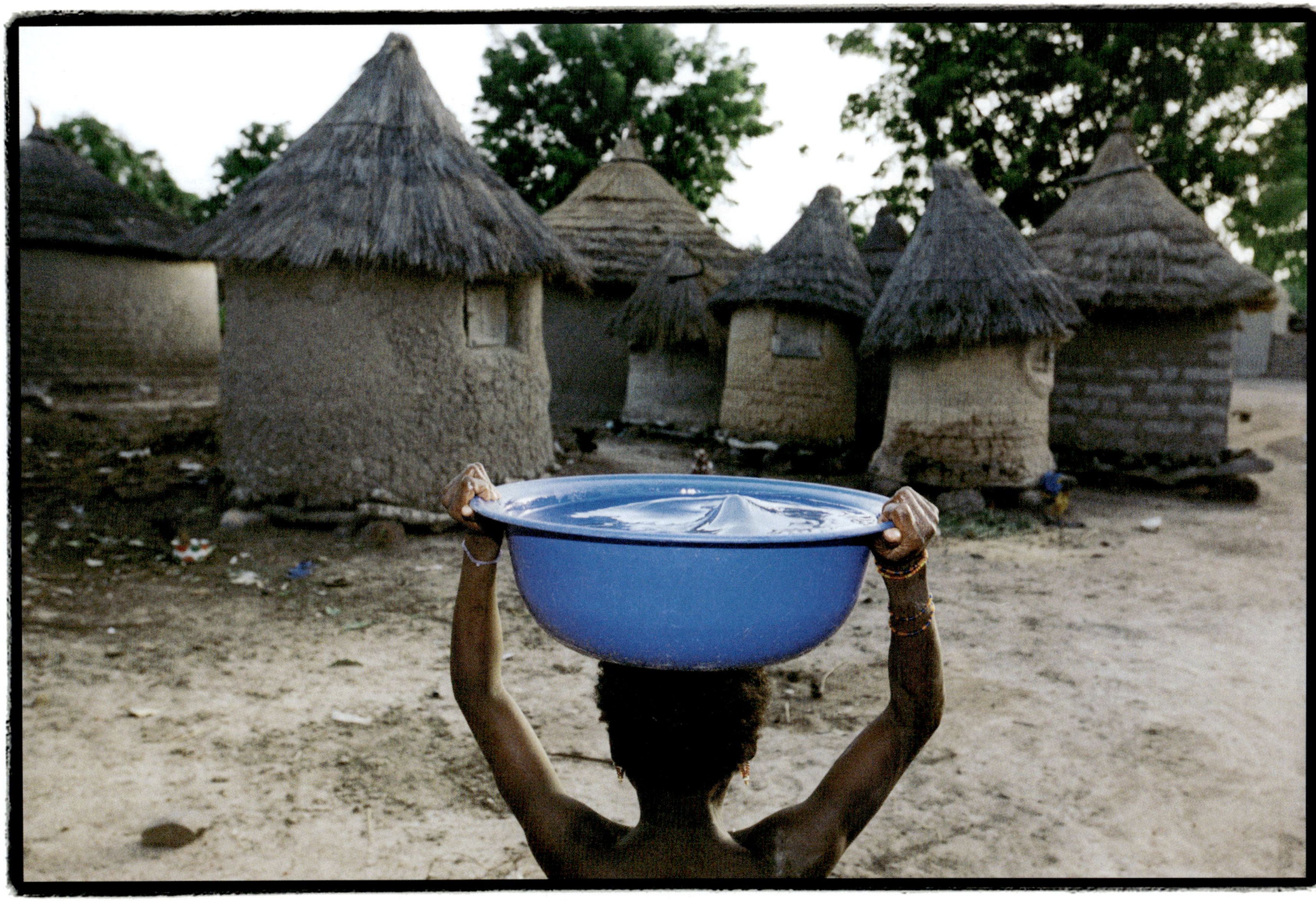

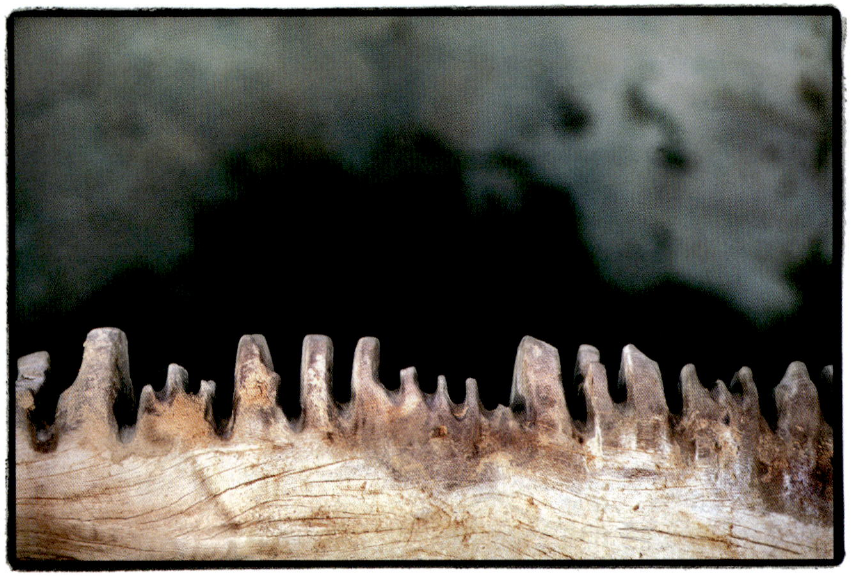

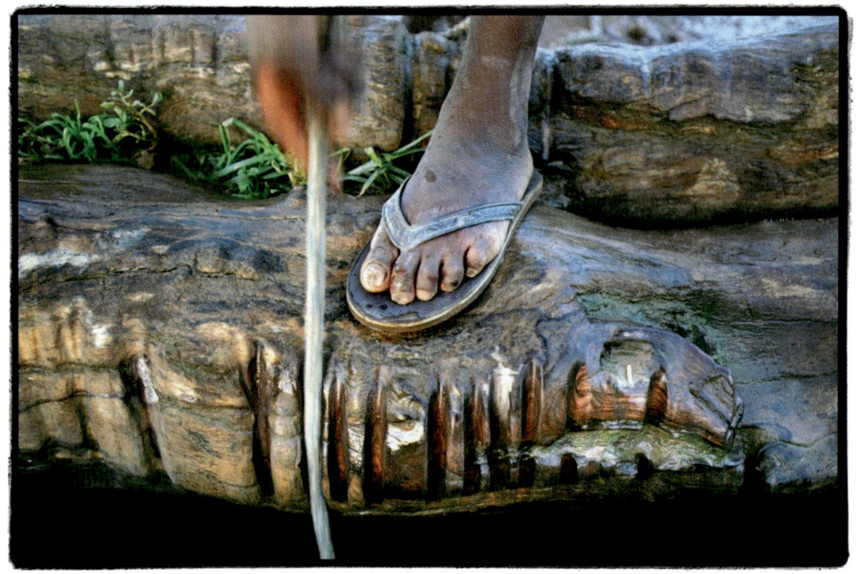

LEFT and MIDDLE Dogon Valley. The deeper traditional wells are fortified with heavy wooden beams. The beams at the top of the wells have deep grooves in them caused by the friction of the bucket ropes.

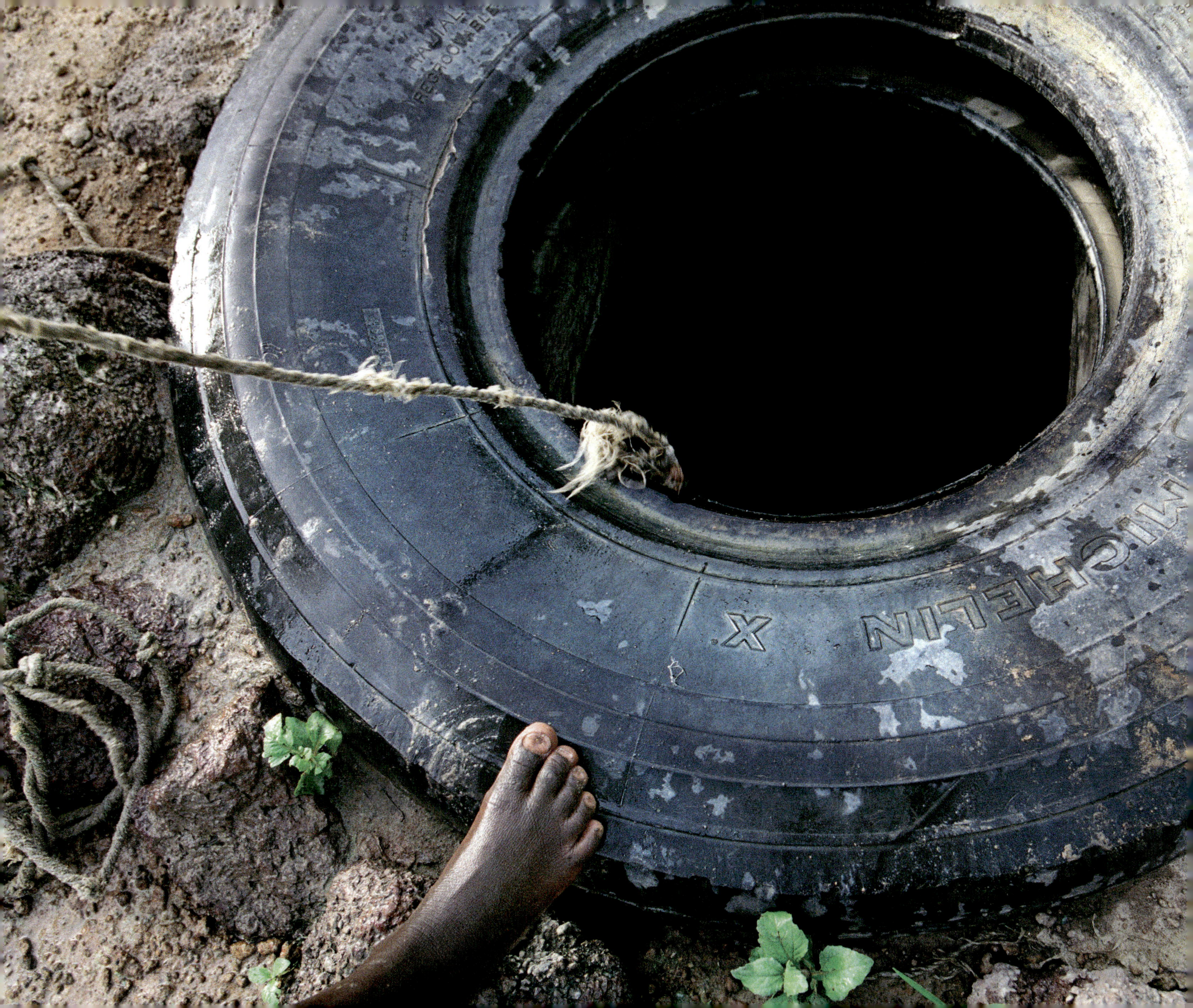
MICHELIN X

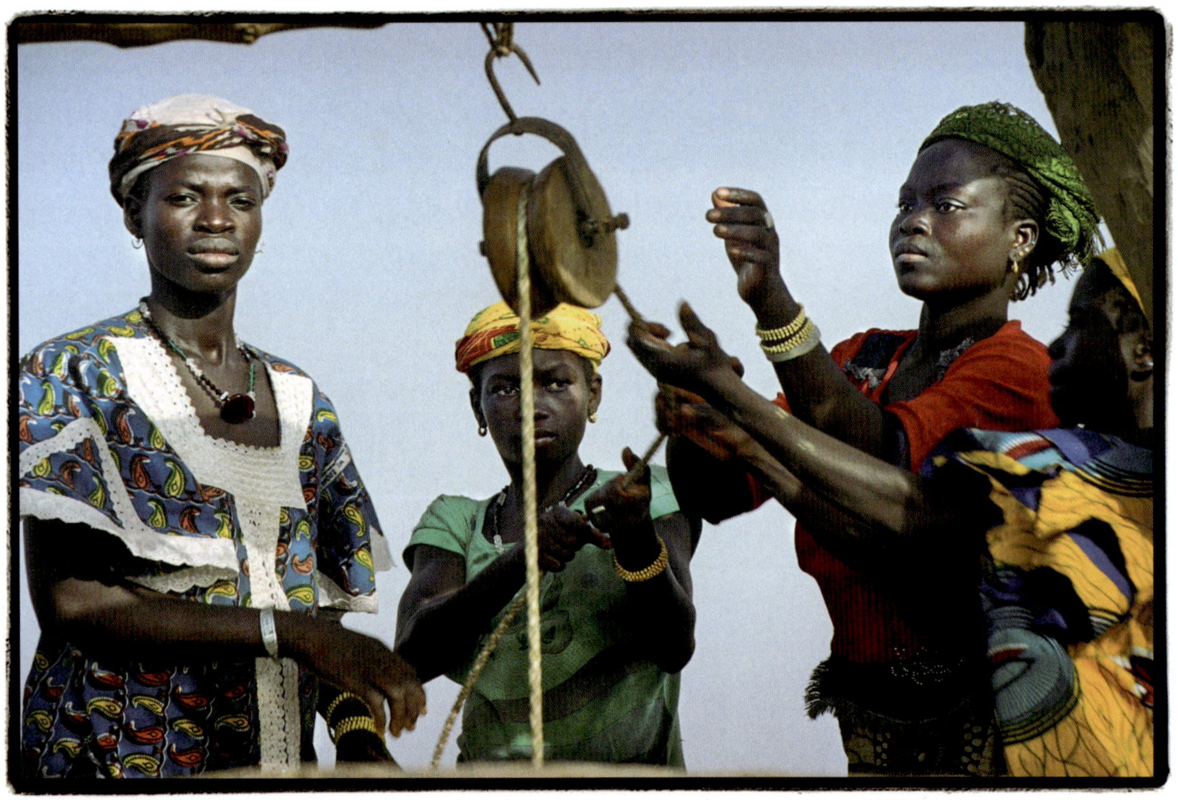

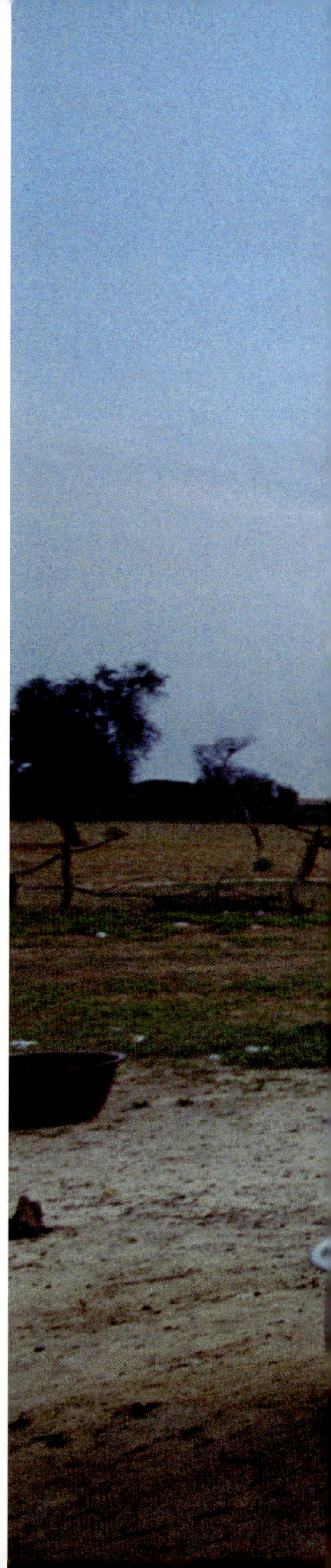

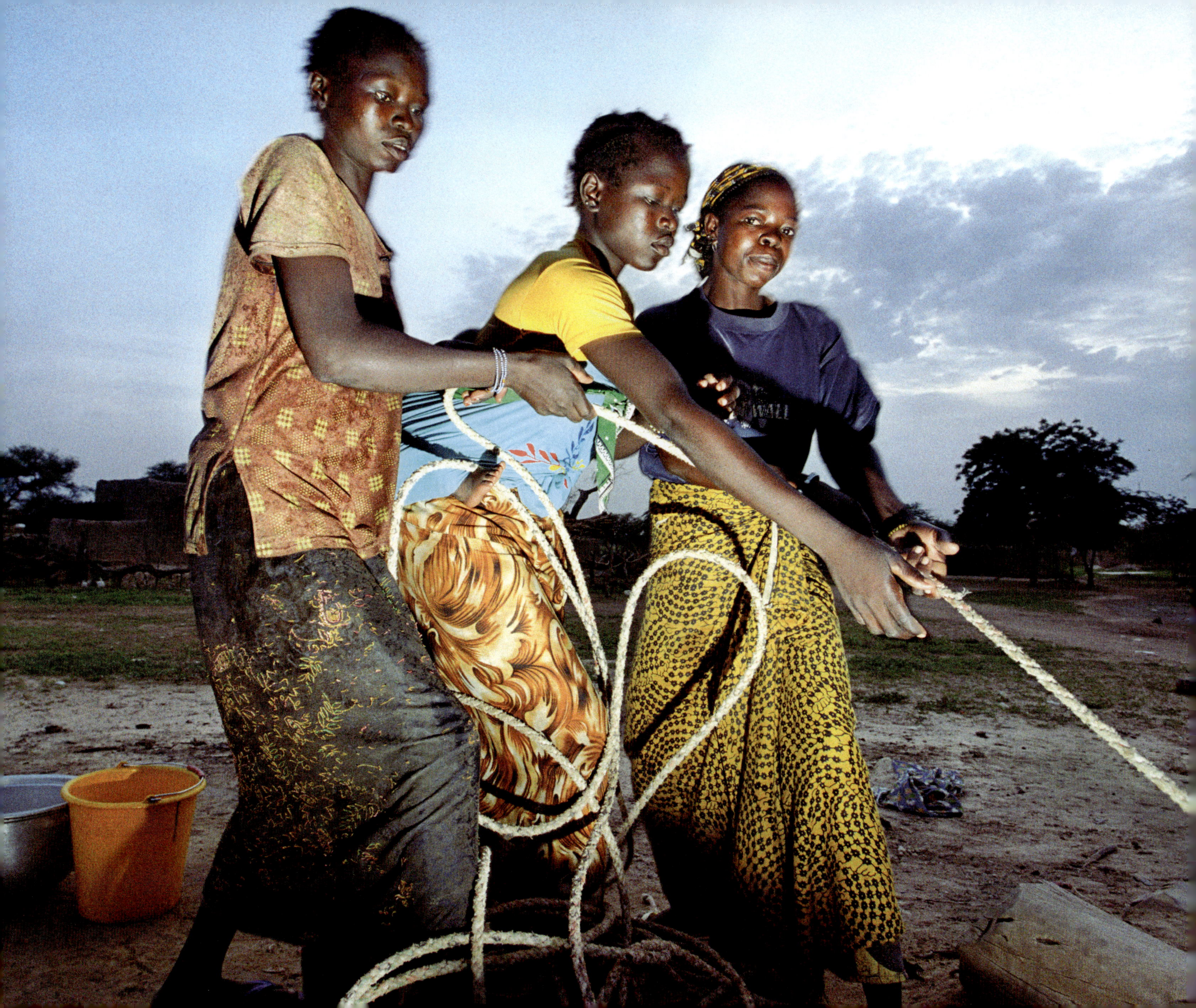

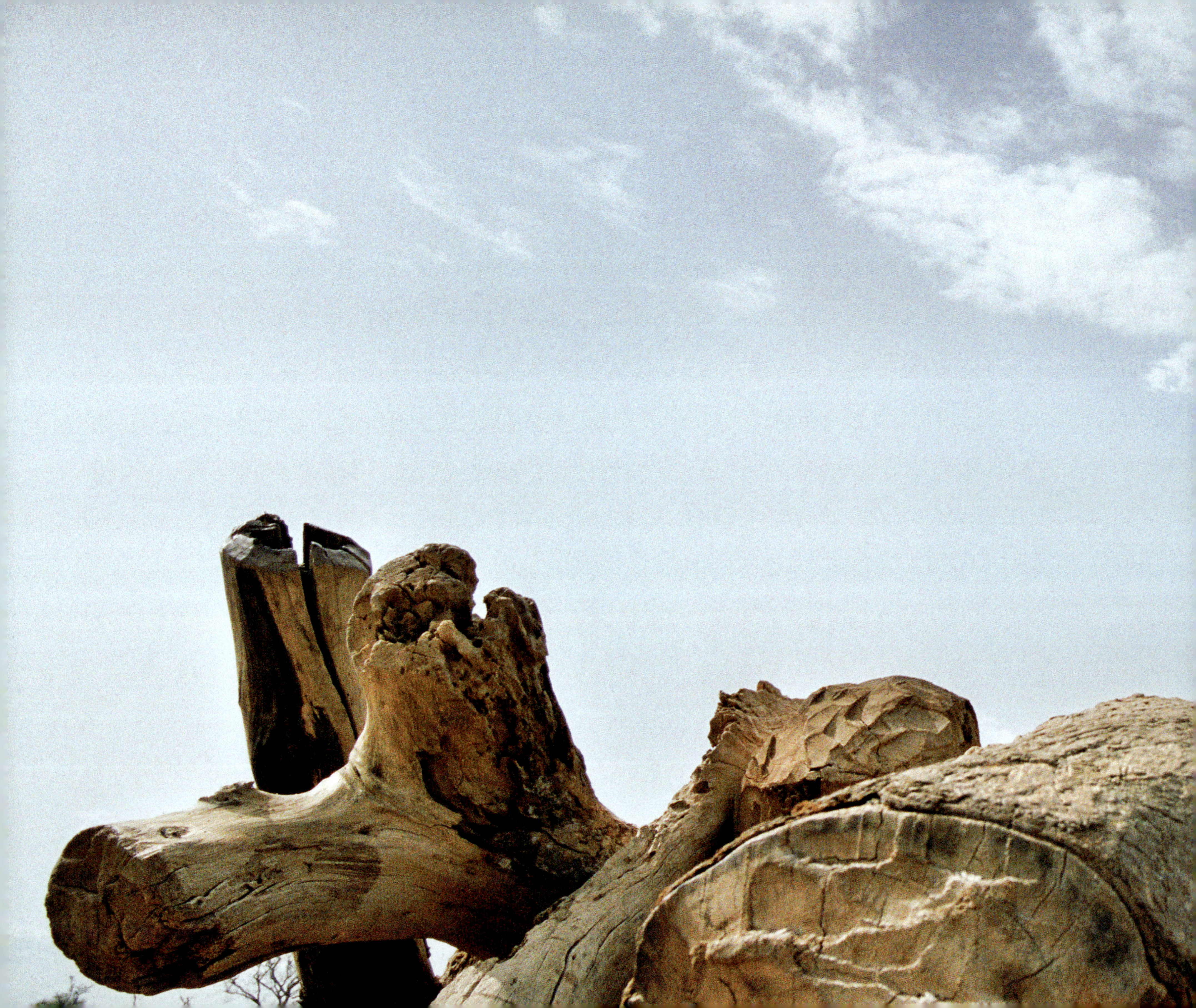

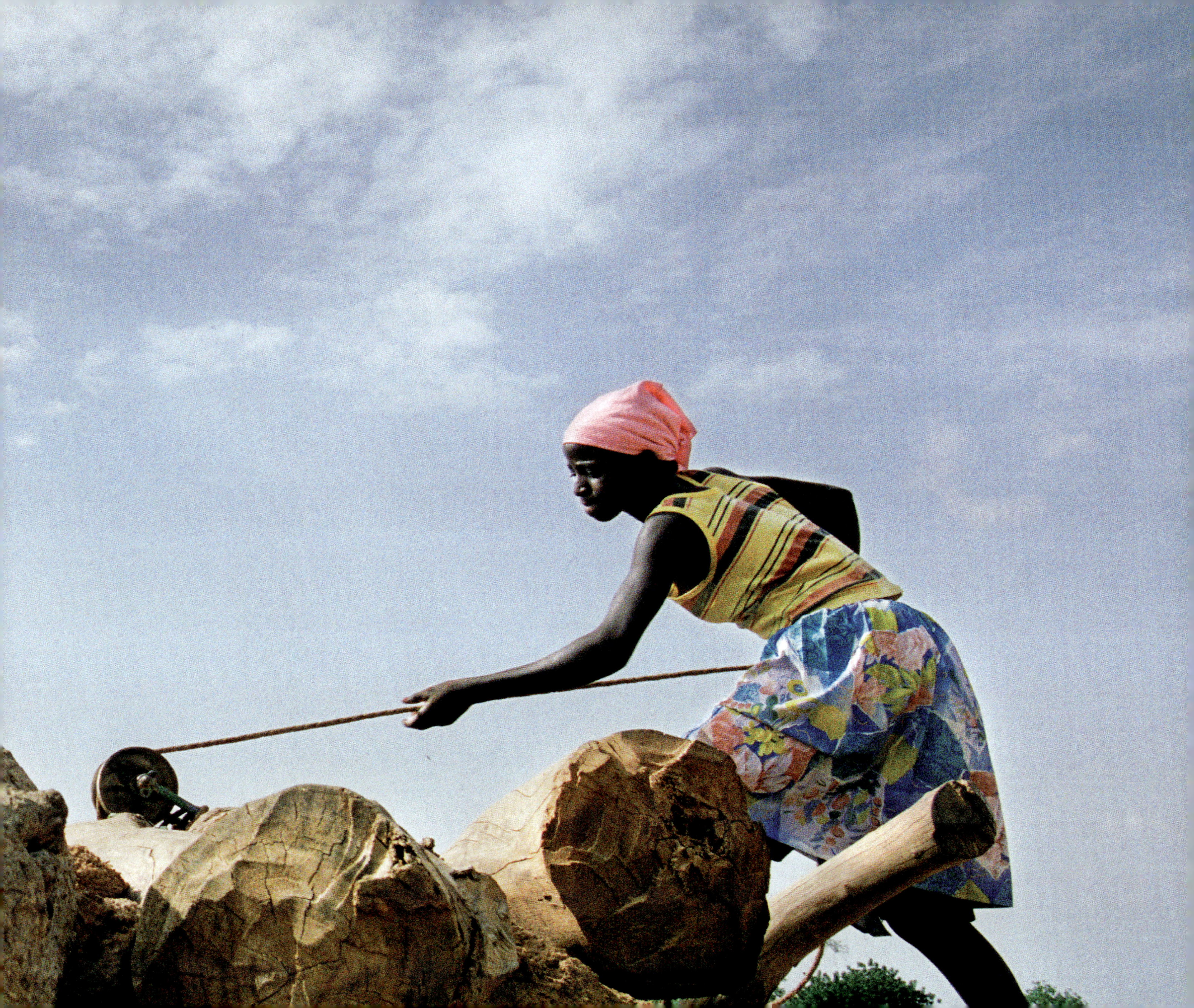

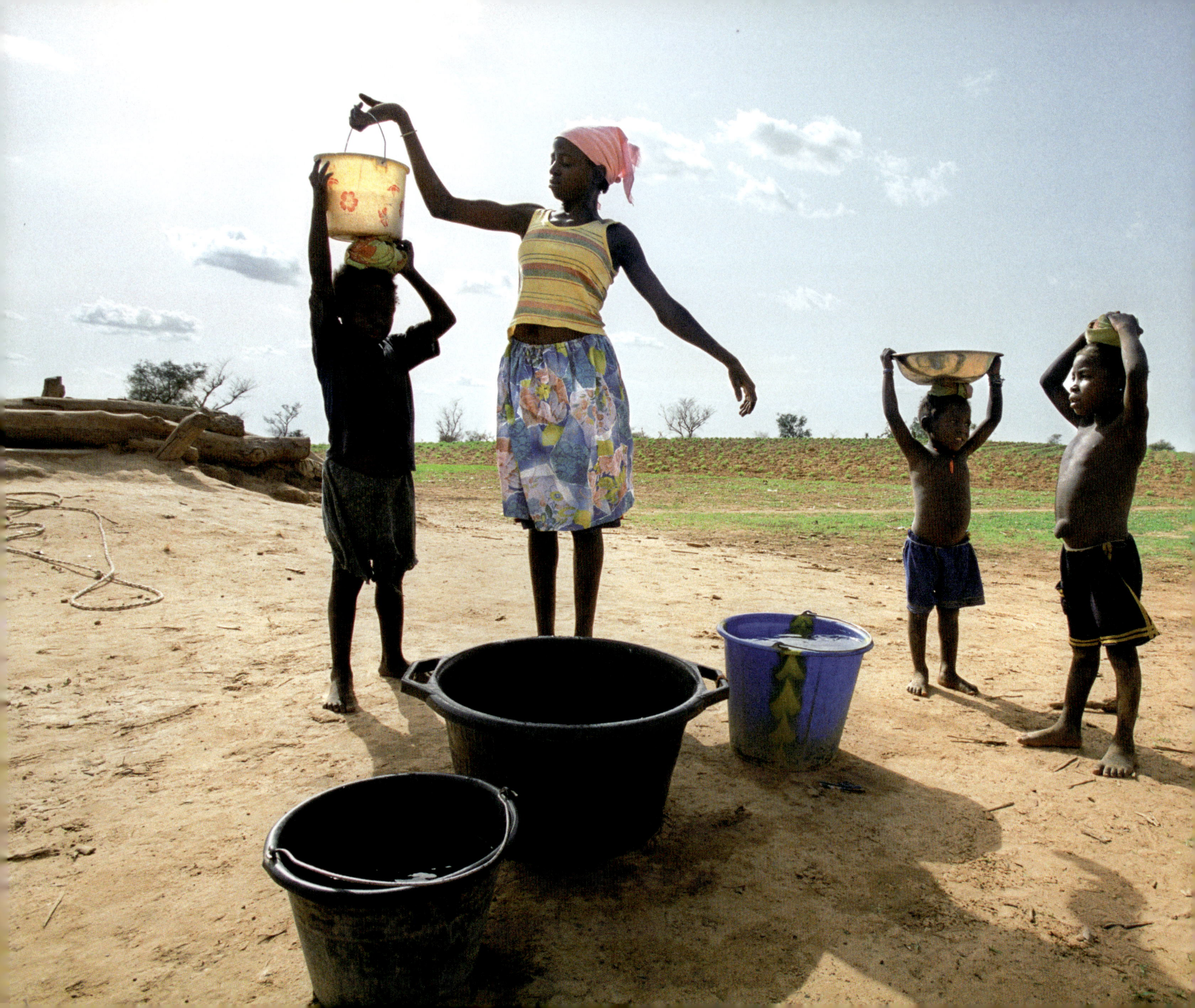

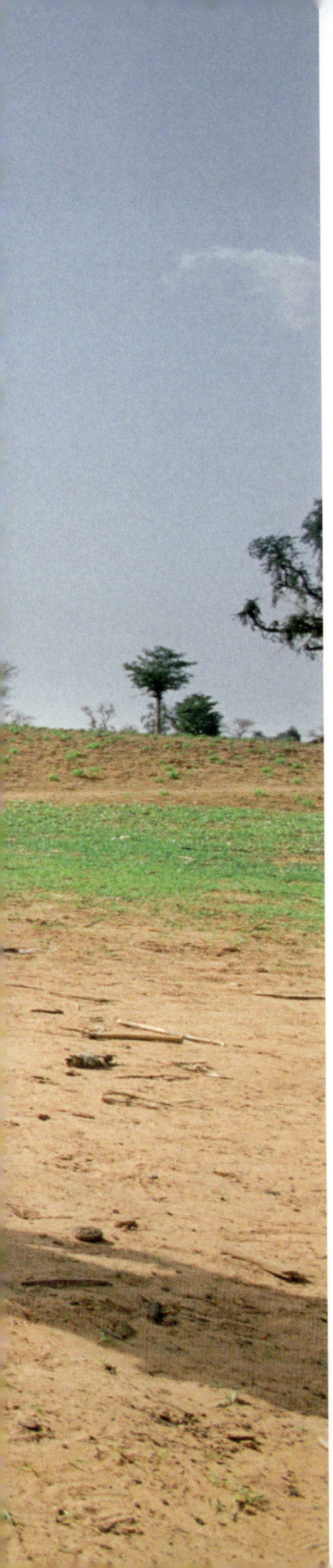

LEFT Dogon Valley. 13 year-old Marietou and her younger sisters and brother are collecting water. Marietou goes to the well five or six times a day. Her sisters, the youngest barely six years old, carry small yet symbolic amounts of water home, in order to practise what will one day become part of their daily routine.

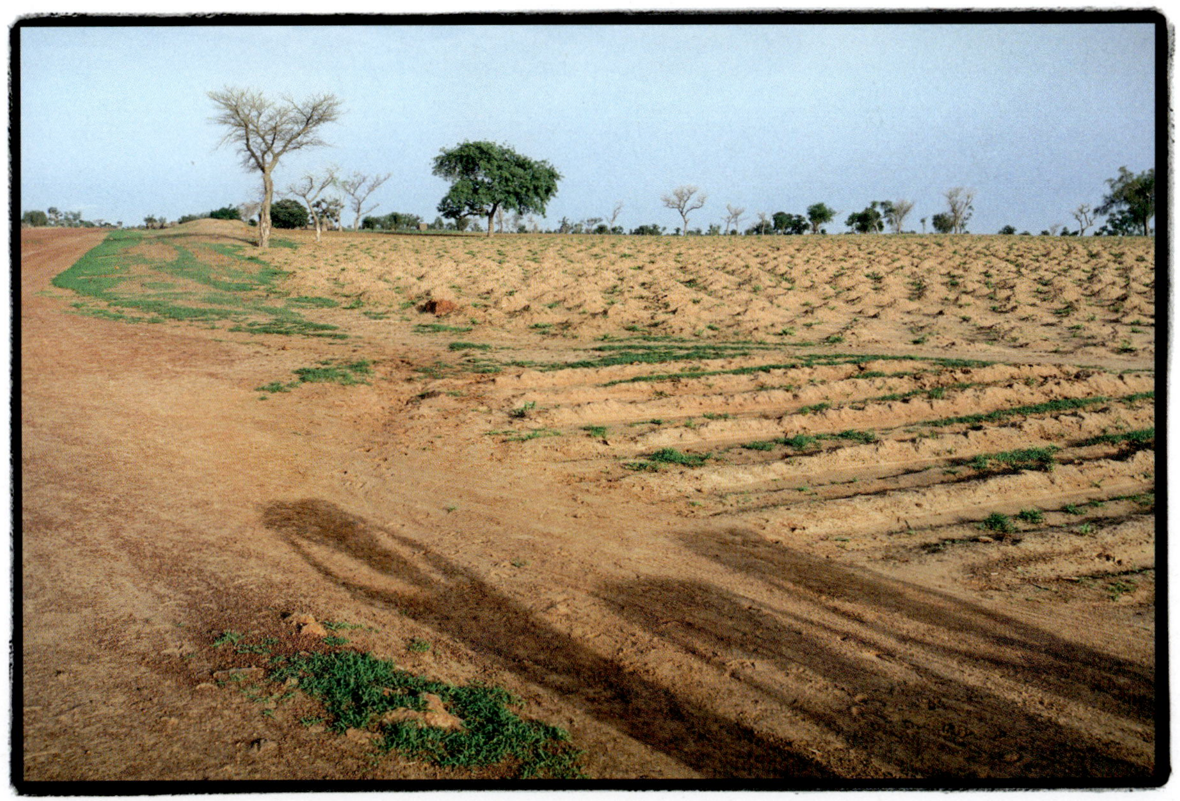

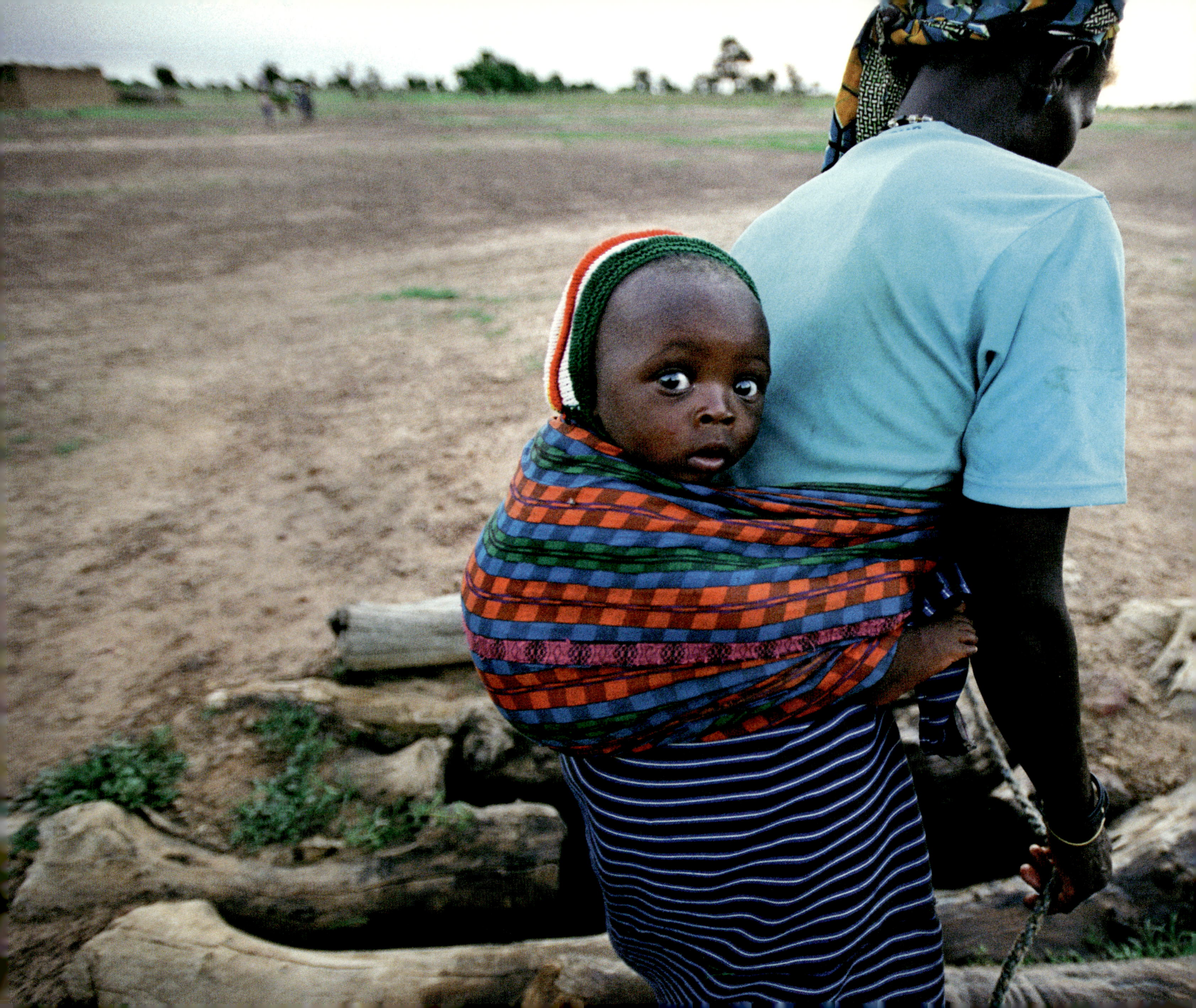

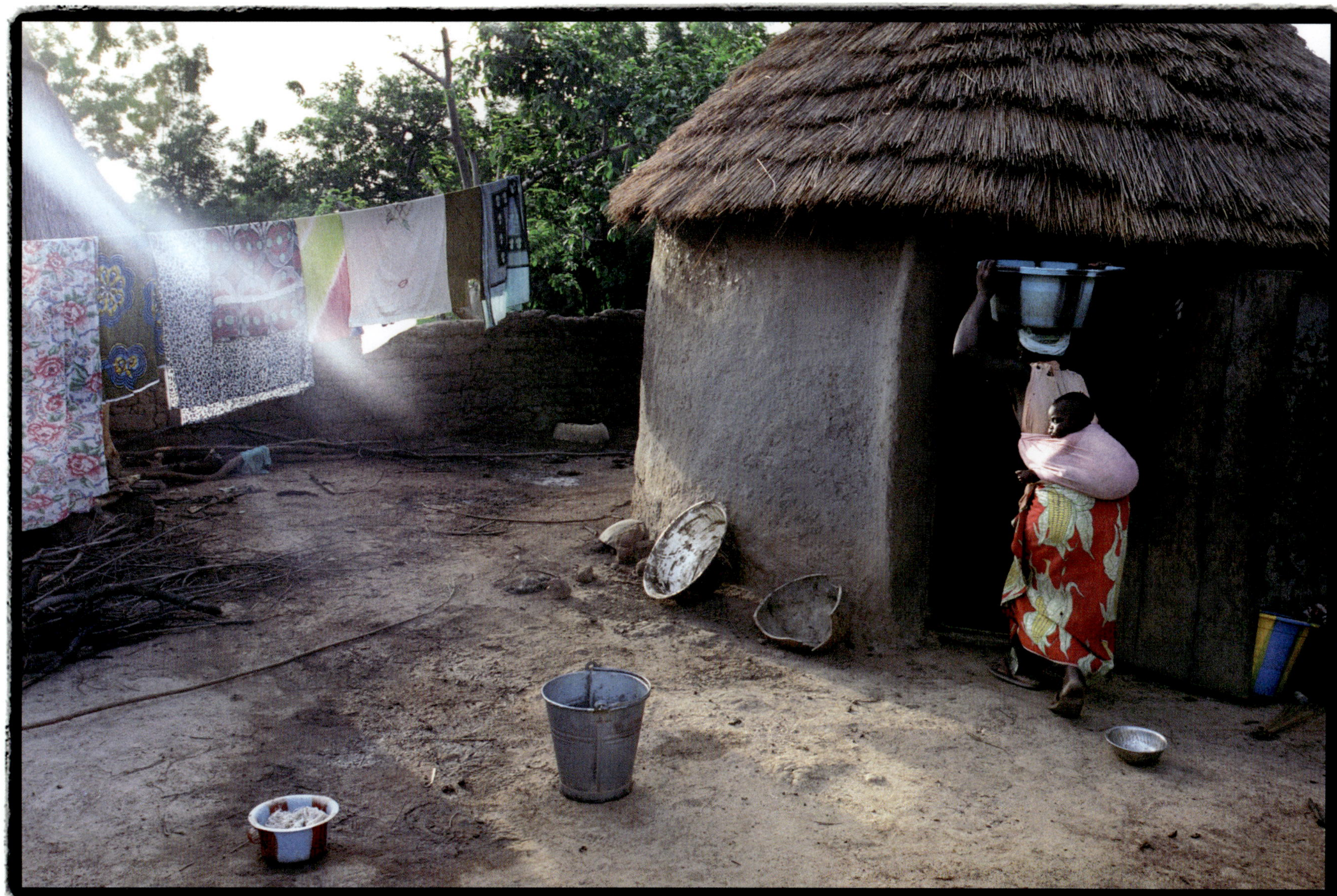

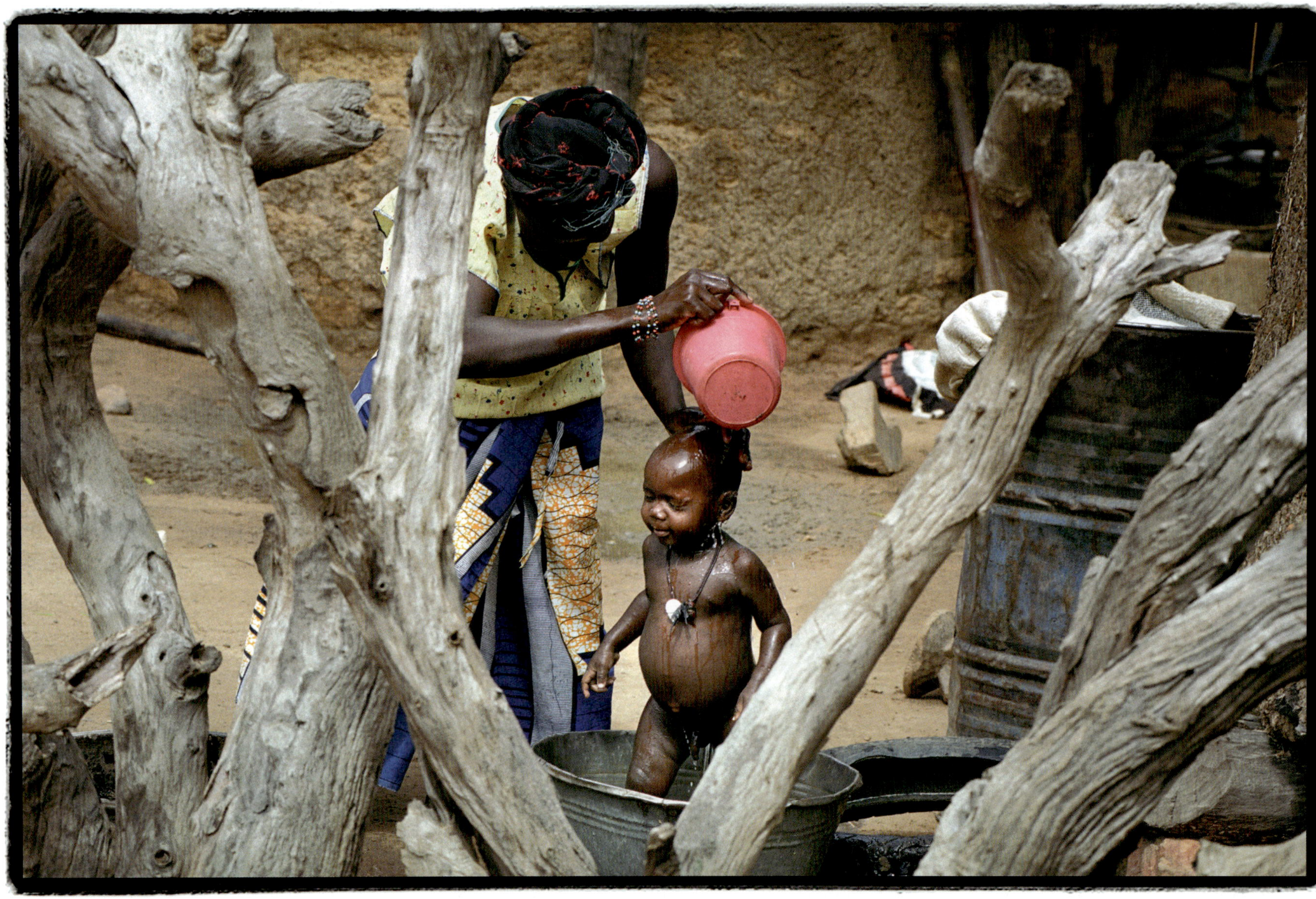

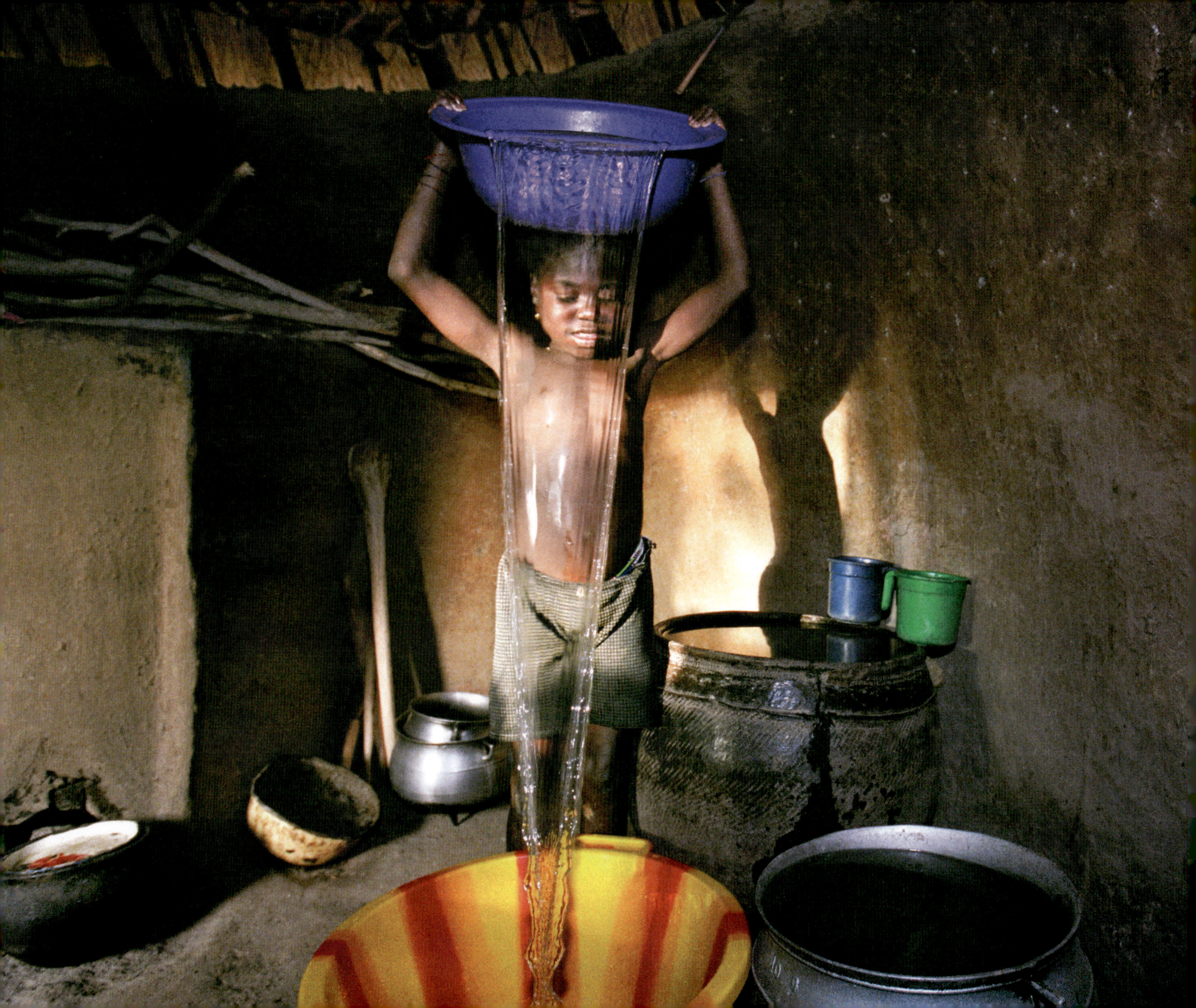

> water-related diseases are affecting mainly the poor in developing countries

> about 2 million people die every year due to diarrhoeal diseases, and most of them are children less than 5 years of age

> Southeast Asia accounts for 62% of the world's burden of dengue

> the state of human health is inextricably linked to a range of water-related conditions

> only 18% of people in Asia and 13% in Africa have access to sanitation linked to a sewage system

> safe drinking water and proper sanitation can reduce the mortality rates of water-related diseases by up to 80%

> on present trends the UN says we shall probably number about 8.9 billion by 2050

> city and urban populations are growing by an estimated 60 million people per year: over 1 million every week

4 HEALTH / SANITATION URBANISATION

> population growth –coupled with industrialisation and urbanisation– will result in an increasing demand for water

> over two billion people lack access to adequate sanitation

> over 50% of the population of developing countries are exposed to polluted water sources

> water-related diseases are the leading cause of death in the world

> a child dies every 15 seconds from drinking contaminated water

> in India, over one billion people do not have access to safe water

> in Third World countries up to 80% of hospital beds are occupied by patients who are ill from drinking contaminated water

> crowded cities are attracting more people, leading to increased water use and pollution: it is a vicious circle

> one out of 10 children born among the urban poor in India is not destined to see their fifth birthday

> dirty water kills 5,000 children a day

> two-thirds of sub-Saharan Africans do not have proper toilets

> one in three urban dwellers (1 billion people globally) live in slum settings with little or no access to clean water, sanitation

> higher temperatures will increase the extent of malaria and dengue fever

> in Bangladesh, the surface water is so badly polluted that many farmers and cities rely on groundwater containing arsenic

> some 2 million tons of waste per day are disposed of in receiving waters

> 1 litre of wastewater pollutes 8 litres of freshwater

> many urban dwellers resort to open defecation or defecation in a bag or wrapping, which is then dumped

> water-related diseases such as cholera, diarrhoea and the guinea worm represent more than 80% of all illnesses in Mali

> about 2 million people die every year due to diarrhoeal diseases, and most of them are children less than 5 years of age

> Millennium Development Goal: Halve the proportion of people who do not have access to basic sanitation by 2015

> in the Kibera slum in Nairobi, 71,000 people make do with just 50 bathrooms and 110 toilets

> the UNDP estimates that nearly half of all people in developing countries are suffering from an illness caused by bad water or sanitation

> when 400 people are confined to share one toilet, basic standards of hygiene fall rapidly

> more than half of the world's major rivers are being seriously depleted and polluted, degrading and poisoning the surrounding ecosystems

> the number of slum dwellers living in informal settlements is expected to double to 2 billion in the next 30 years

> over 1 million people die from malaria every year

> a large proportion of slums do not exist in official records and therefore receive no government services

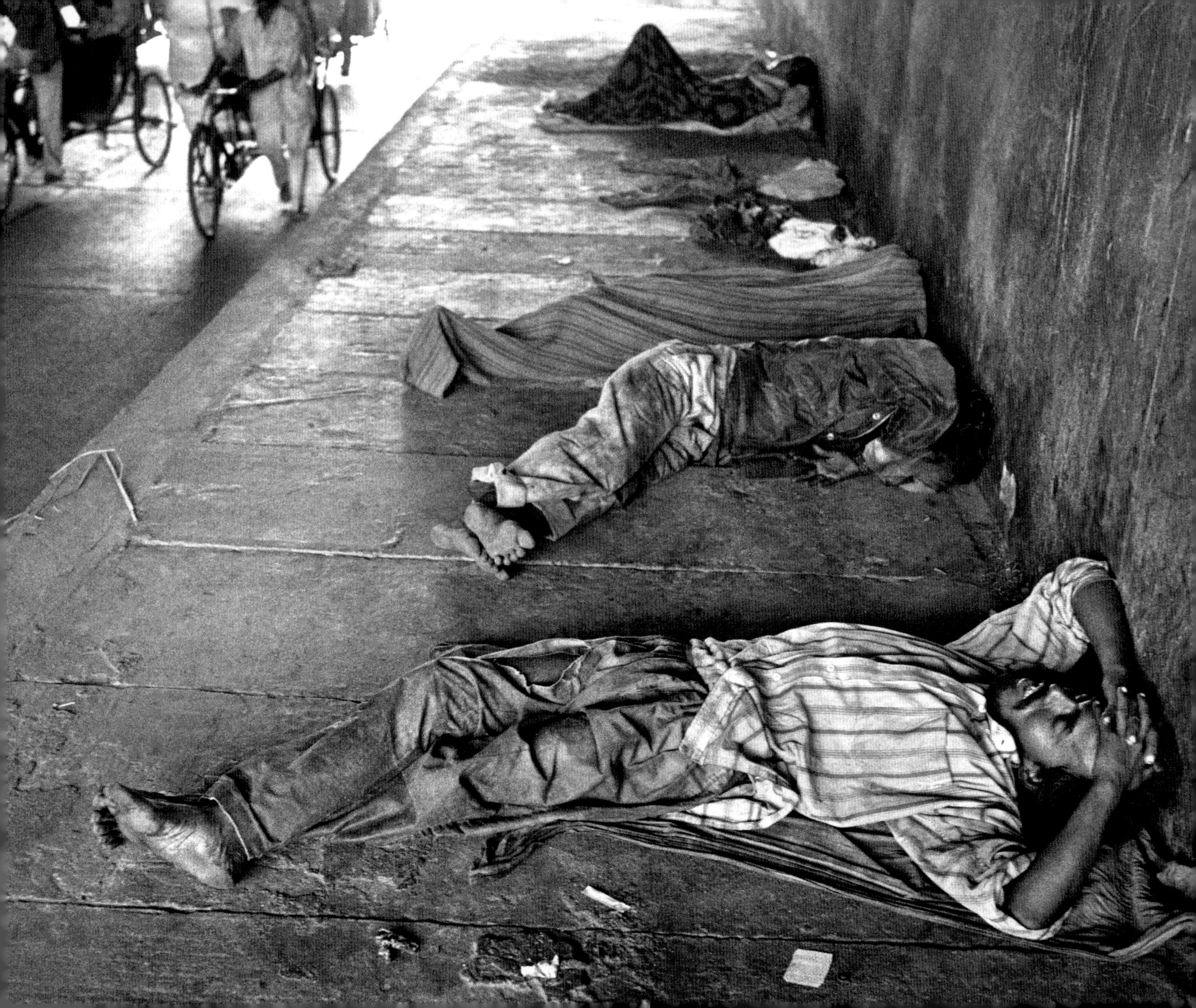

India
From rural poverty to urban misery

India's main water sources are under threat. Many farmers have free access to ground water and grow thirsty cash crops for export, such as cotton, instead of water-prudent food crops for local needs. This has resulted in over-exploitation of ground water and fast-falling water tables. Community wells have dried up or the water has become too saline for any use except washing. Rainwater is also less reliable today due to drought and desertification. In Punjab, once the most prosperous agricultural region in India, vast stretches of land have turned into desert.

For many of India's 700 million farmers, agriculture has become unsustainable. In Gujarat and Rajasthan alone, 50 million people face a water famine. Thousands of farmers in the state of Maharashtra have committed suicide in recent years. They were in despair after a vicious circle of crop failures due to drought, and growing indebtedness for the purchase of new seeds and pesticides. Finally, they took their lives by swallowing some of the pesticides, leaving their widows and children with the debts.

Millions of farmers, desperate to keep their families fed, have already left the land in search for a better life in the city. The urban population in India has risen ten times over the last century and is now the second largest in the world after China. According to the government, half of the population will live in urban areas by 2025. The driving force behind this movement is the concentration of economic opportunities in the urban areas. But the majority of the migrants are illiterate and belong to lower castes, and thus have access only to low-income jobs. They are automatically excluded from the safer, higher-valued sites in the cities and have no option but to live in overcrowded slum areas.

Growing urban poverty comes with environmental and social ills. Most slums lack basic public services such as clean water supply, proper drainage and electricity. In India, these services are provided through weak governance systems in which corruption and patronage are widespread. In the 1,100 slums of New Delhi, tap water is either non-existent, or the water is running only a few hours a day. Wherever a toilet complex is available, it is often so poorly maintained that people prefer to save the rupee per visit and defecate in the open. Without these key services, slum dwellers often suffer from skin disorders, water-borne diseases and tuberculosis. This is not exactly what they expected when they left the land for the city.

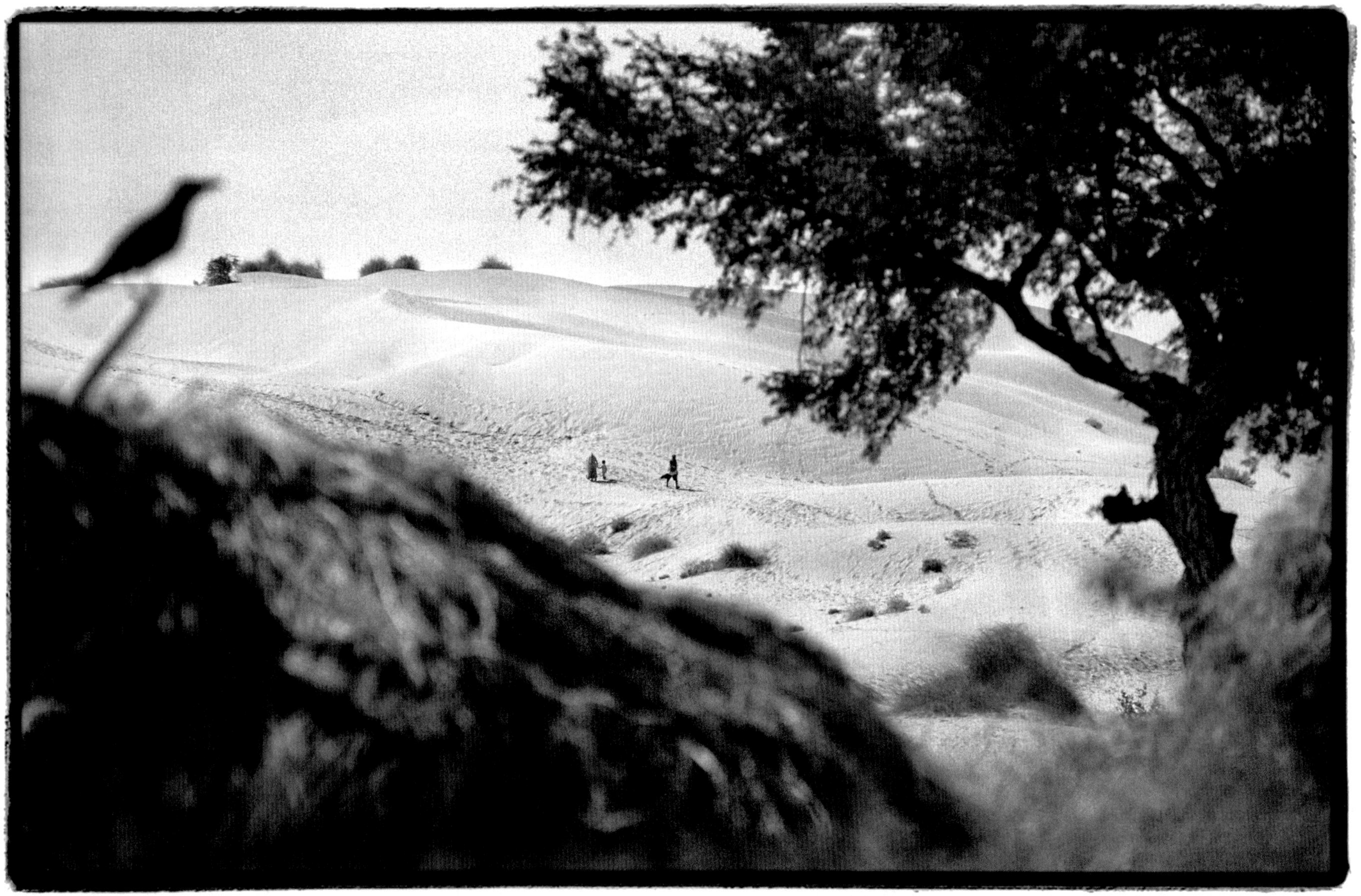

LEFT Rajasthan. Sand dunes between Pokaran and Jaisalmer. The Great Indian Desert or Thar Desert is characterised by its massive sand dunes, excessive heat, dust storms and an average annual rainfall of less than 250 mm. The Thar Desert covers about 65% of Rajasthan state.

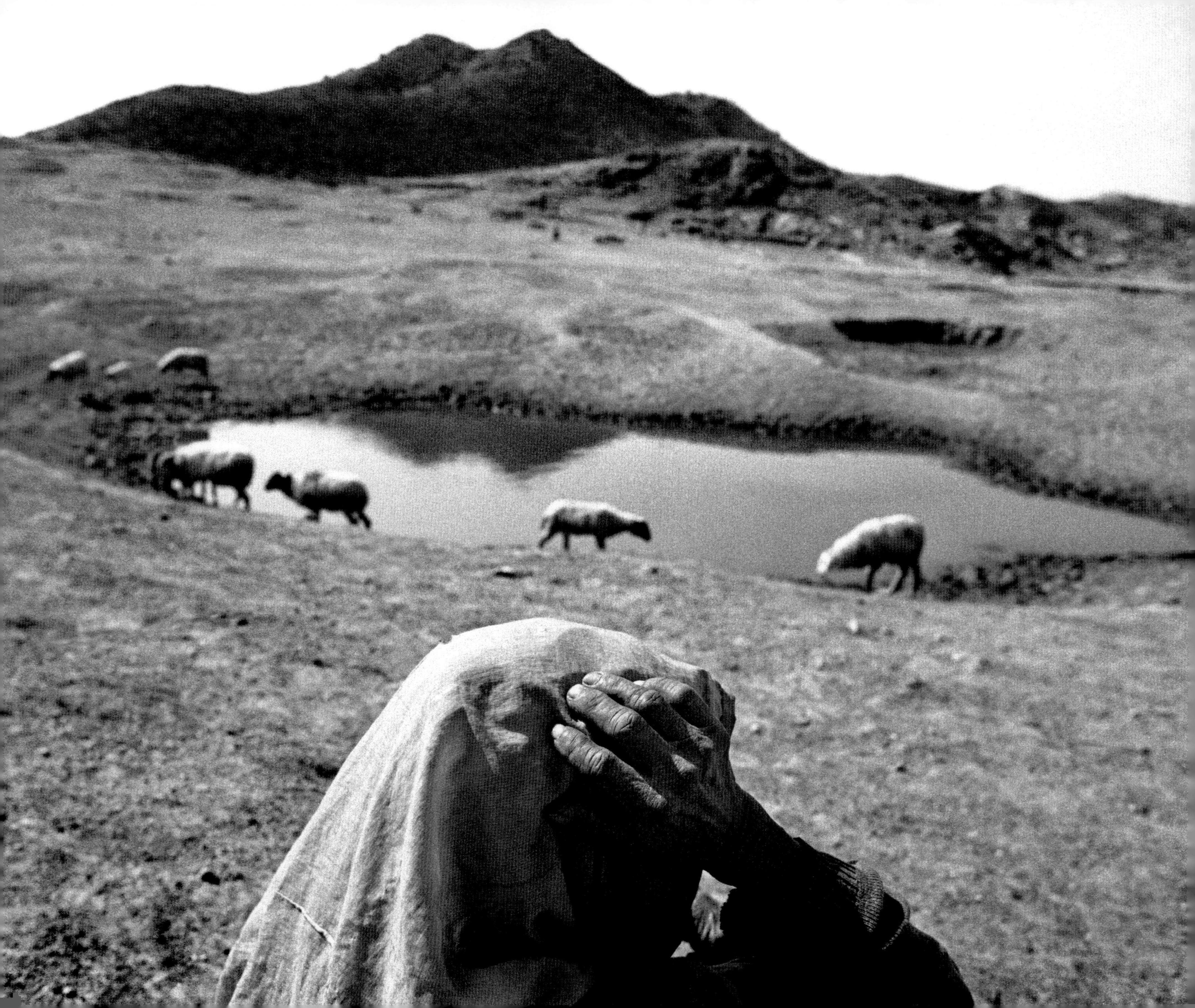

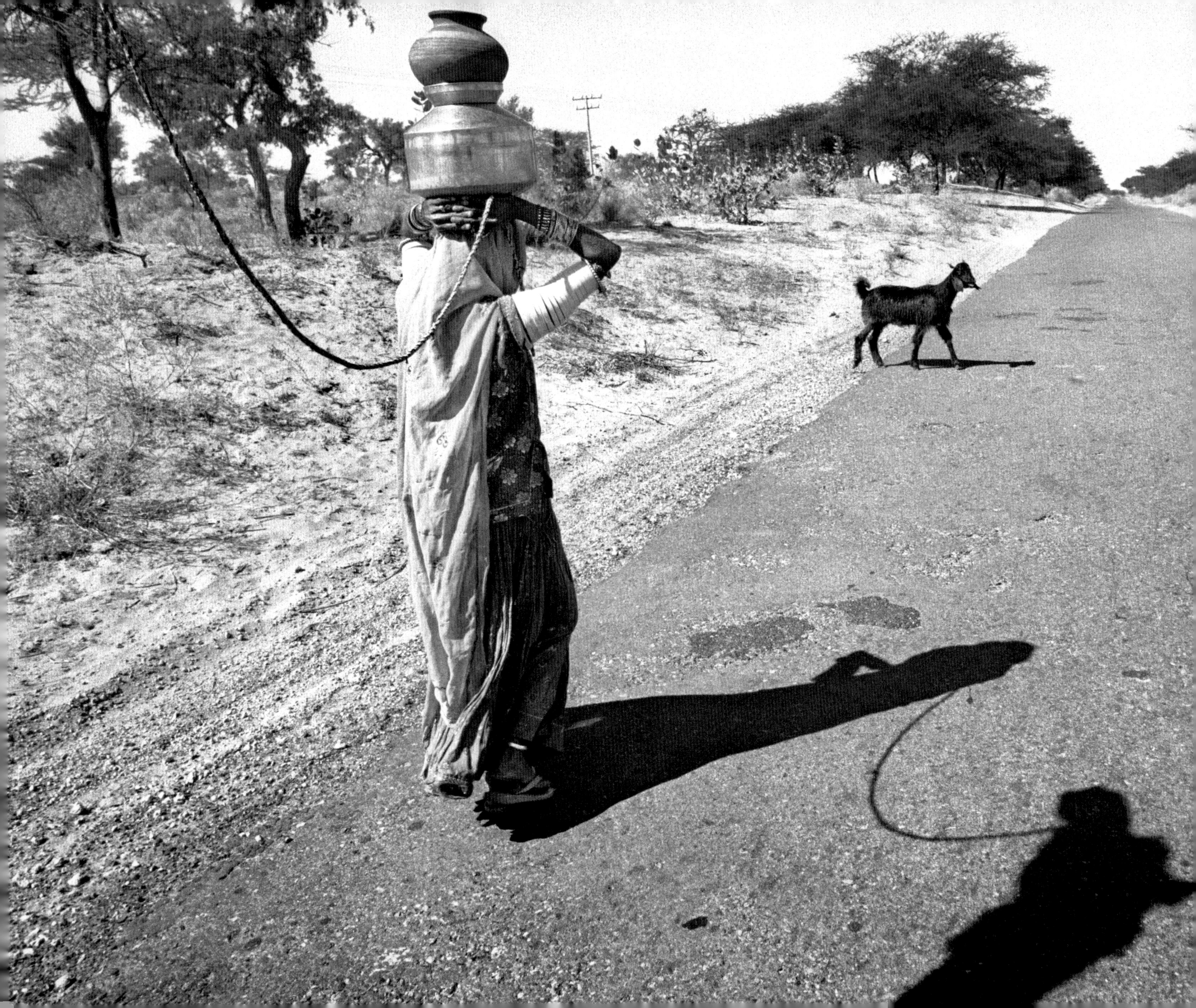

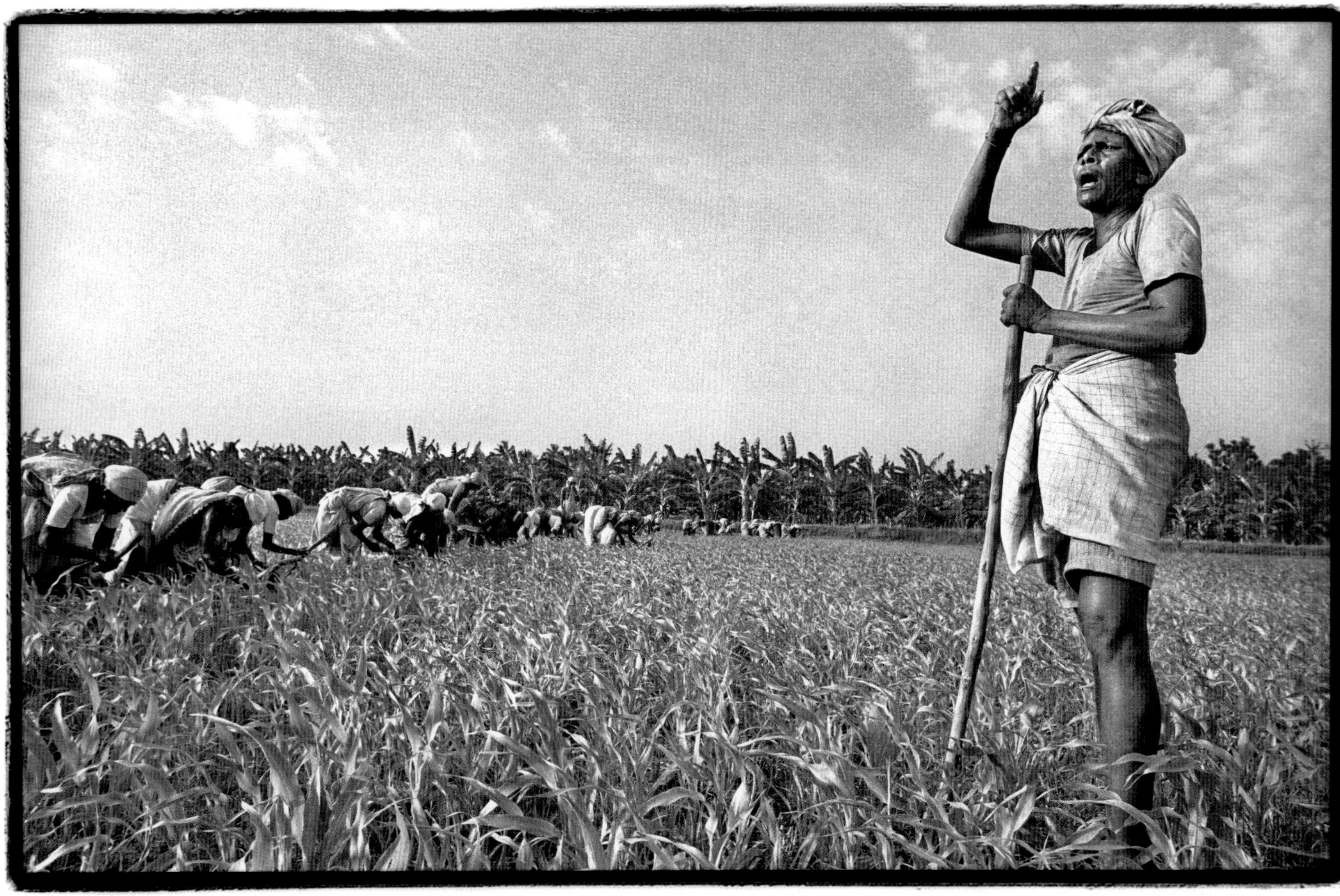

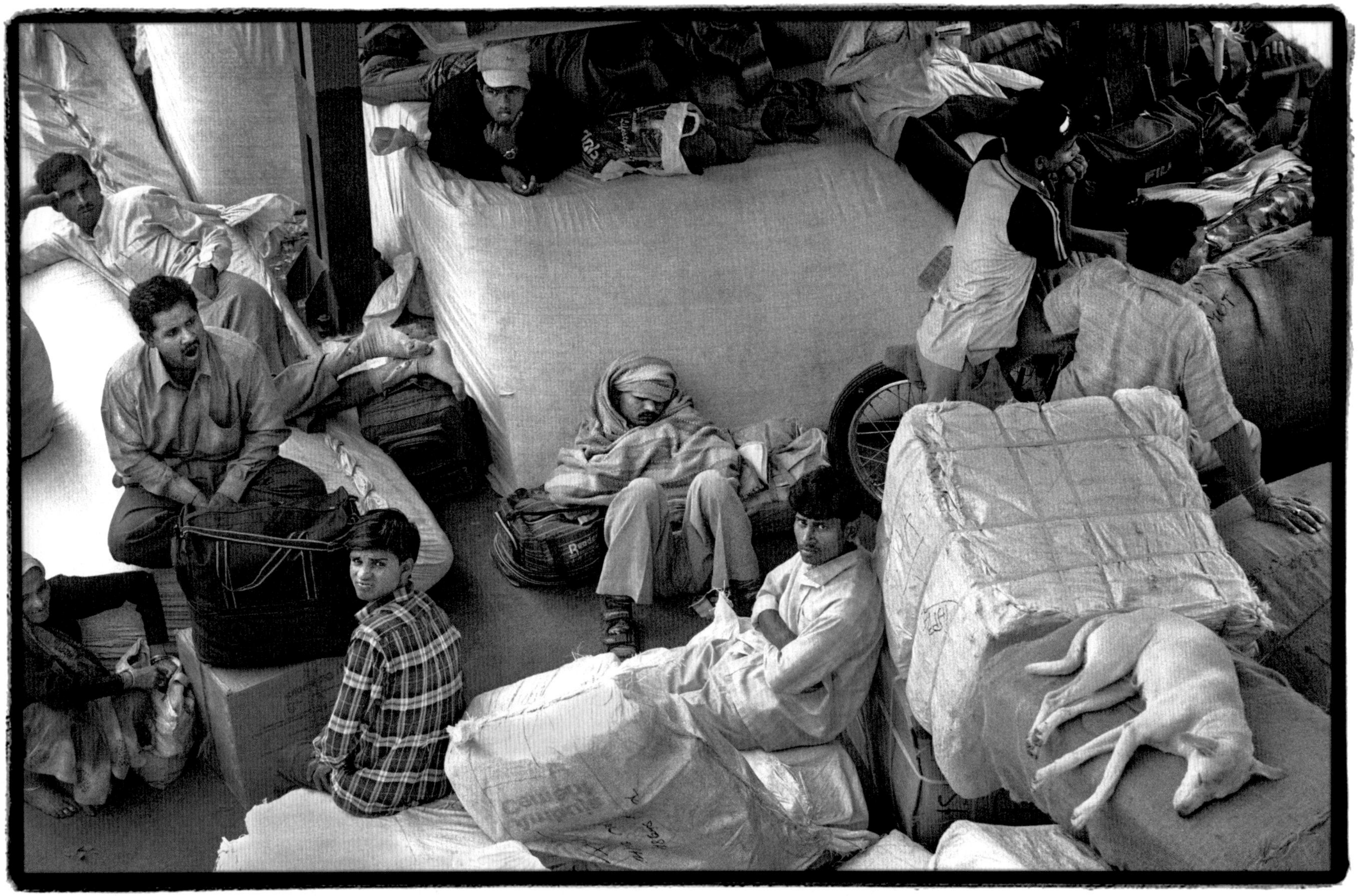

LEFT Delhi Railway Station. Delhi's relatively high per capita income and better living standards attract people from rural areas.

RIGHT Delhi. Slum behind Delhi's World Health Organization building in the city centre. A few hundred families live here on a small plot between the building and two railway lines. They have one sanitary block which is poorly maintained. People prefer to defecate next to an open sewer that separates them from the fancy WHO building.

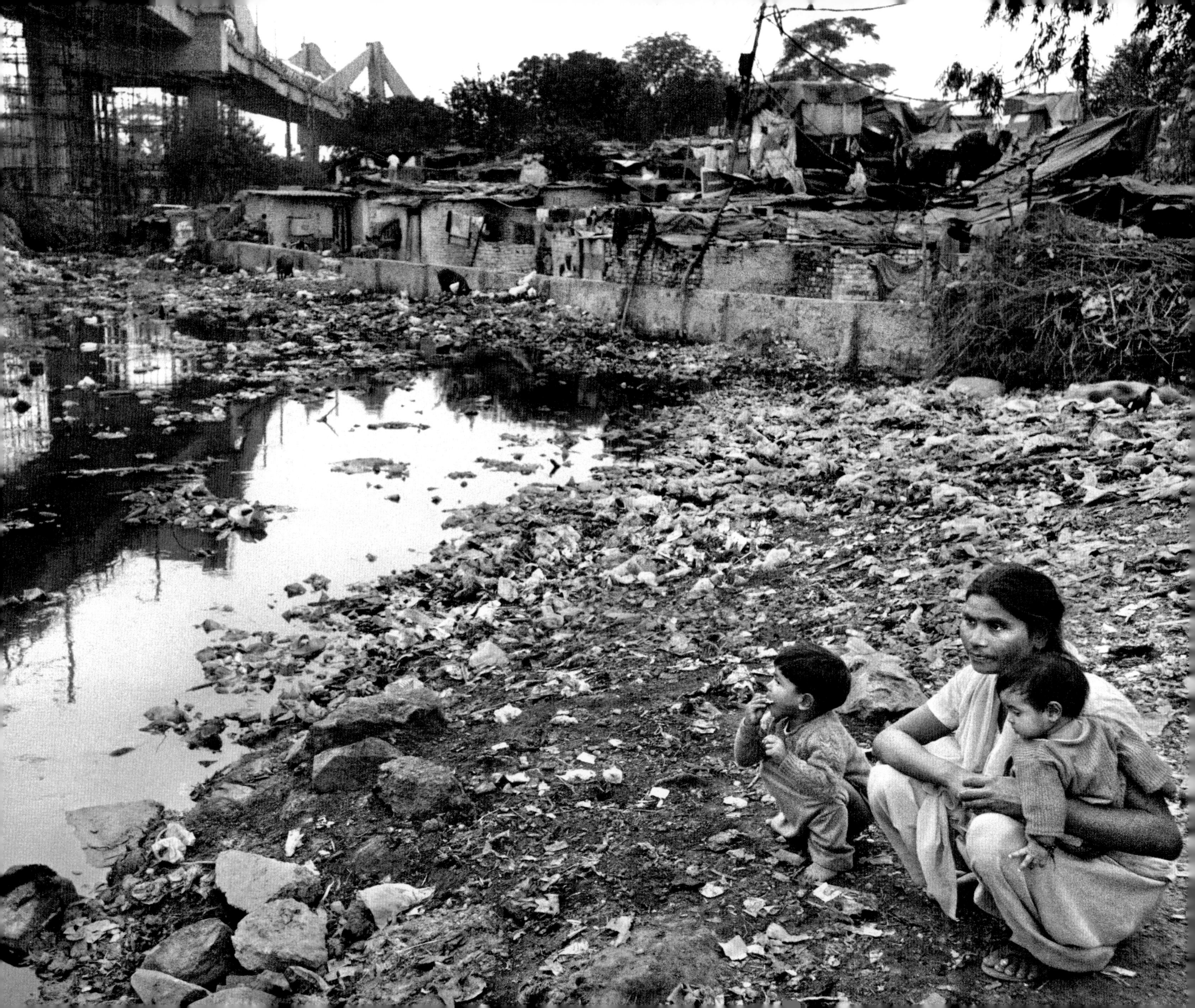

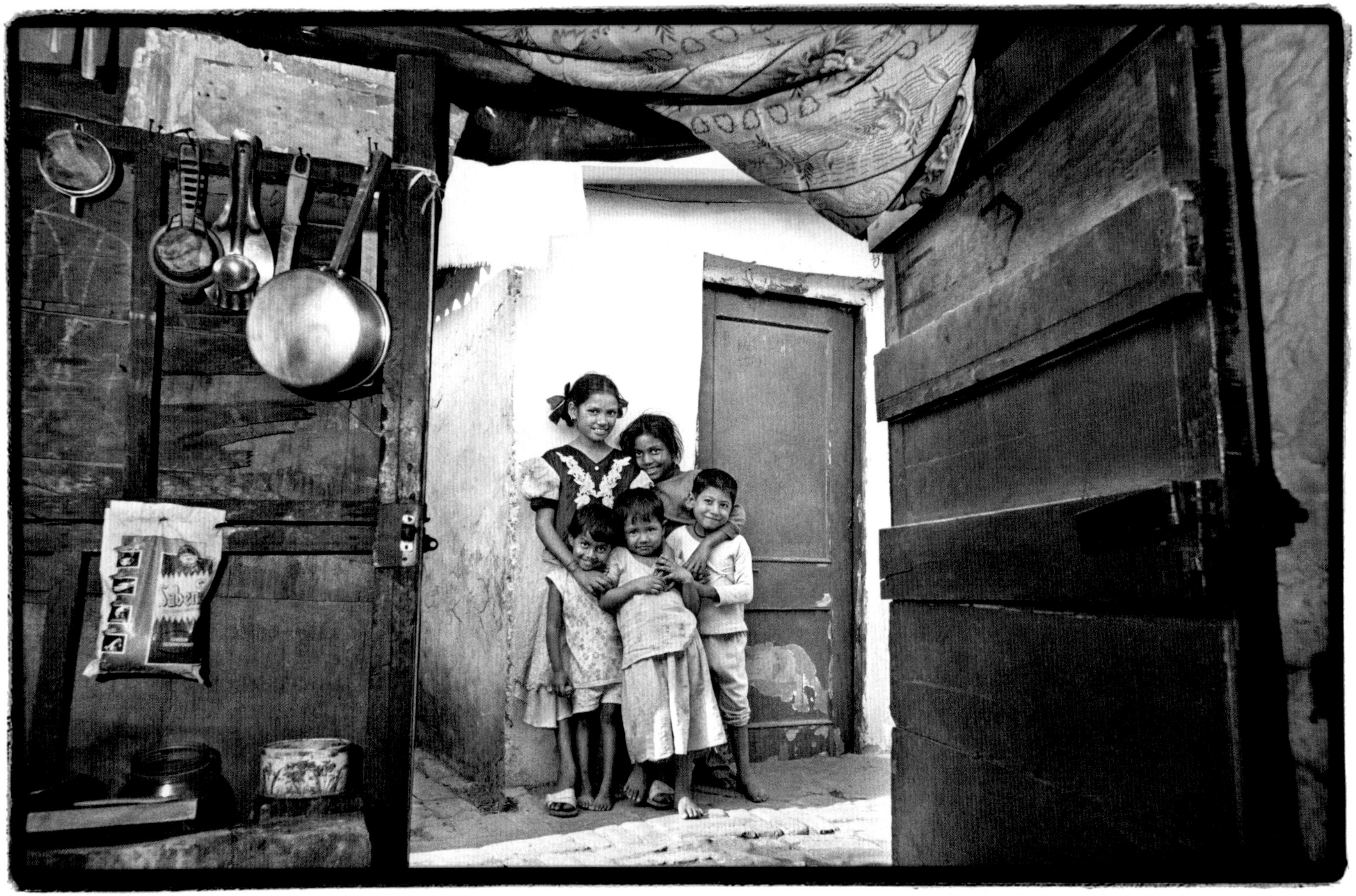

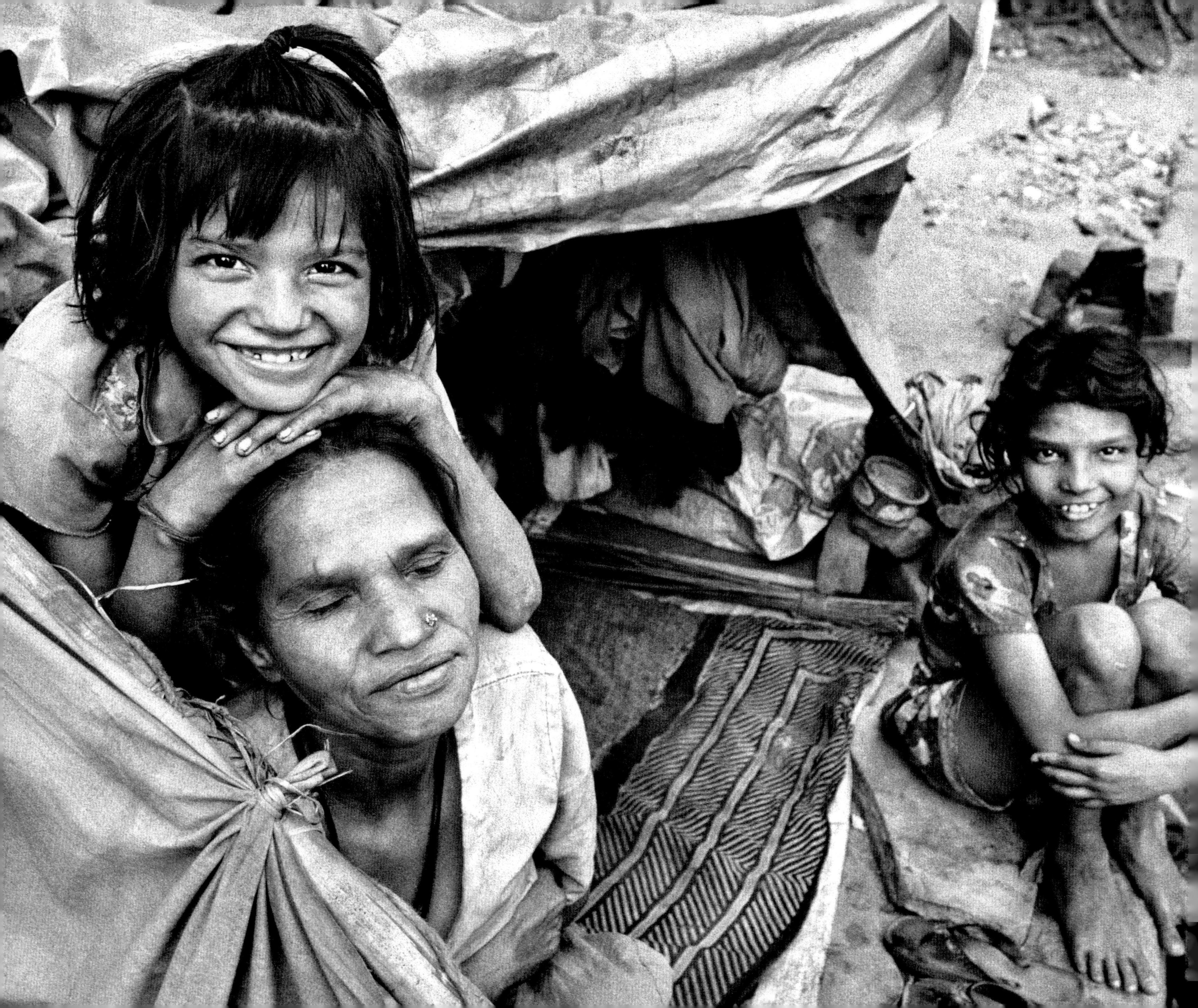

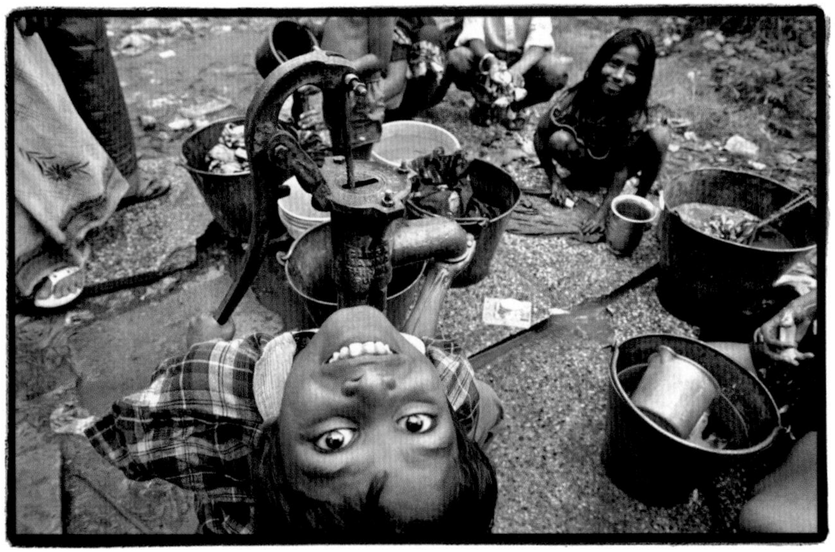

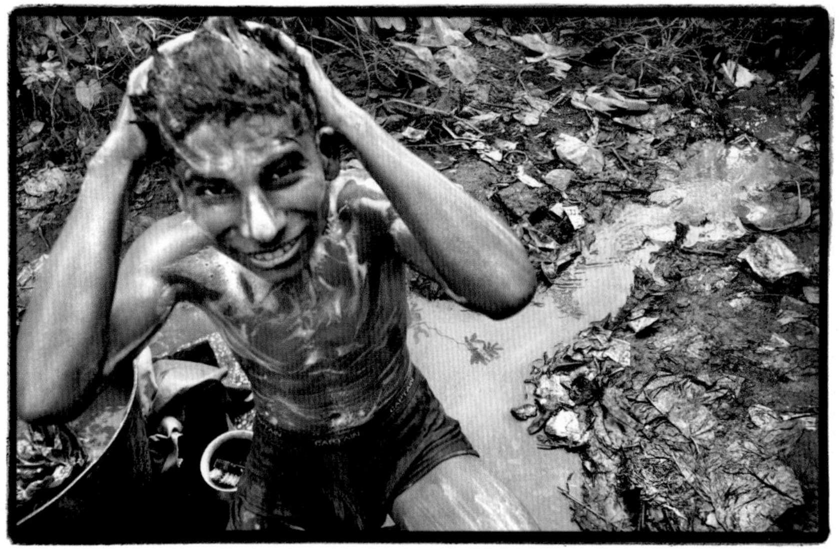

Delhi. The city that will host the 2010 Commonwealth Games cuts a sorry figure as far as its sanitation is concerned. Most of its sewer lines are silted and not functioning.

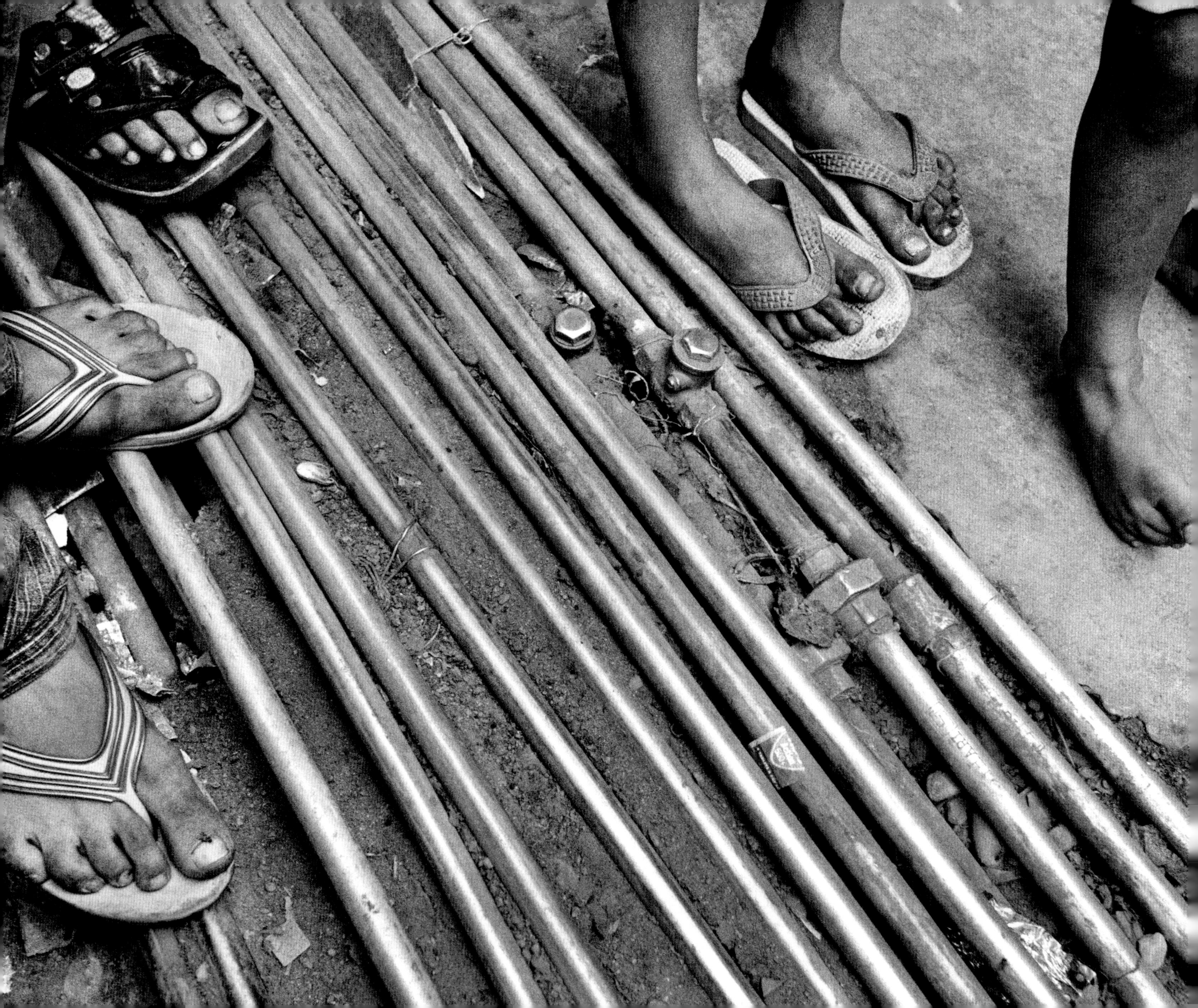

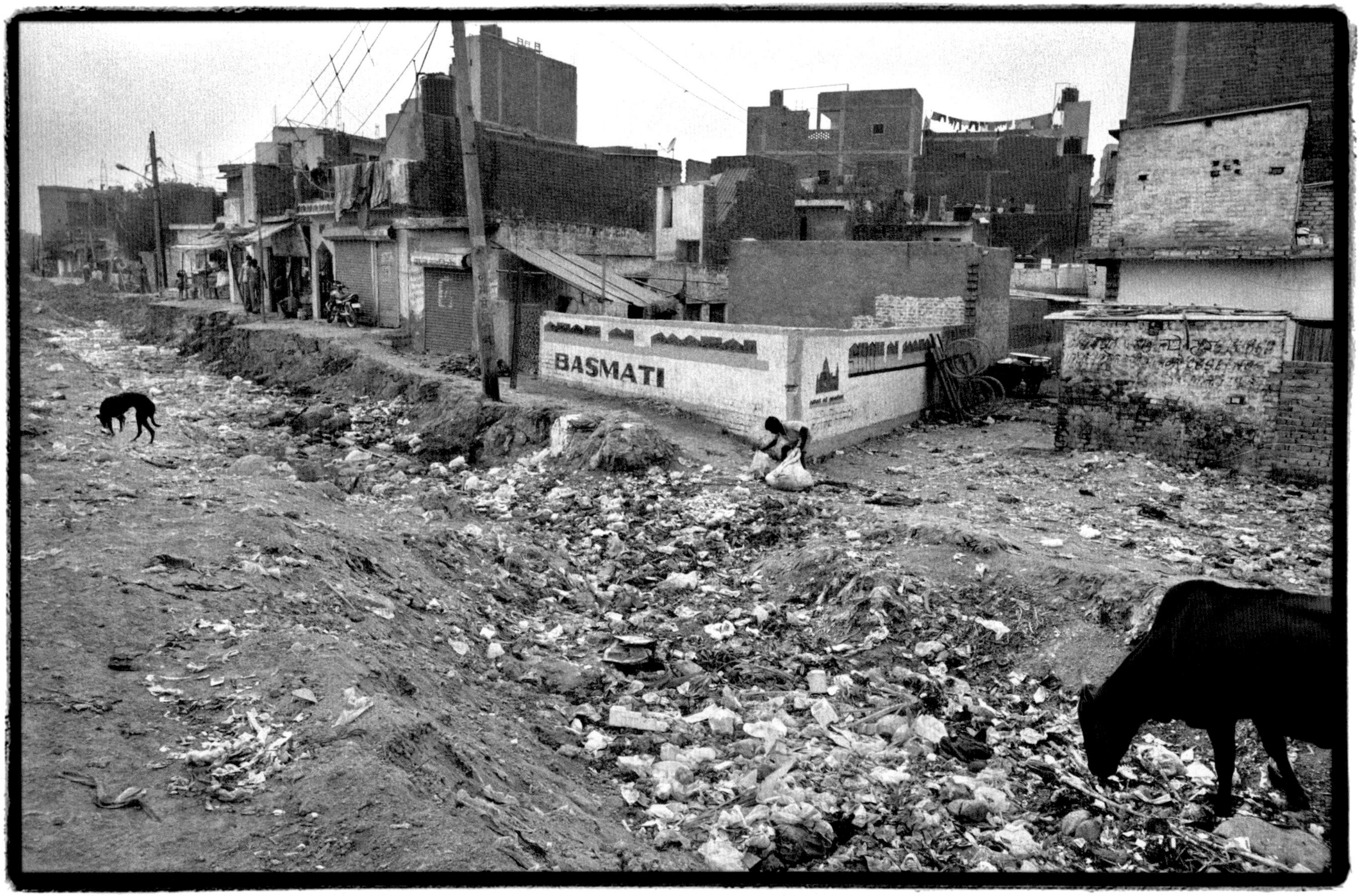

RIGHT Delhi. Less than half of the city's waste is processed by sewerage plants. The remaining filth is dumped into the already polluted Yamuna river.

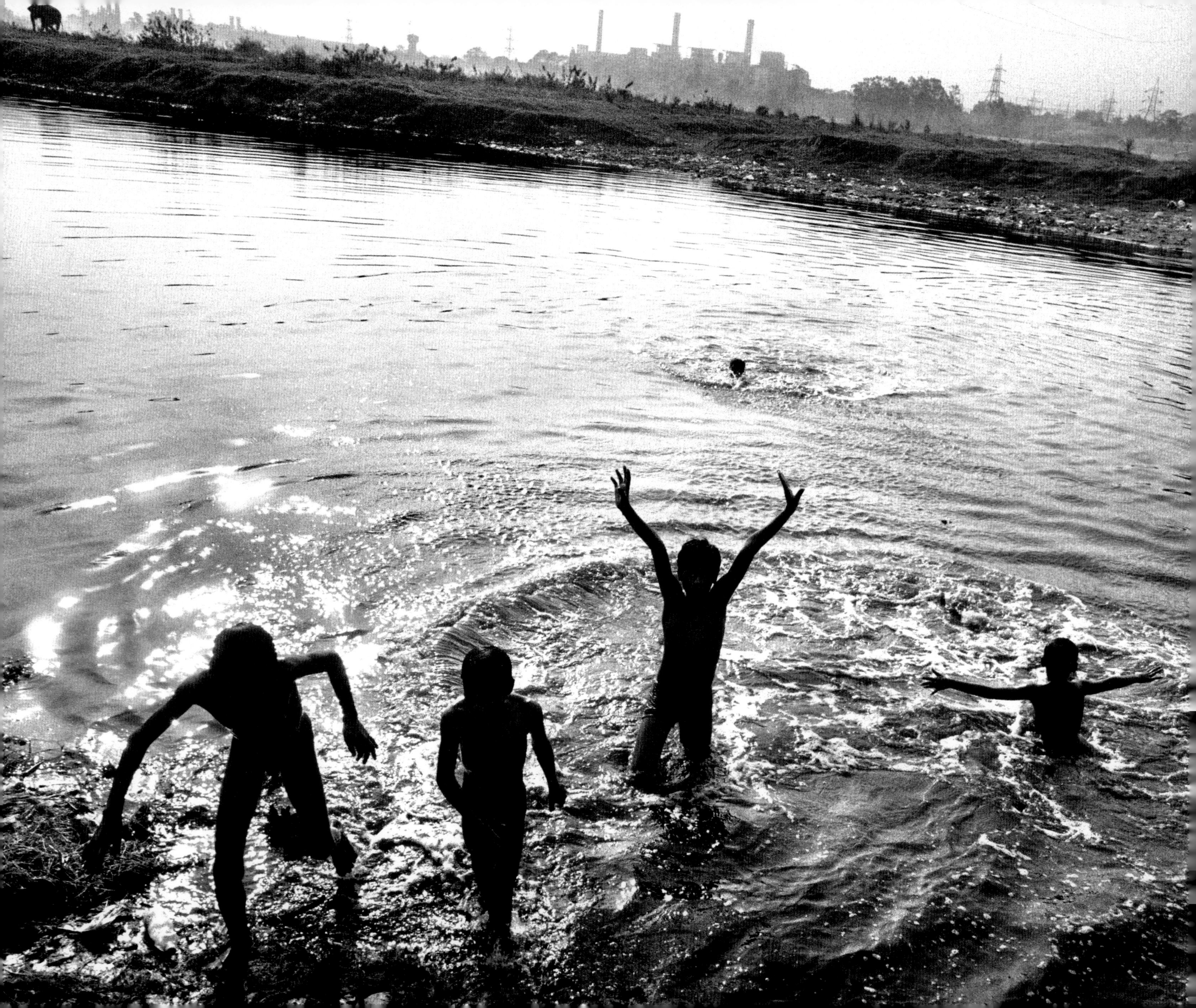

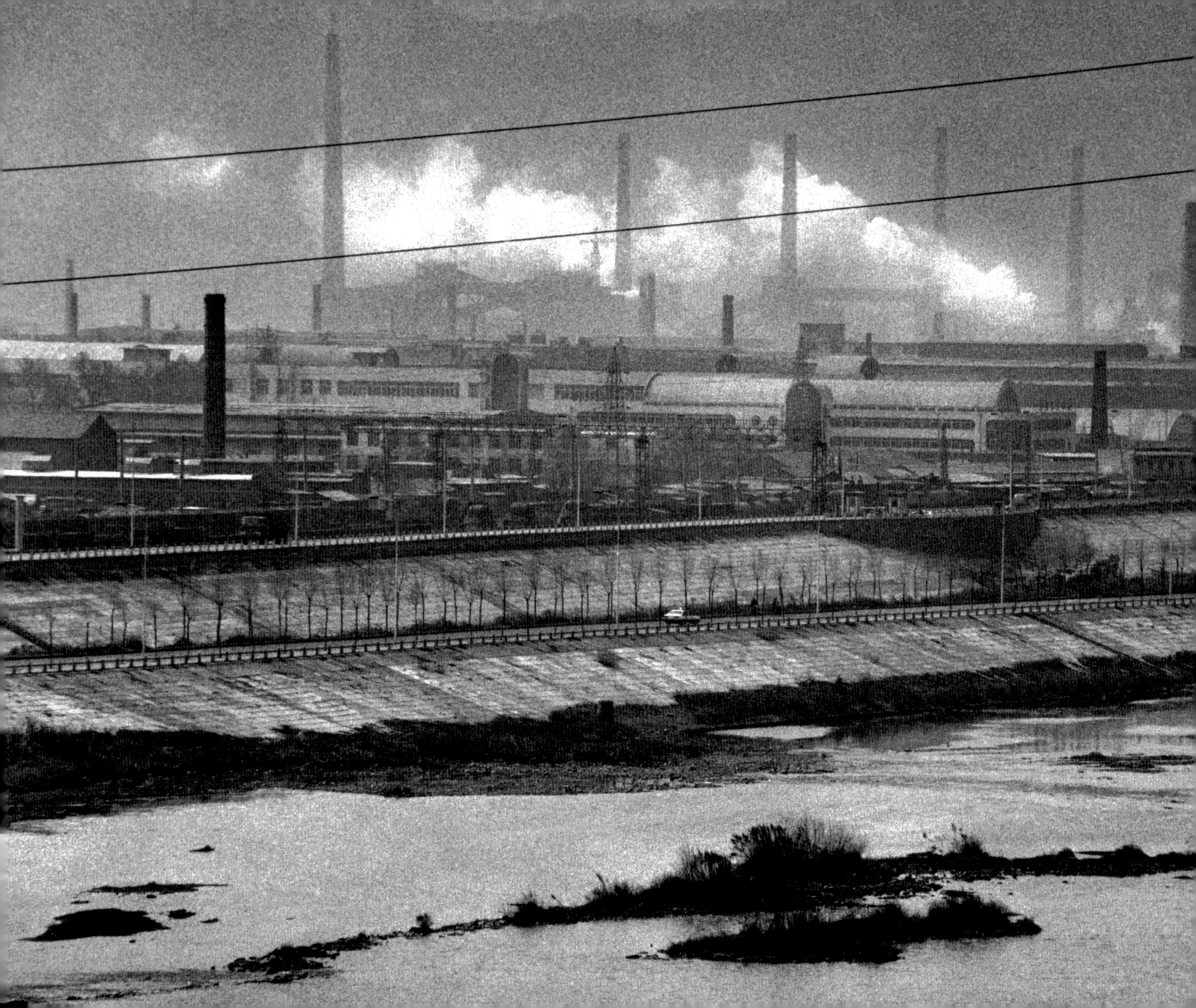

China
Pollution poisons progress

With an average growth of more than 9% per year in the past two decades, China is among the world's fastest growing economies. Its demand for energy is increasing and will continue to do so until it reaches nearly 15% of the total world energy consumption by 2025. The average income has gone up and poverty is being reduced.

However, fossil fuel emissions have also increased ninefold over the last 50 years. According to the World Bank, China has 16 of the world's 20 most polluted cities. The main pollution source is coal, as it supplies three-quarters of all primary energy and is mostly burned untreated, without pollution controls.

China already has the world's highest emissions of sulphur dioxide, while emissions from rocketing car ownership are only just becoming an issue. Respiratory infection is now the number one childhood disease and one of the leading causes of death overall in China.

Per head, China's water resources are among the lowest in the world and are concentrated in the south. The north and west experience regular droughts. Even when/where water is not scarce, it is rarely clean. The Ministry of the Environment labelled over 70% of the water in five of China's seven major river systems as "unsuitable for human contact". In 2005, the authorities announced that the Yellow River was so polluted that its water was not safe for drinking. It is estimated that 300 million people in China do not have access to safe drinking water.

The government finally enacted the Cleaner Production Promotion Law. It established demonstration programmes for pollution remediation in ten major Chinese cities, and designated several river valleys as priority areas. In order to reduce the air pollution, it introduced a tax on high-sulphur coals, established coal-free zones and made plans to construct natural gas pipelines.

The pollution is not only damaging human health, air and water quality, but ultimately also the economy. Adding it all up, the World Bank concludes that pollution is costing China an annual 8 to 12% of its annual gross domestic product in direct damage, such as the impact of acid rain on crops, medical bills, lost working hours from illness, money spent on disaster relief following floods and the implied cost of resource depletion.

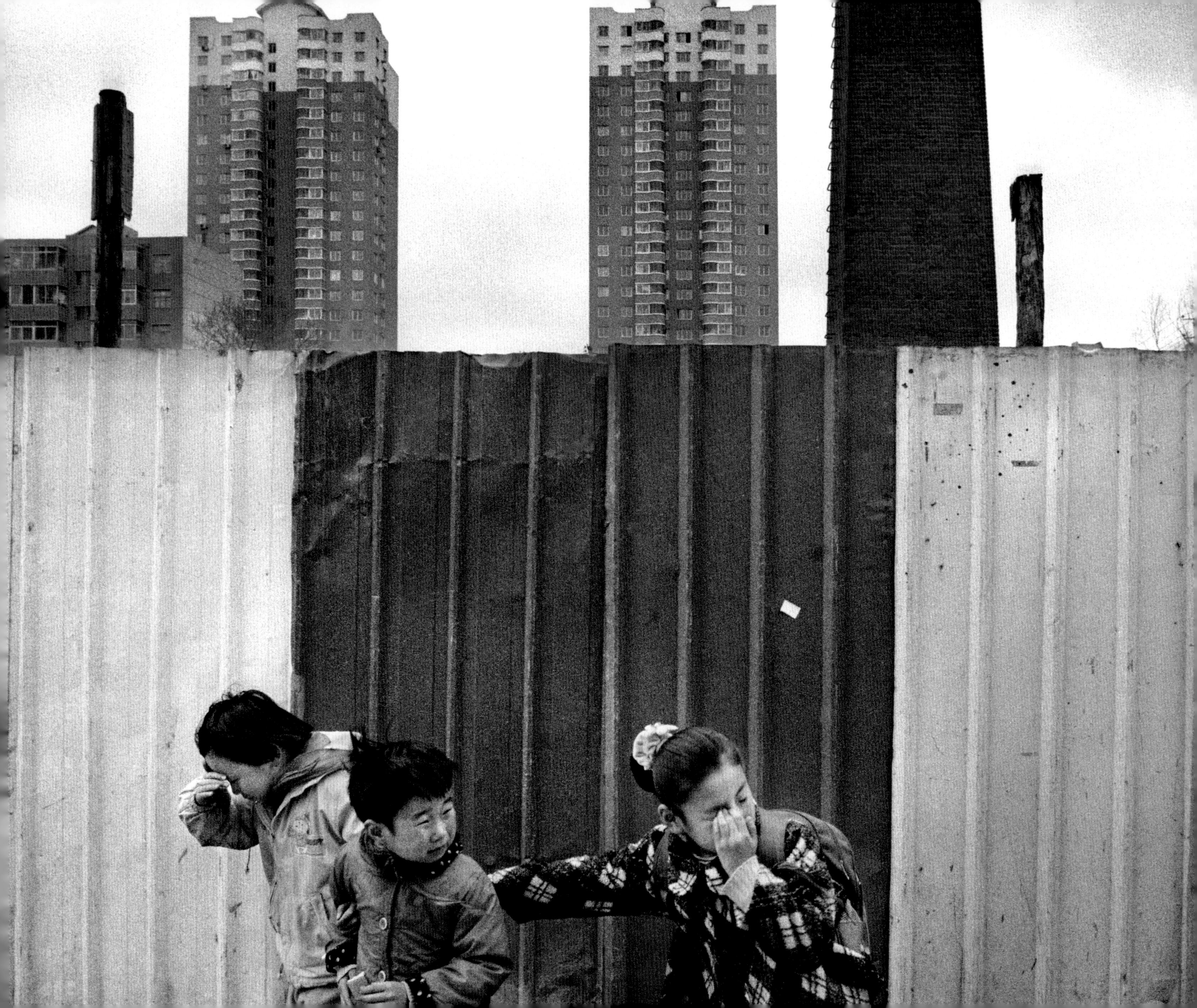

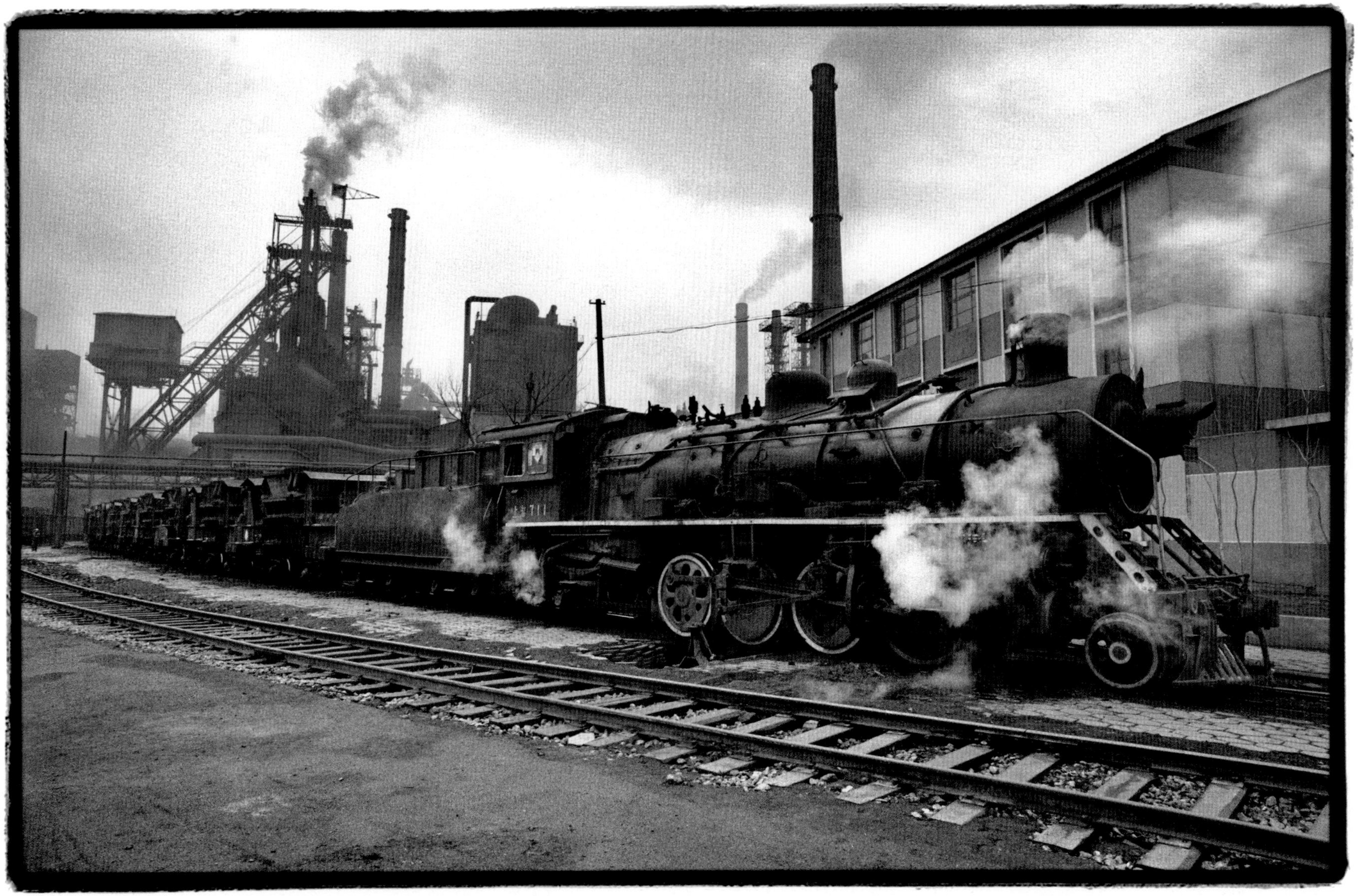

LEFT Shenyang. Children are trying to shield their eyes from dust that is blown up by the wind. Shenyang is one of the main heavy industry bases in China and has become one of the most polluted cities in the country.

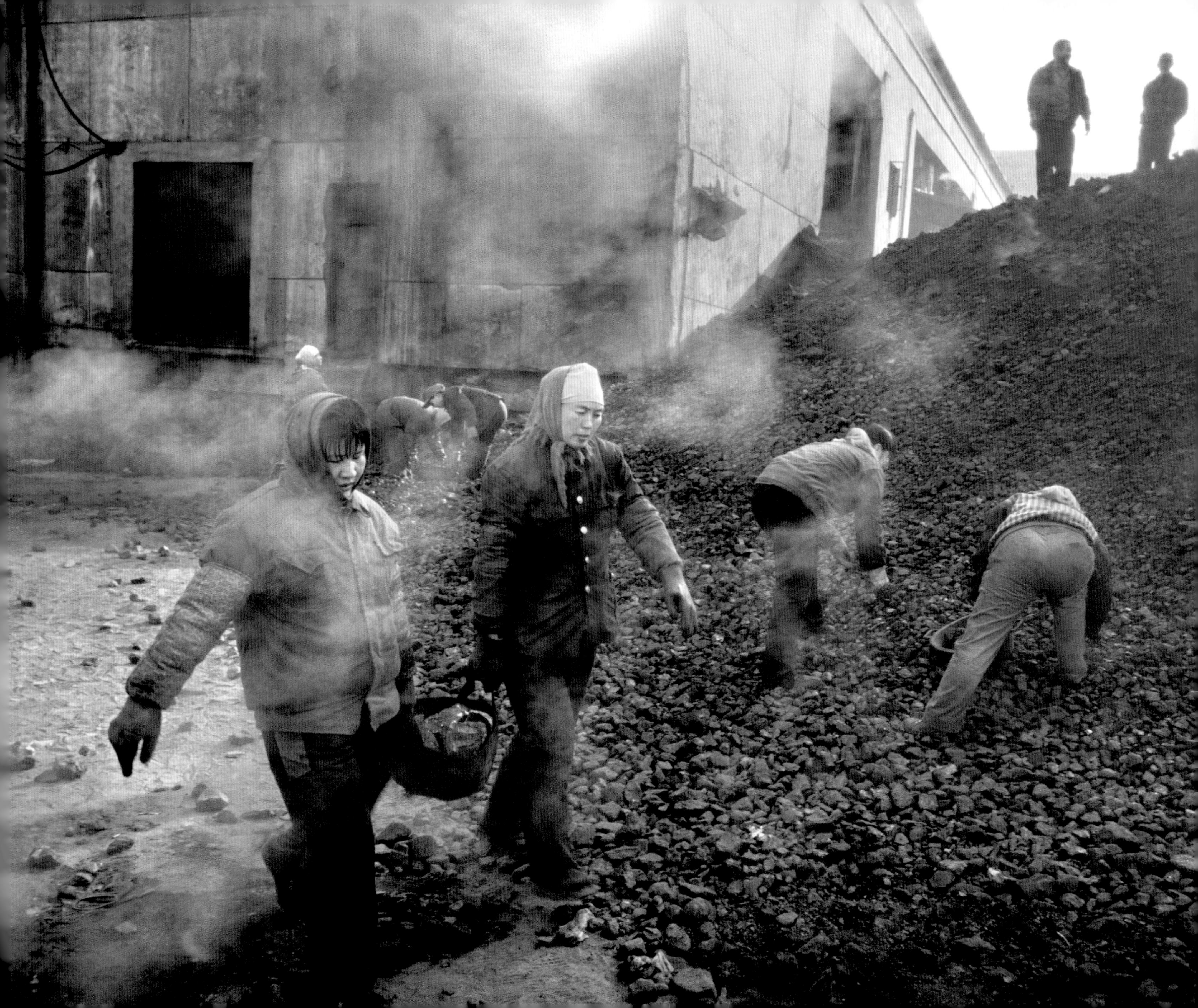

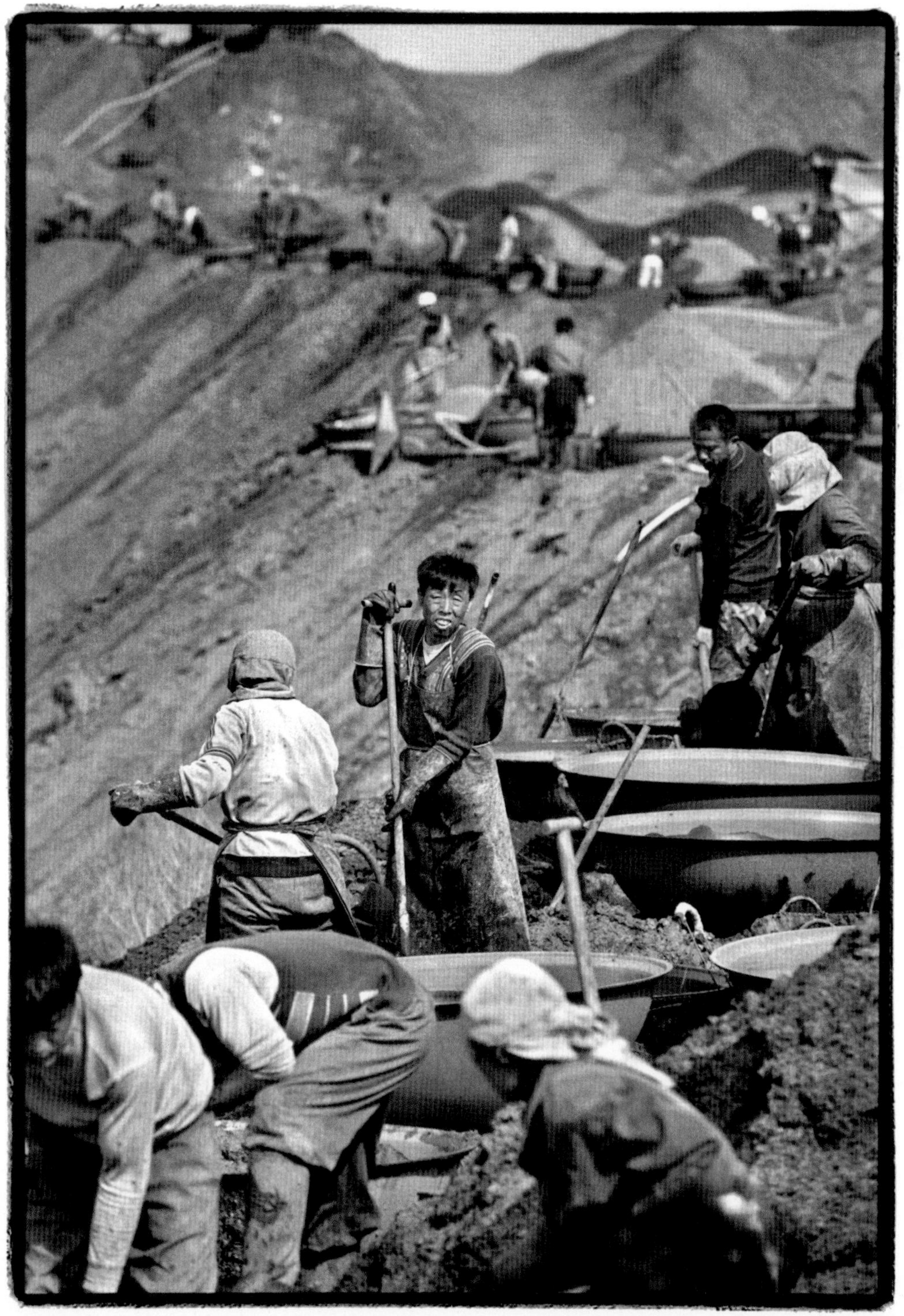

Liaoning Province, Fushun. Workers in the open-pit coal mine of Fushun, one of the biggest in China. Although many anti-pollution measures were introduced, the air pollution in Fushun remains dangerously high. Many industrial processes in this province are old-fashioned, and the industry lacks the money to adopt anti-pollution measures.

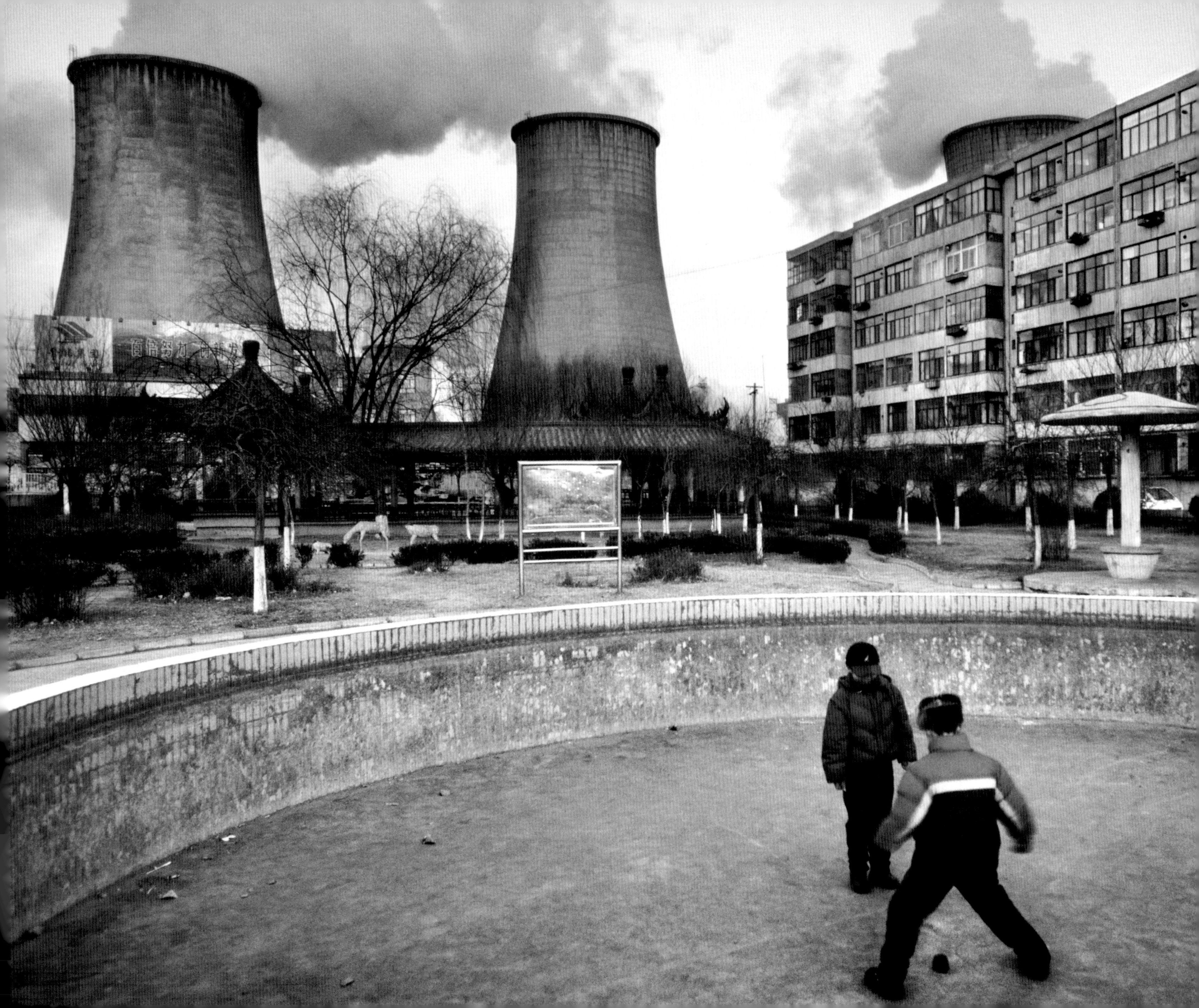
百倍努力

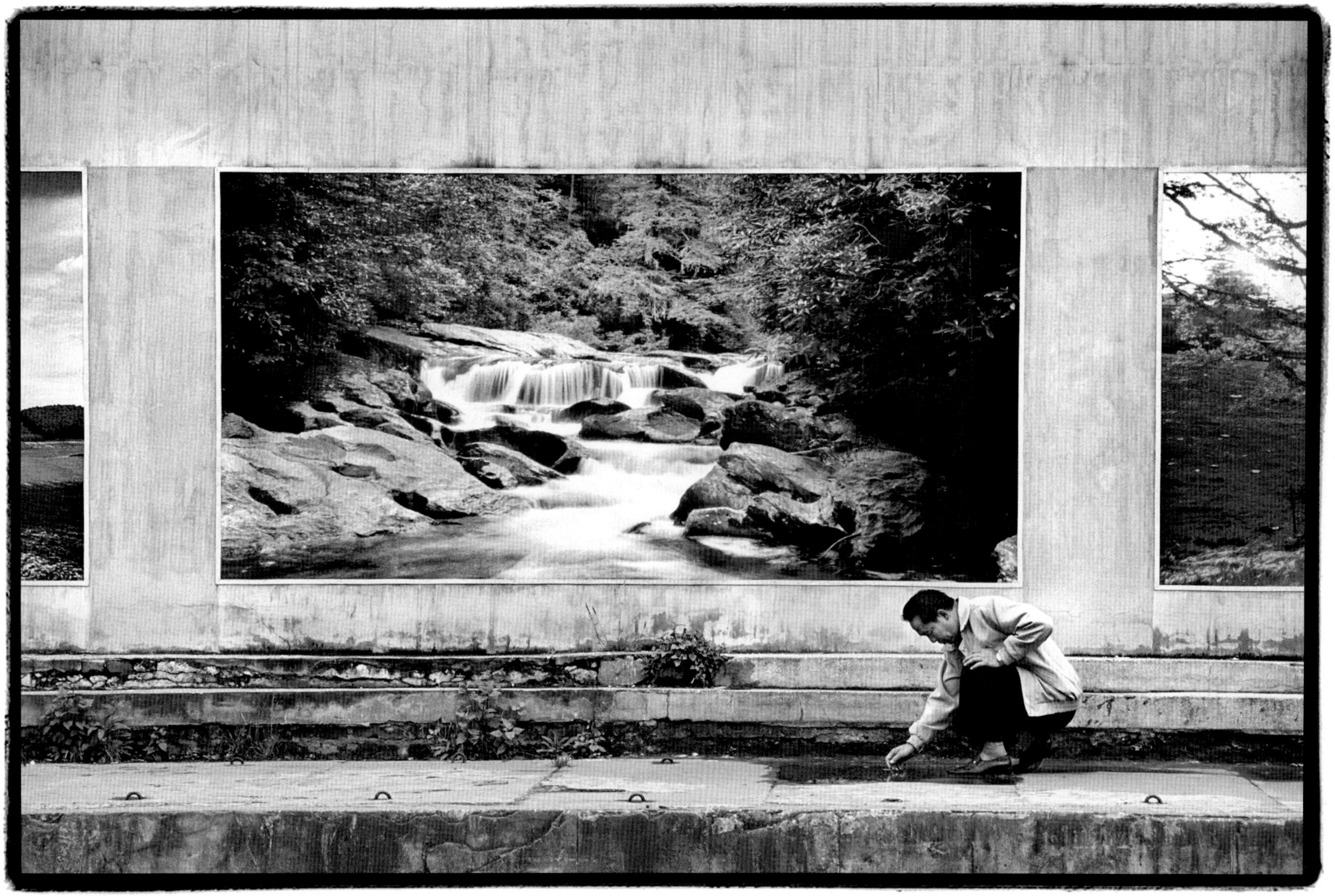

RIGHT Yunnan Province, Kunming. A man walks along a wall with images of beautiful natural landscapes. In some cities, these artificial images are the only natural environment one may see. Subject to deforestation, land degradation, soil erosion and diminishing biodiversity, China is one of the most environmentally impaired nations in the world.

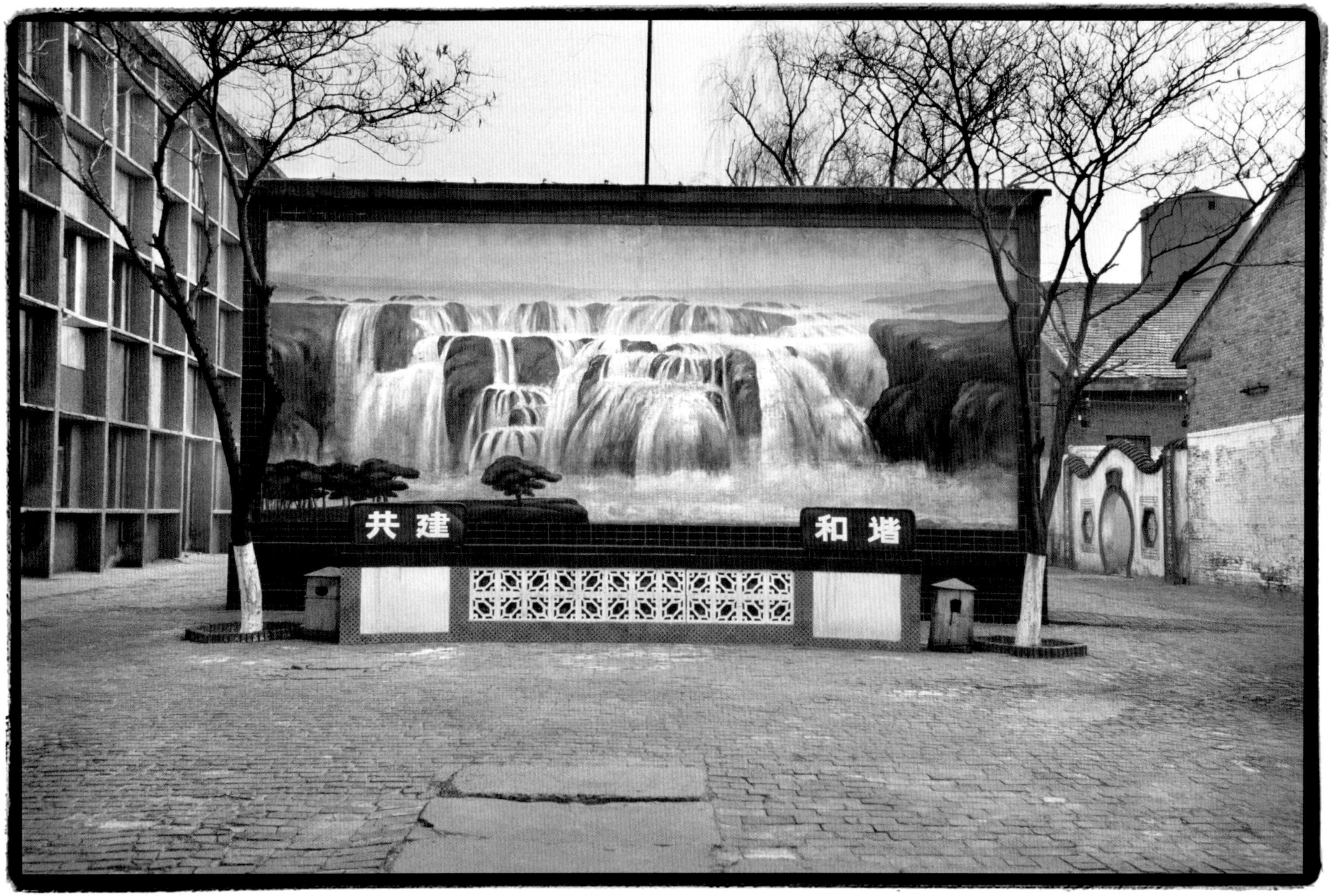
共建
和谐

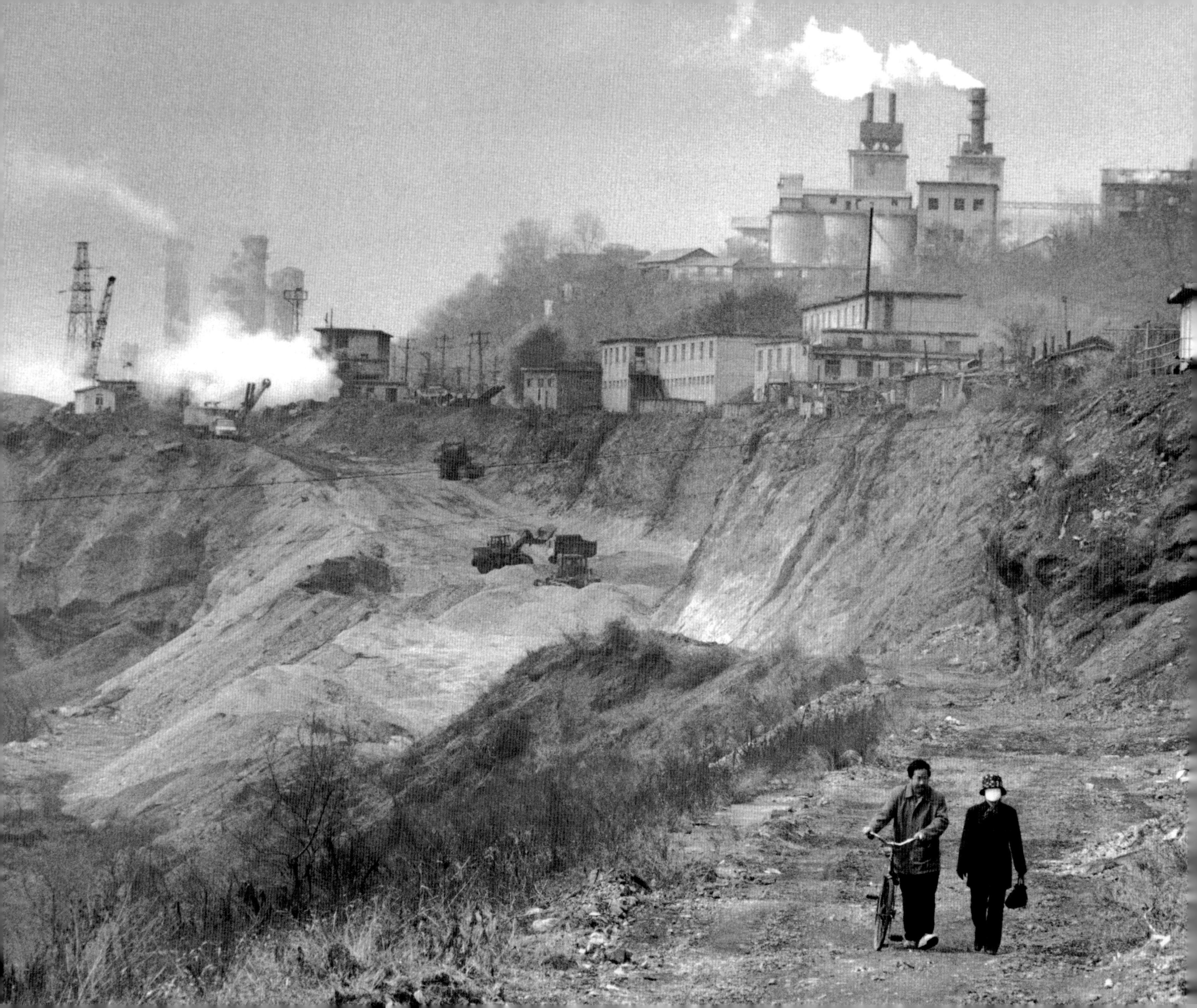

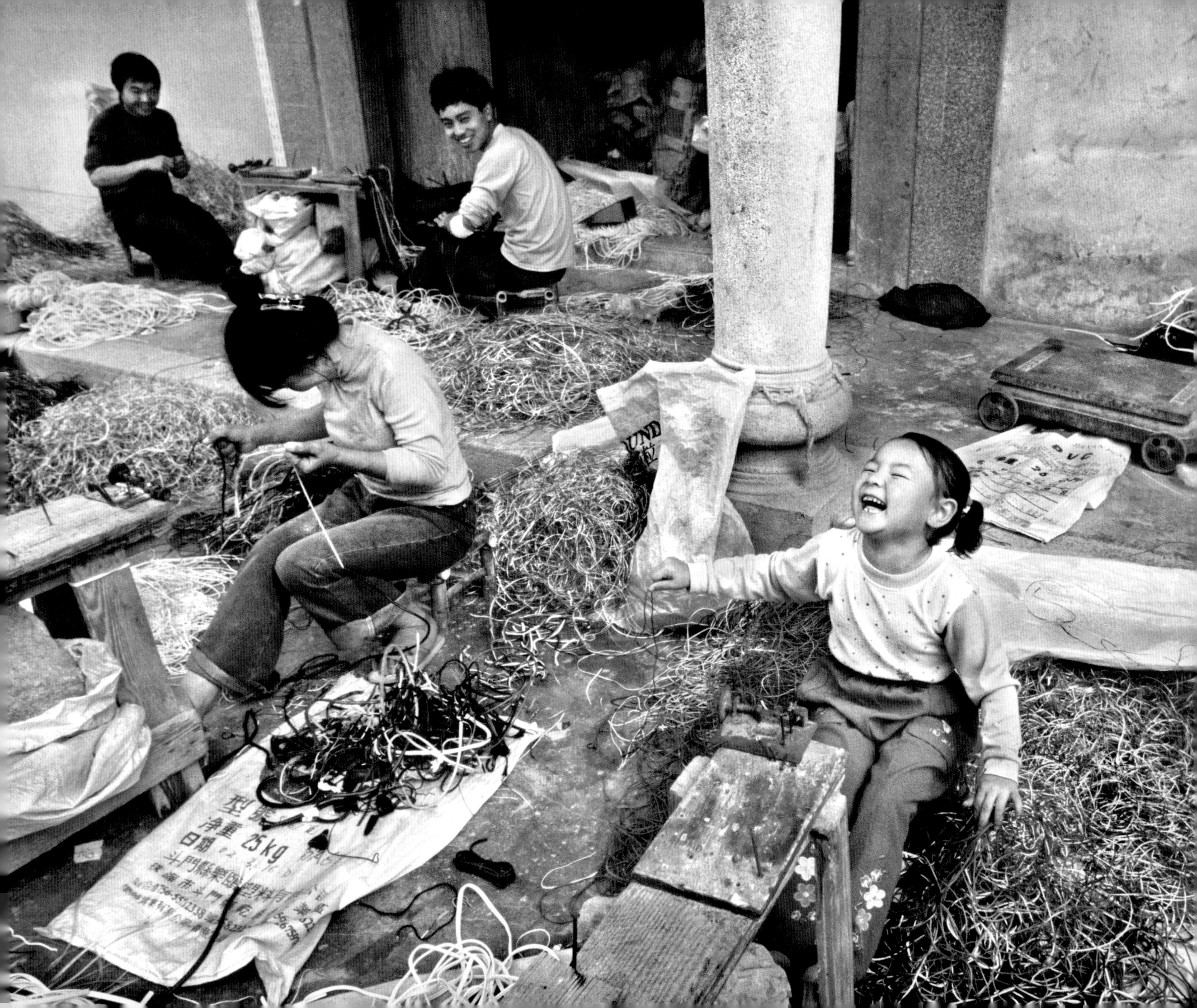
净重 25kg
日期

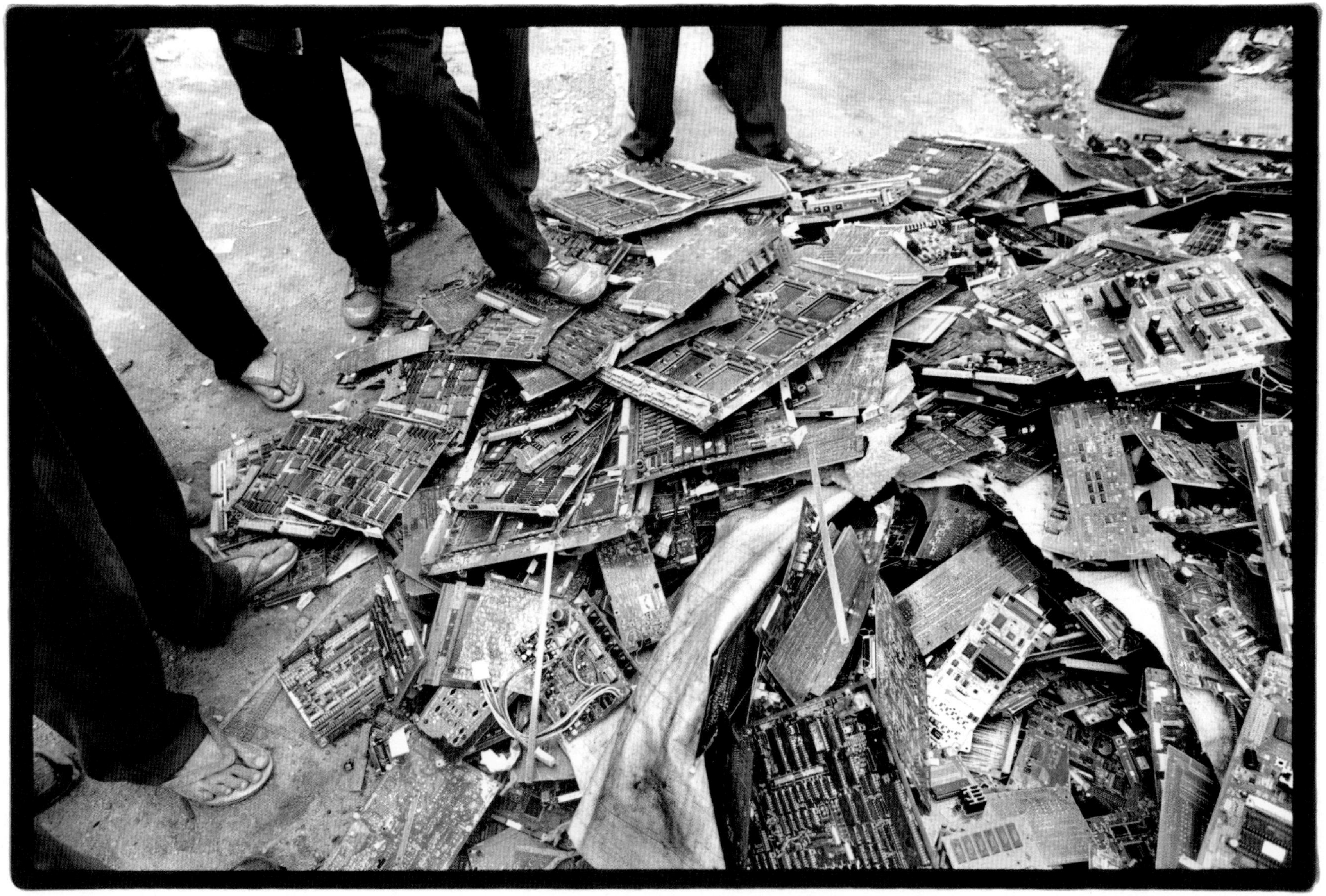

RIGHT Guangdong Province, Guiyu. Piles of circuit boards for sale in a street. Once a poor farming region, Guiyu has become a dumping ground for millions of tons of computer waste, imported primarily from the United States, Japan and South Korea.

LEFT Guangdong Province, Guiyu. Recycling of electric wires. During the night, the villagers burn the waste they cannot recycle. The emissions and ashes contain high levels of pollutants such as lead, beryllium and mercury.

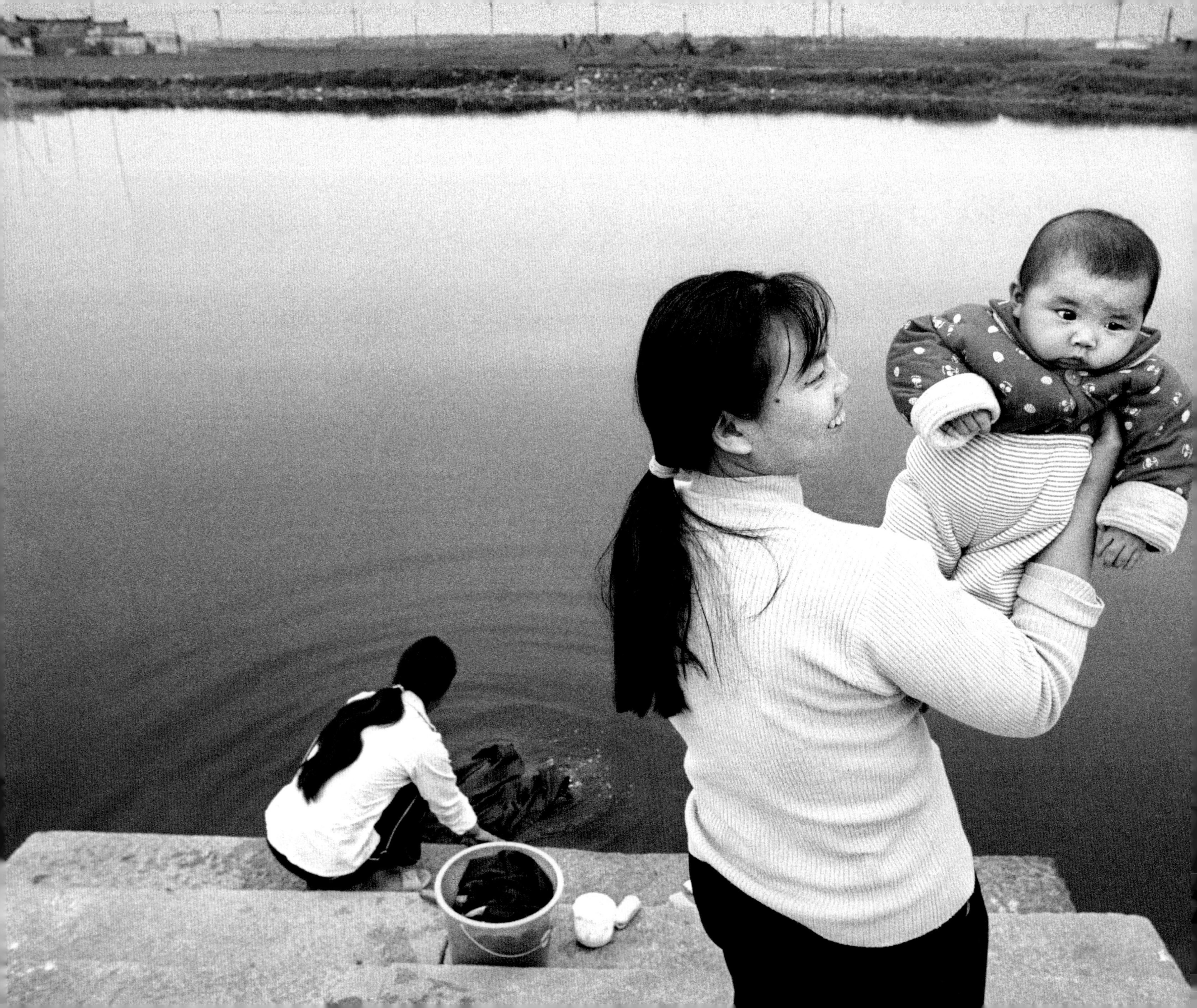

RIGHT Yunnan Province, Kunming. Environmental campaign poster.

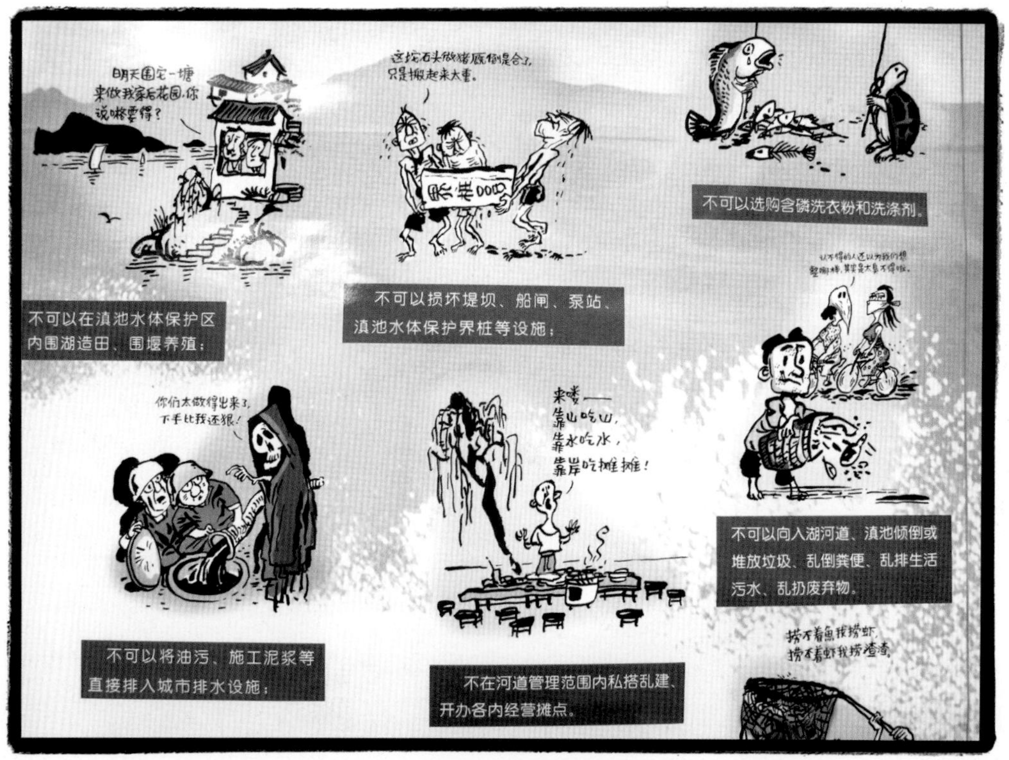

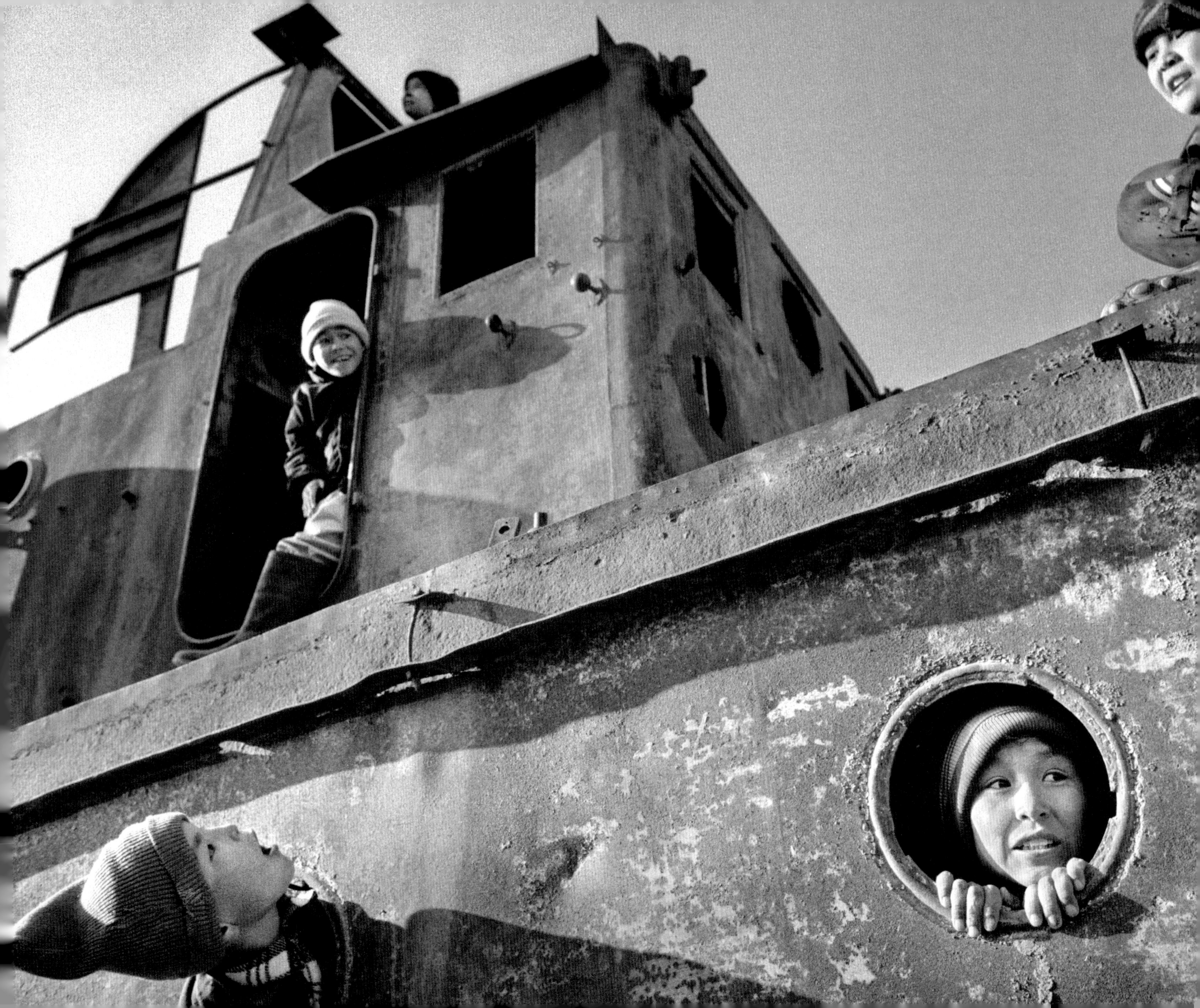

Aral Sea
The missing sea

Karakalpakstan is a region few people can locate unless they are water experts. It is the forgotten region of Uzbekistan where one of the worst natural calamities of the previous century took place: the disappearance of the Aral Sea.

A dry bed scattered with rusting shipwrecks and sea shells is all that remains of a once bustling harbour of what used to be the world's fourth largest inland sea. In 1918, the Soviet Union decided to divert the Amu Darya and the Syr Darya, two rivers that fed the Aral Sea, in order to irrigate the surrounding desert and make the large-scale cultivation of cotton –the 'white gold' of Central Asia and its agricultural product of the future– possible.

By 1960 the sea had begun to shrink. Within one generation, the water receded over 100 kilometres. By 2004, the sea's surface was reduced to a quarter of its original size, and had split into two smaller seas, the Small Aral Sea and the Large Aral Sea. Over an area as large as Belgium and the Netherlands combined, the immense reservoir of fresh water was replaced by a poisonous silty desert.

Thousands of tons of pesticides and fertilisers, used in the cultivation of cotton, are being spread by the wind and have left local people vulnerable to eye irritations and major health problems. A combination of poverty, malnutrition and drinking water contaminated by the toxic sea salts, has increased the incidence of bronchitis, cancers, liver and kidney diseases 30-40 times in the last twenty years alone.

The veteran fishermen keep dreaming of the sea tides and gentle waves forever gone, and send their children to more prosperous regions or abroad.

There is some hope, however, for those whose lives stay intertwined with the fate of the sea. The construction of a new dam and a series of dykes, completed in 2006, has seen the waters of the Small Aral Sea rising once again. Furthermore, the dry Aral Sea bottom has been opened for oil exploration. Will the 'black gold' save Karakalpakstan?

LEFT and RIGHT Muynak. Sergei Bakajoesjin Lipatovitsj (66) was 19 when he first sailed on the waves of the Aral Sea. Eight years later he became captain of the "Swan." Today Sergei's "Swan" is rusting on the dry bed of the bay, between the wrecks of a fishing fleet that used to comprise 300 ships. "The tide began to turn in 1964," says Sergei. "After that autumn the tide never returned and the water continued to subside until the harbour completely dried up in1981," the year when Sergei became harbour-master. "We continued to work, until we had to travel 30 kilometres to reach the water. The water was too far, and the salinity was too high for fish to be able to survive." Nowadays the sea has receded over 100 kilometres, leaving him unemployed.

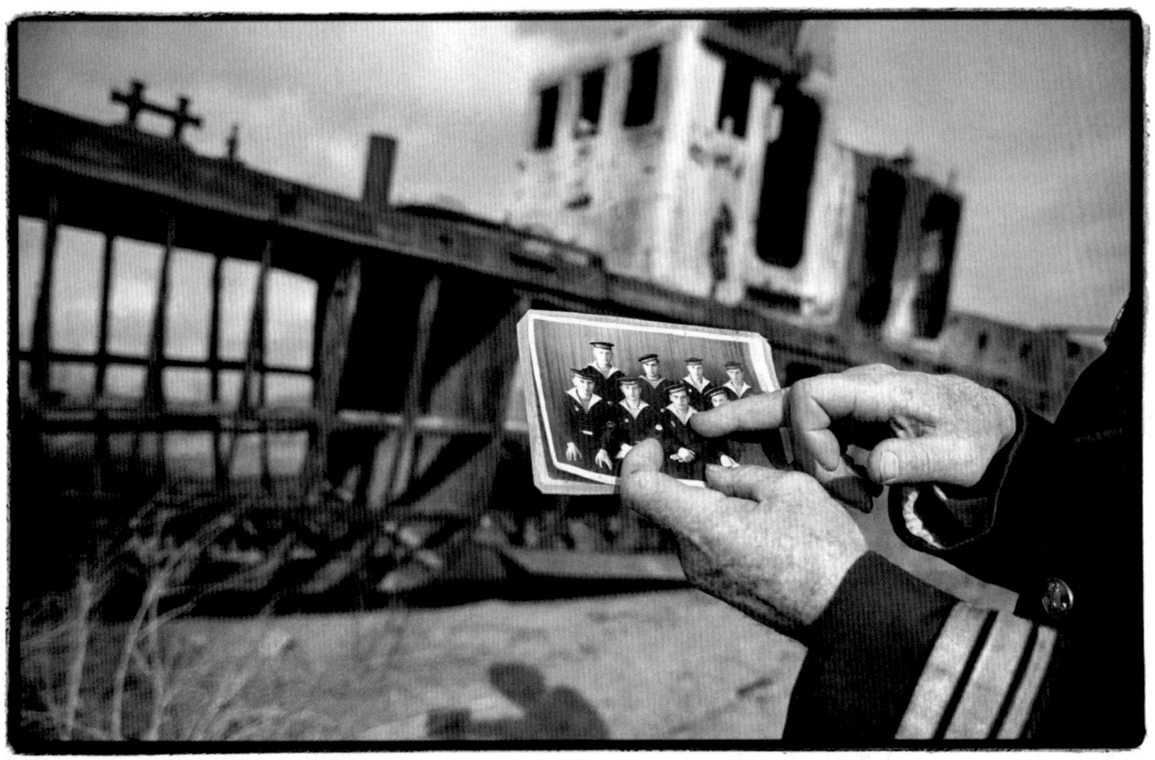

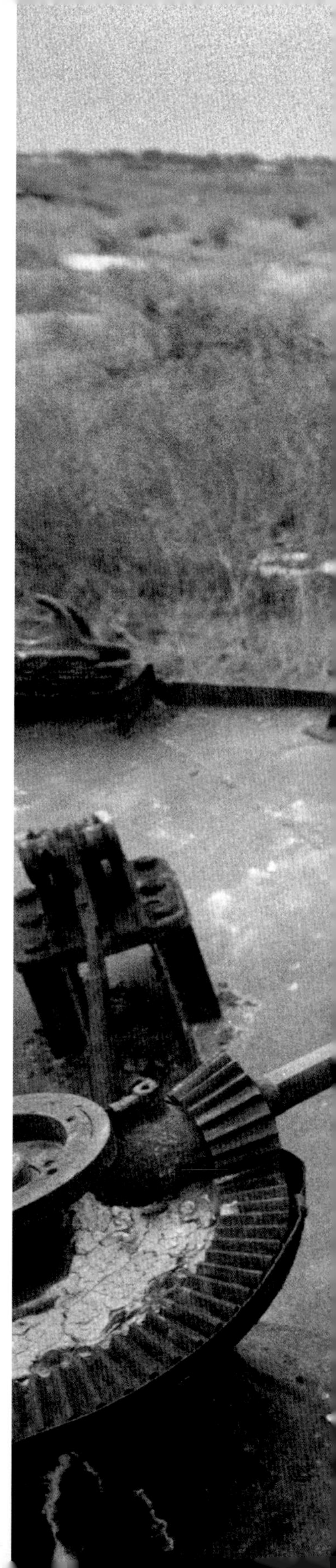

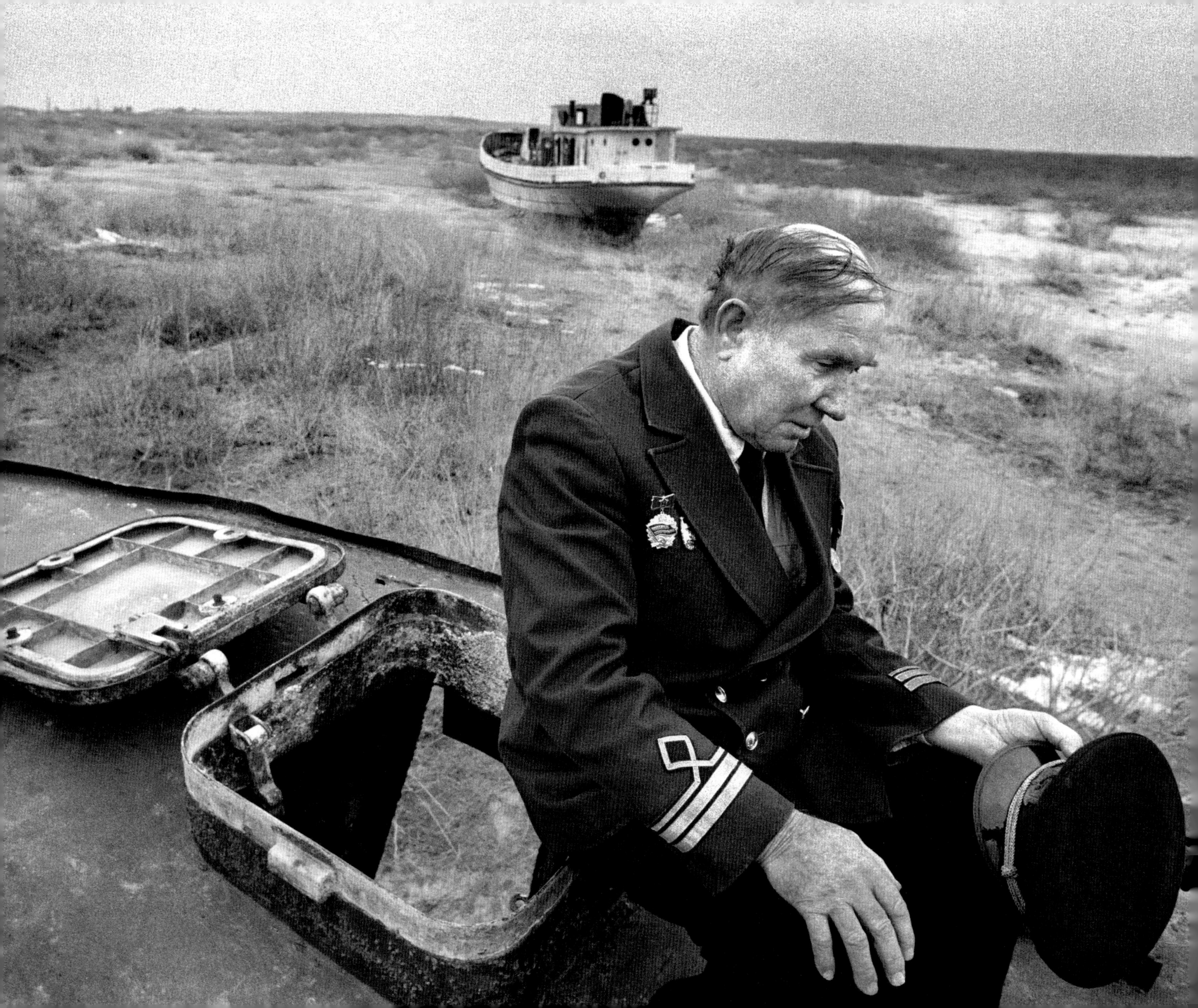

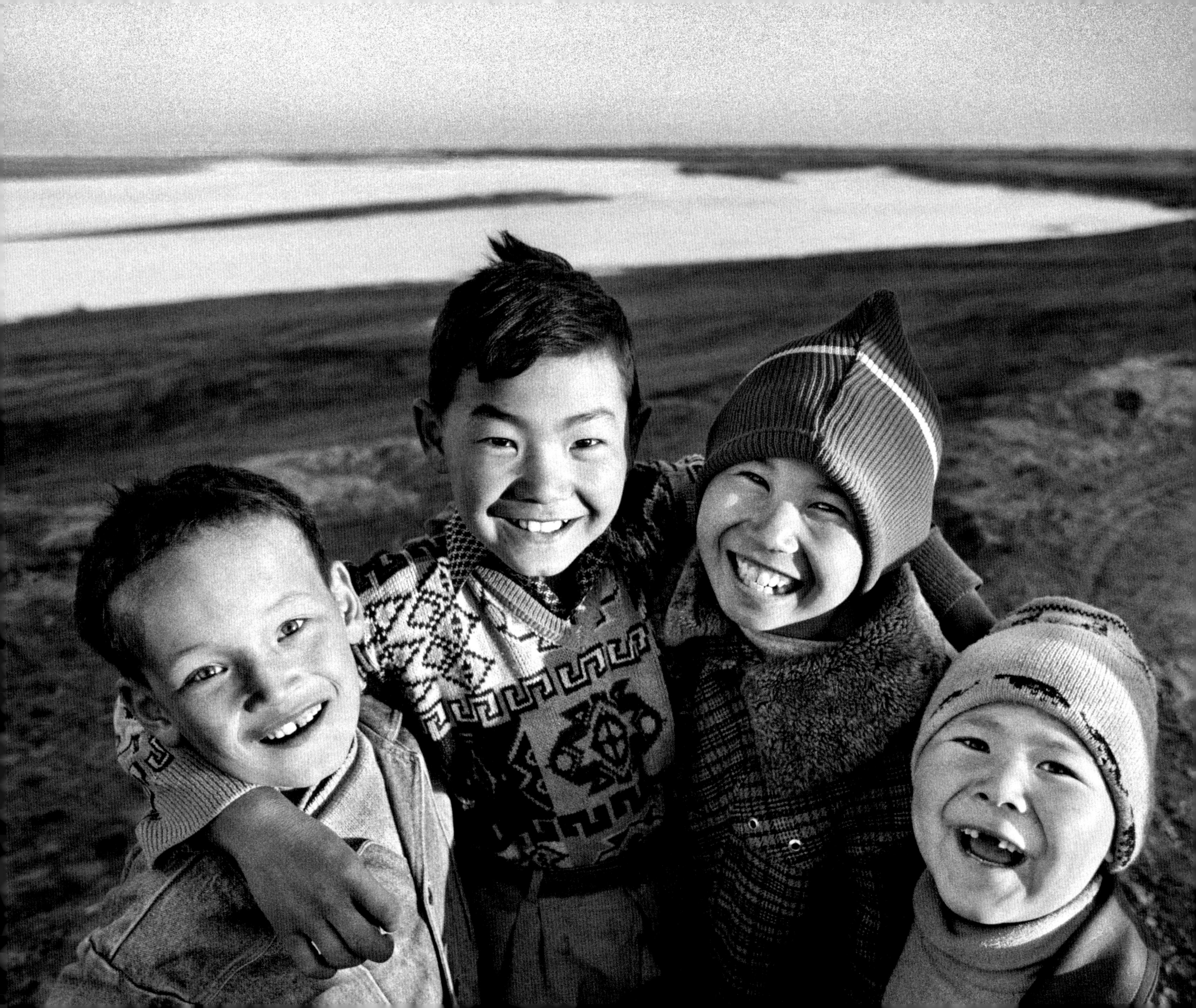

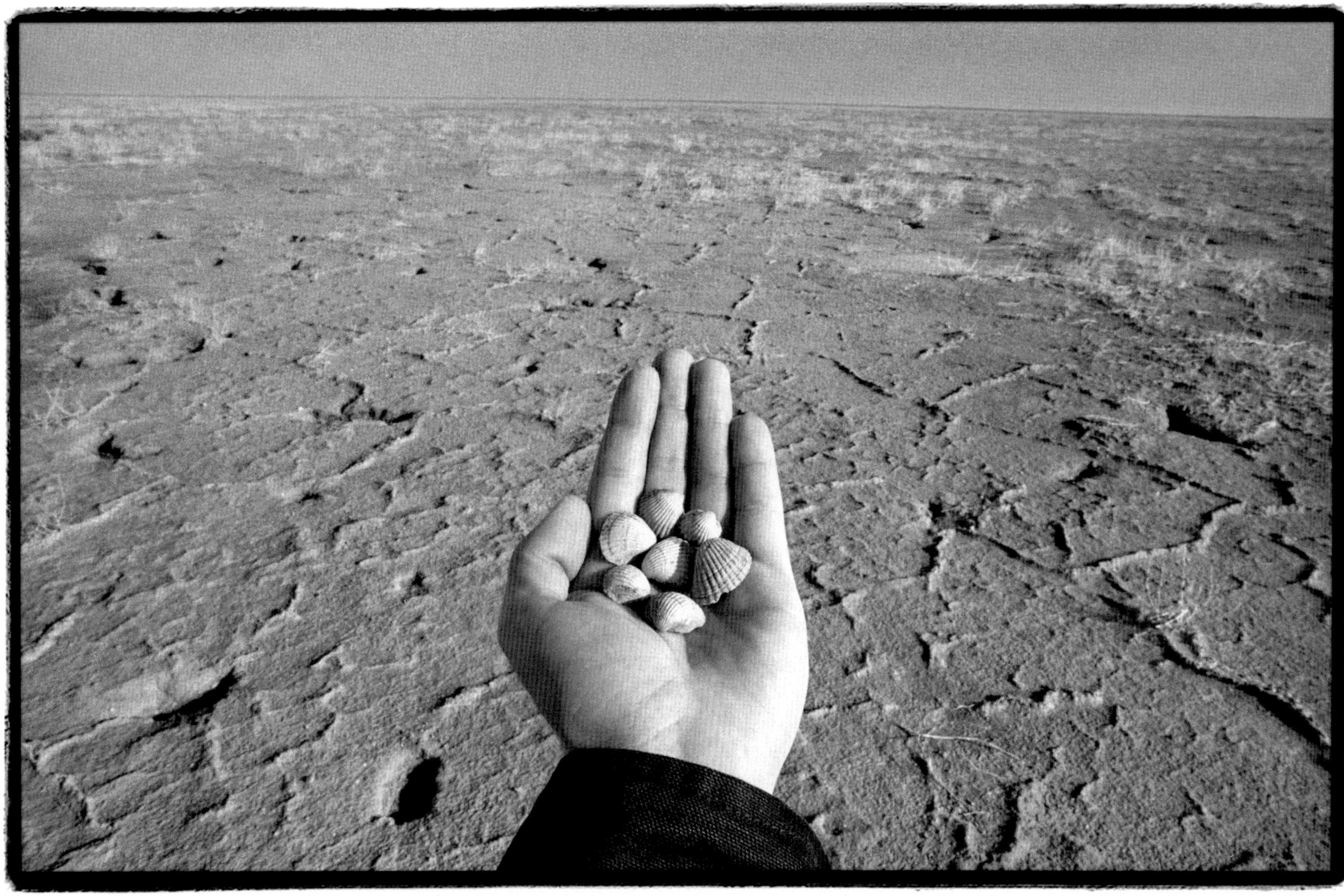

LEFT Muynak. These children are no longer familiar with the Aral Sea. With few prospects, many families have left the area.

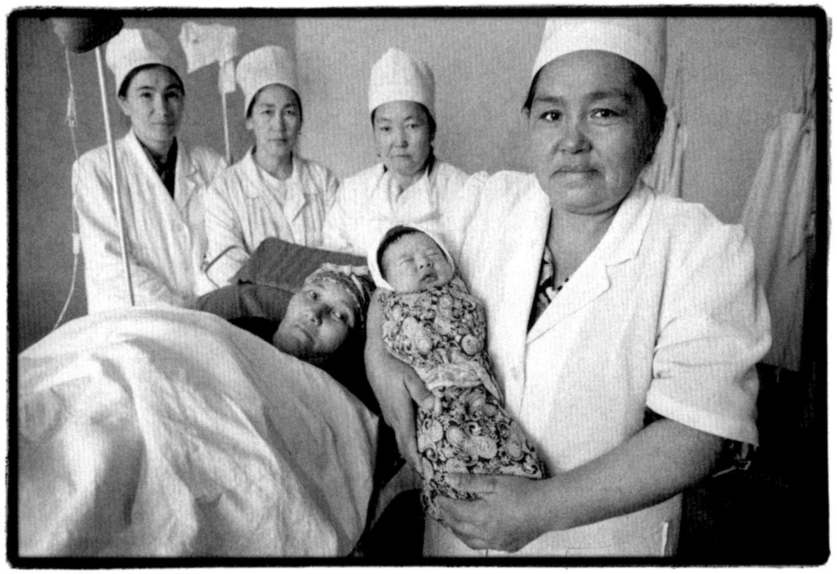

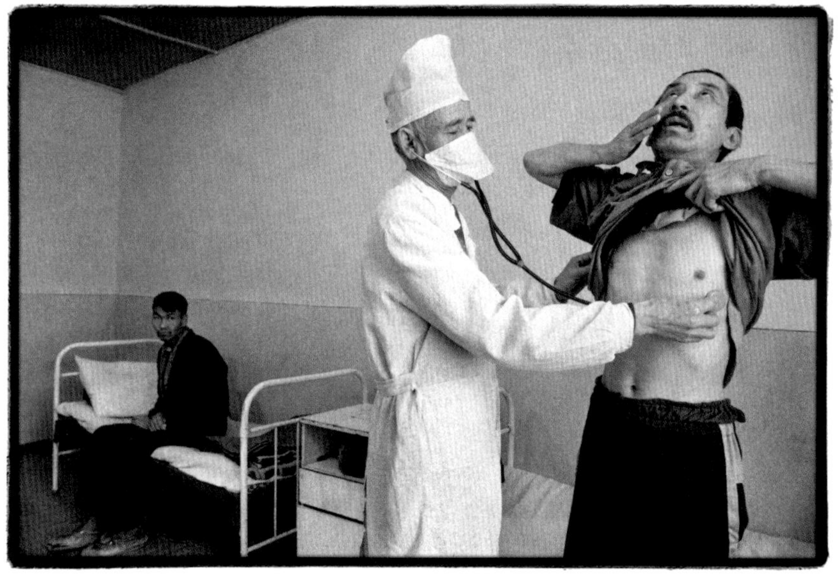

LEFT Muynak. A tuberculosis epidemic is raging in Muynak. The impoverished community is a breeding ground for the tubercle bacillus. The 50 beds in the sanatorium are all occupied. With over 250 TB patients among a population of 100,000, Muynak has the highest rate in the former Soviet Union.

RIGHT Muynak. Ainaev Musa (73), with a photograph of himself in a rowing boat. Musa caught his first fish in the Aral Sea when he was 13. His last fishing expedition dates back to 1980. What remains of his boat he can no longer say. He knows that their vessels were loaded on huge trucks to be taken ever further away from Muynak until, in the end, there were no more fish to catch. His three sons are all unemployed. One of them worked in the building industry, but since the Aral Sea disappeared, nobody has wanted to build a house. Many people have already left the area.

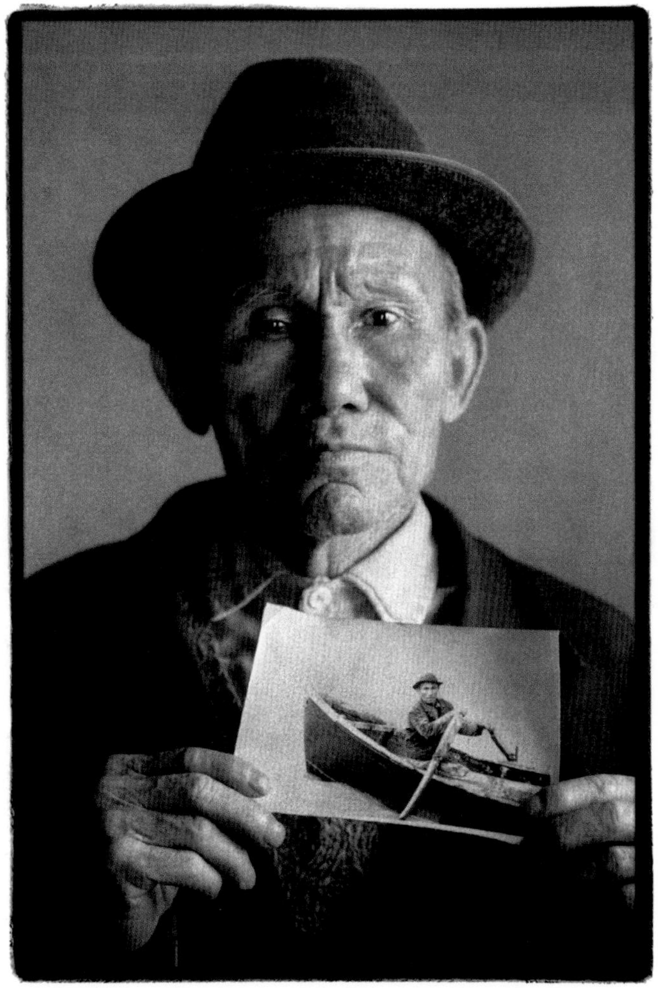

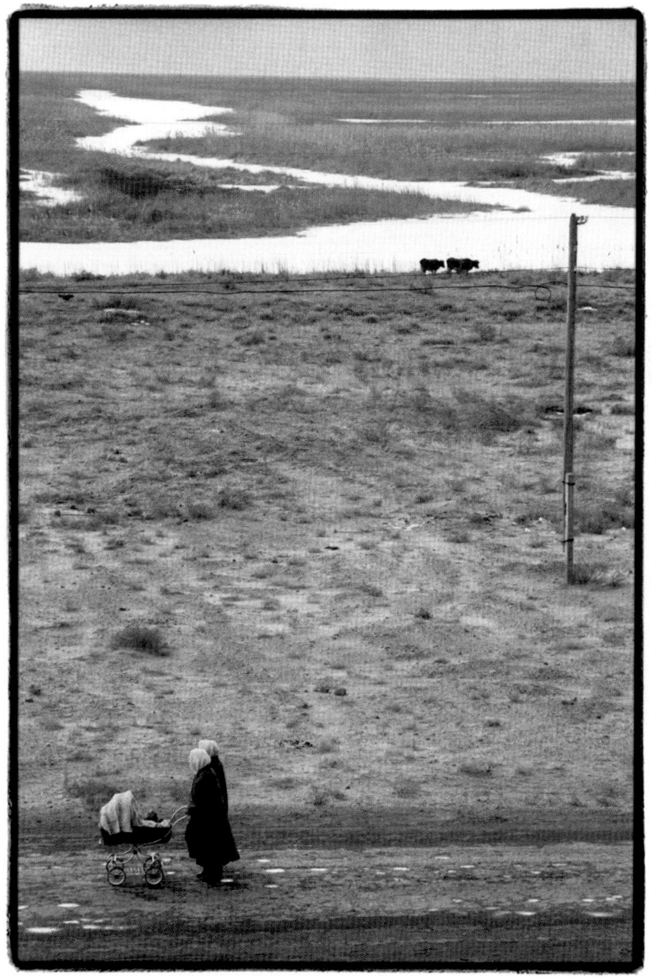

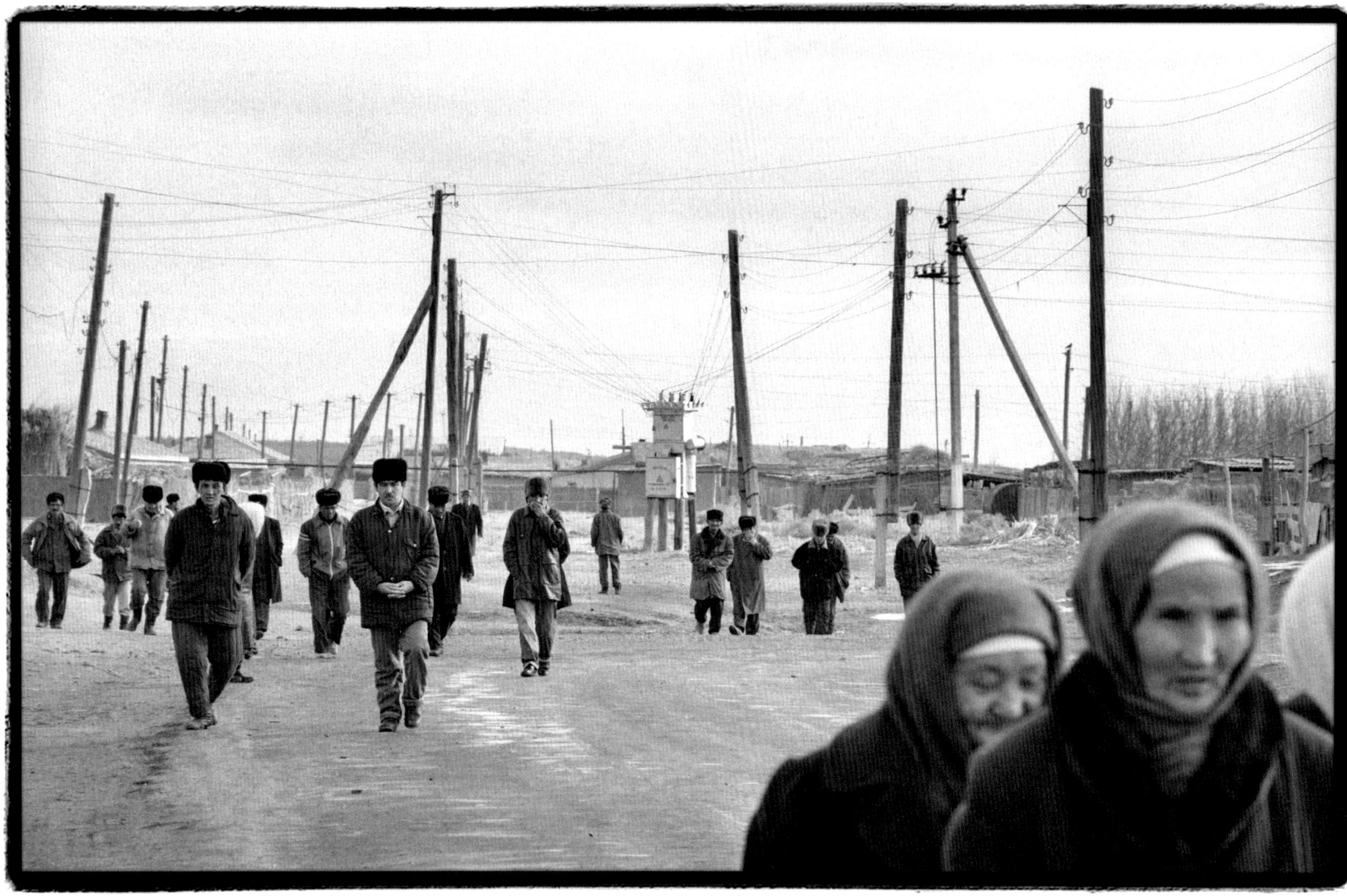

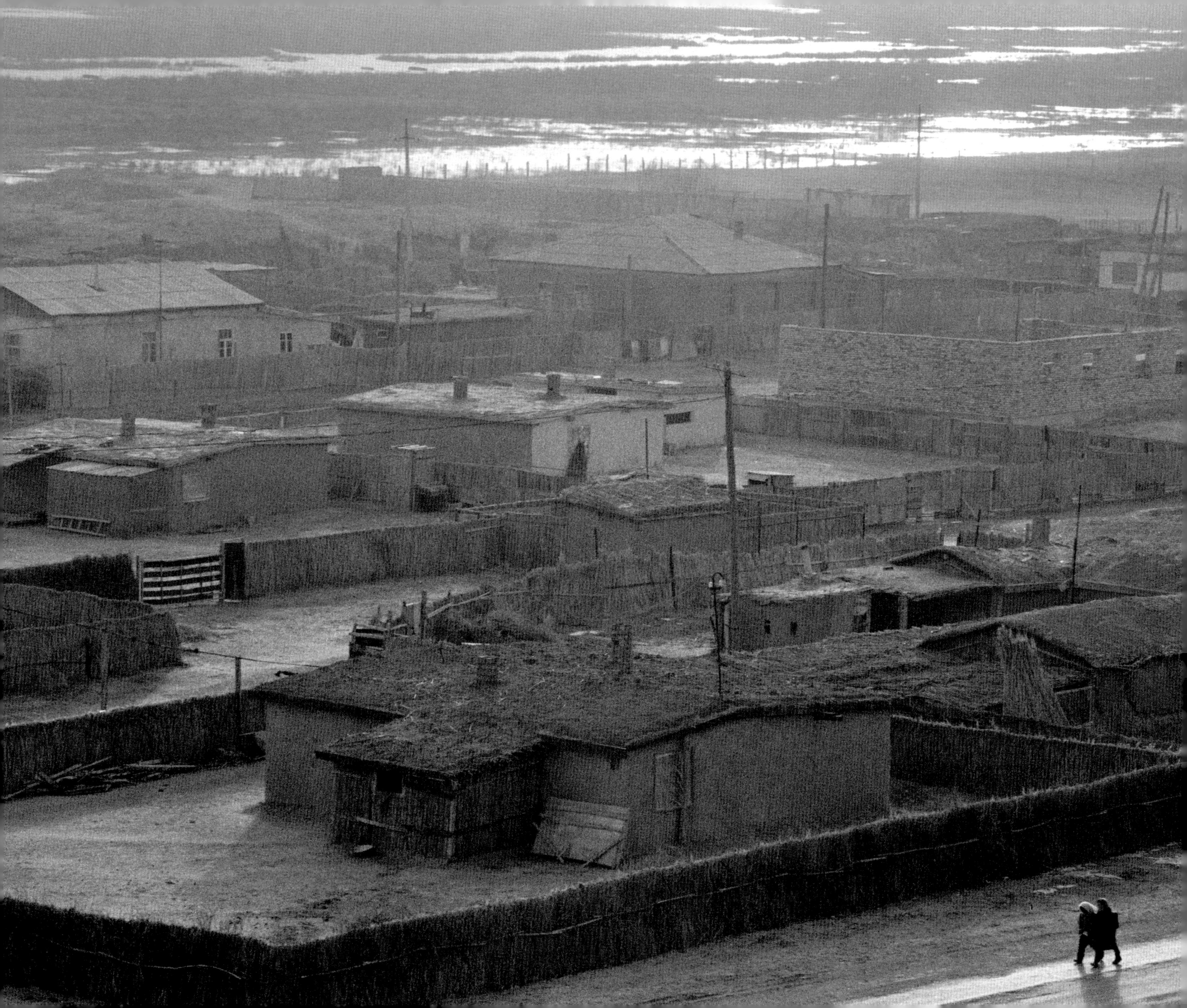

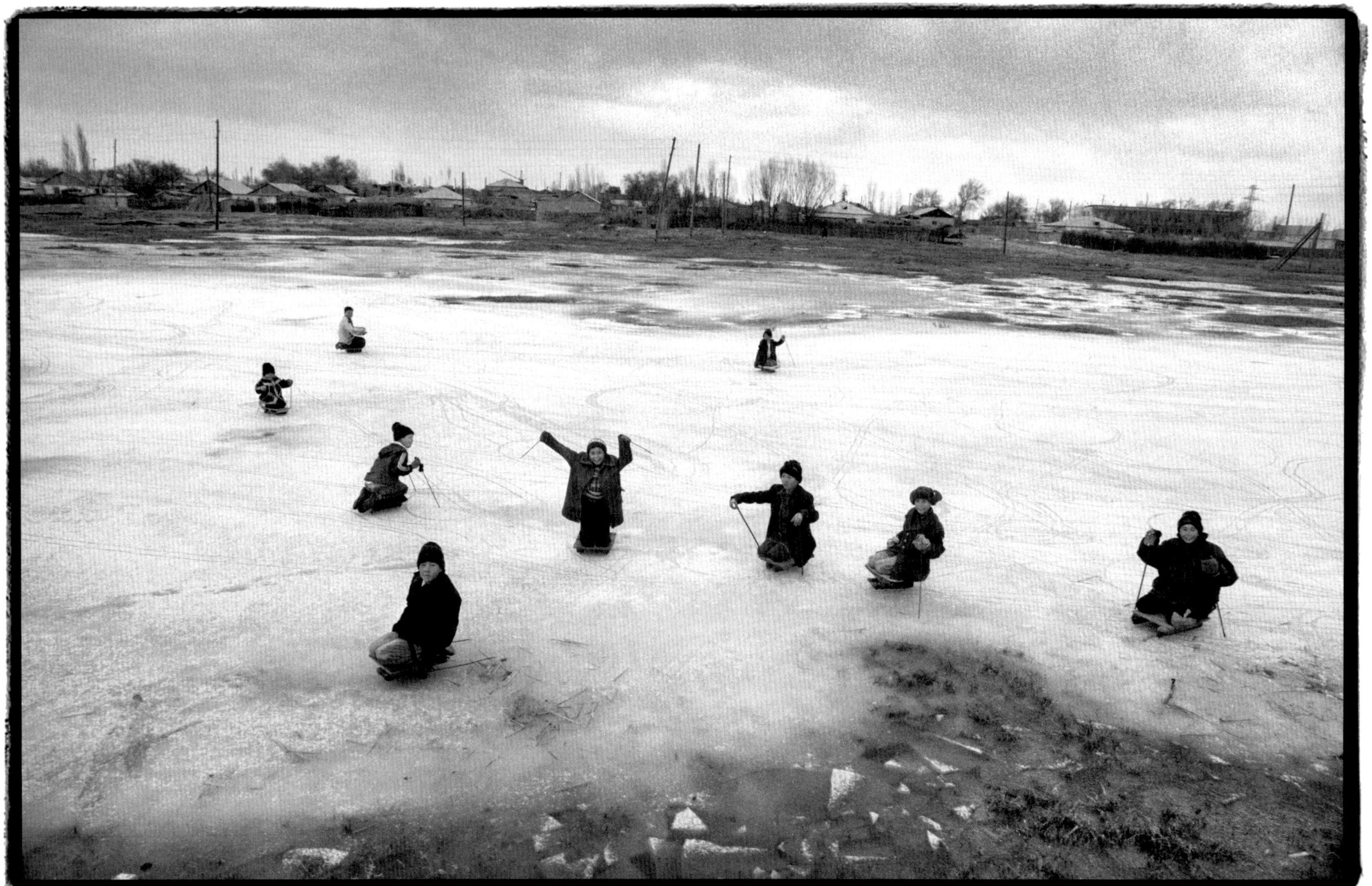

RIGHT Muynak. For the fishermen who stayed, this artificial lake may be their only source of income. If they are lucky, they can catch snake fish here, one of the few species that still survive in the salty water.

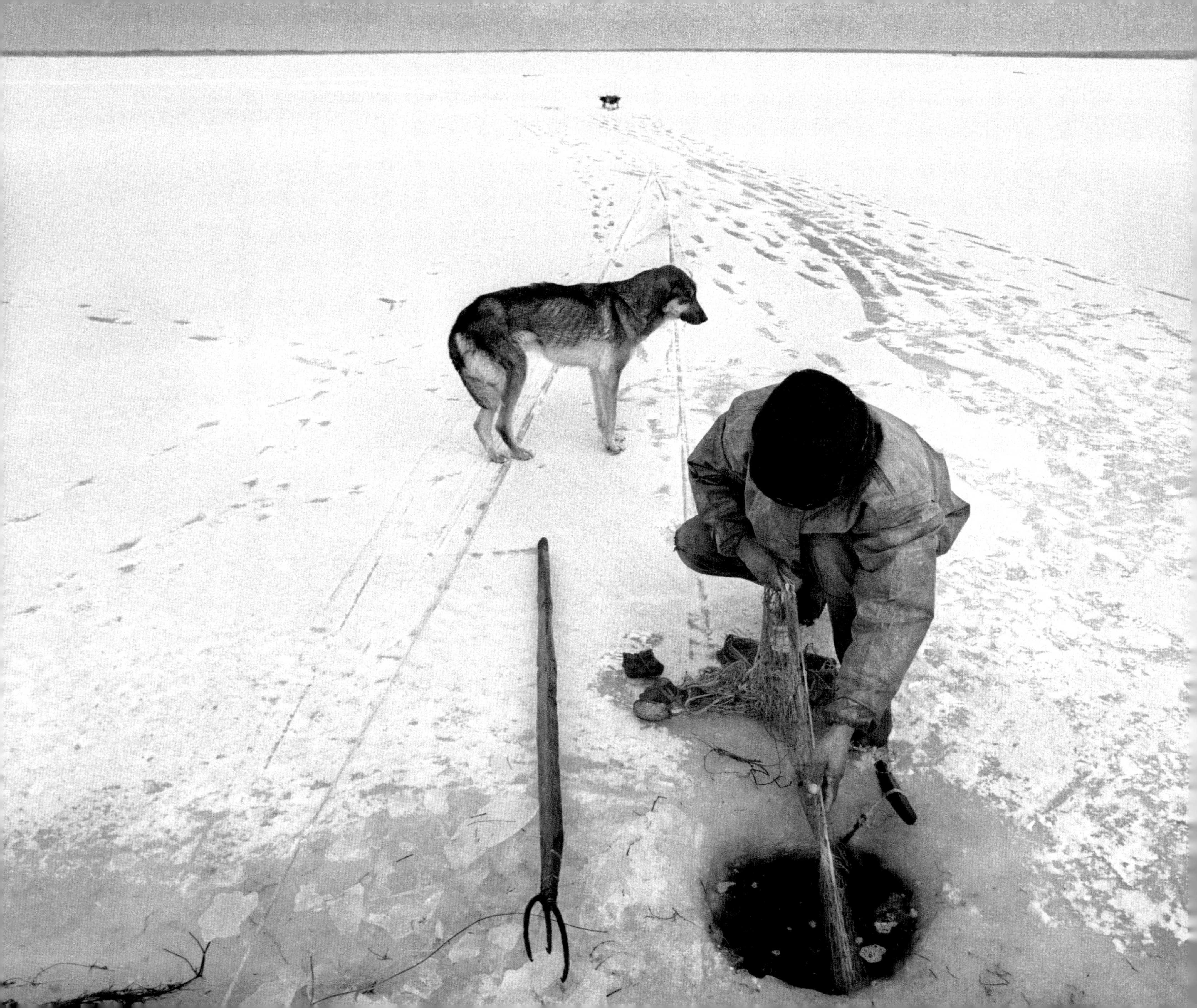

> water scarcity directly affects agriculture and threatens the livelihood of billions

> environmental degradation and desertification are threatening food security

> the consumption of water is growing twice as fast as the world population

> water shortages have created conflicts between nomadic and sedentary tribes

> while 86% of the Nile finds its origin in Lake Tana in Ethiopia, millions of Ethiopians depend on food aid to survive

> Palestinians are prohibited from pumping water

> Boutros Boutros-Ghali: the next war in the Near East will not be about politics, but over water

> the link between environmental degradation, water scarcity and violent conflict is a serious threat

5 AGRICULTURE & CONFLICT

> in the Middle East, 5% of the world's population survives on 1% of the world's fresh water

> competition for water may turn violent in several communities in the Horn of Africa

> whoever controls the water supply in the Near East will dominate economic and political events

> the Nile is a matter of life and death to Egypt, as the country has almost no other source of water

> the Nile is vitally important to the survival of 160 million people in 10 countries

> crop production will be threatened by increases in competition for water and land losses due to desertification and sea level rise

> Cairo said it was ready to use force to protect its access to the Nile

> Israeli settlers in the West Bank use several times more water than their Palestinian neighbours

> agricultural activities consume most water on a worldwide basis

> agriculture uses up to 70 times more water than what is used for drinking and domestic purposes such as cooking, washing and bathing

> the total amount of water used for crop production each year could double by 2050

> irrigation allows wastage on a prodigal scale, with water evaporating before it can do any good

> Middle Eastern leaders have stated that water is the factor most likely to lead to war

> Israel has been increasing the amount of water it takes from the Jordan river as a bargaining chip in peace negotiations

> irrigation accounts for over 70% of all water withdrawals

> crop yields are likely to decrease due to an increase in temperature

> the impact of adverse climate changes on agriculture is exacerbated in Africa by the lack of adapting strategies

> climate change will affect the world's food supply

> climate change will bring increased competition between agriculture and urban as well as industrial users

> fast climate change assessment is vital for safeguarding the world's food supply, UN warns

> instead of encouraging farmers to adapt to the climate by dry farming, for instance, governments are bringing water to them

> 10 million hectares of productive land are lost every year to the desert

> agriculture accounts for about 35% of Africa's gross national product, 40% of its exports, and 70% of its employment

> land used for millet cultivation has become barren sand dunes in east Niger and northeast Nigeria

> poor crop yields are forcing Mexicans to risk death by illegally fleeing to the US

> per capita water availability in the Middle East has fallen by 62% since 1960 and is expected to fall by another 50% in the next 30 years

> growing 1kg of potatoes requires only 100 litres of water, whereas 1kg of beef requires 13,000 litres

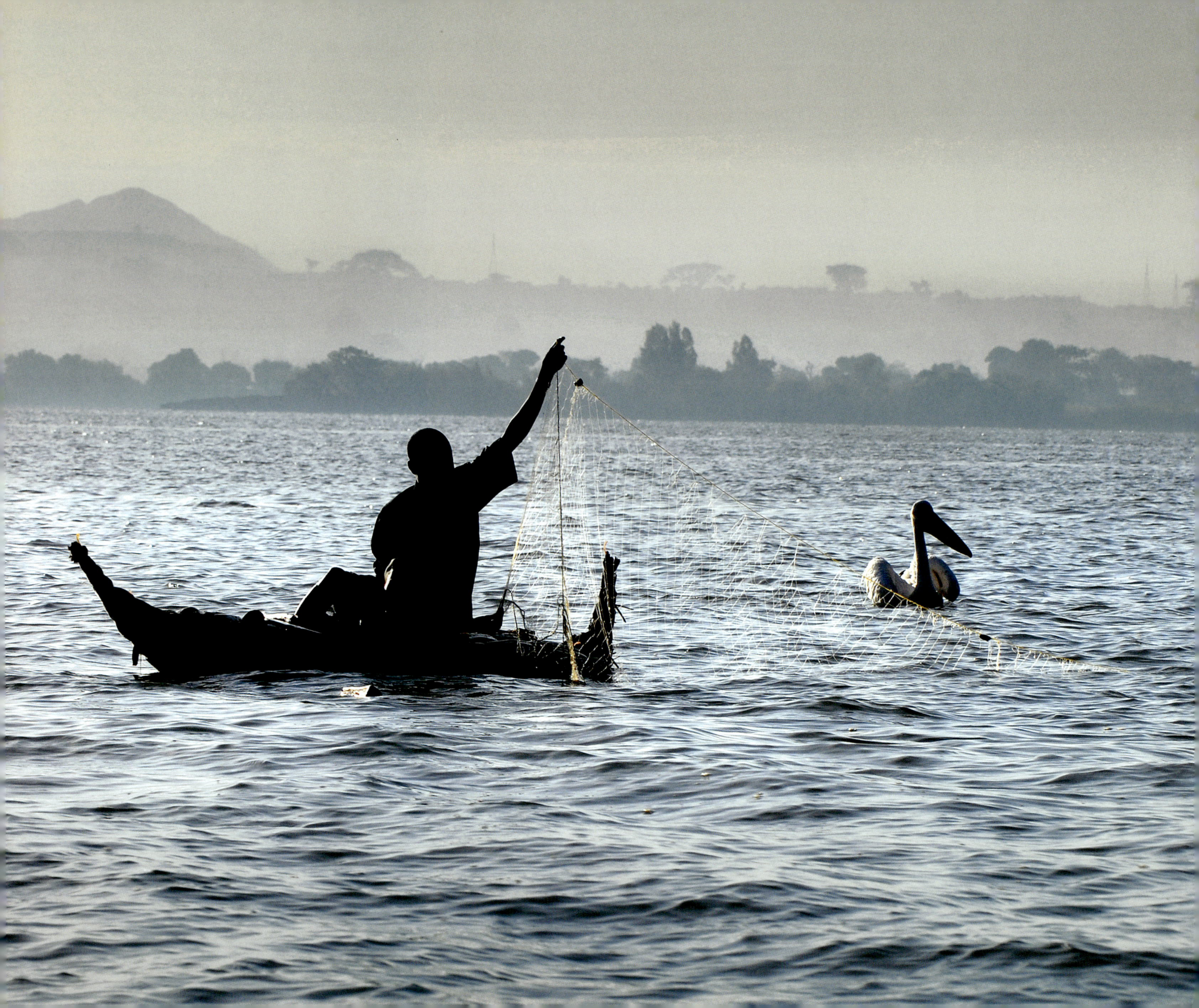

Ethiopia & Egypt
The Nile treaty

The Nile is vitally important to the survival of 160 million people in ten countries that share its basin. To Egypt in particular, the river is a matter of life and death, as the country hardly has any other water resources. A 1929 treaty between Egypt and colonial Britain (that represented Uganda, Tanzania, Kenya and Sudan) said no work would be done on the river that would reduce the volume of water reaching Egypt. The land of the pharaohs sees every change in this treaty as a strategic threat.

Due to population explosion and climate change, tensions have been rising and neighbouring countries question the colonial treaty. Ethiopia, where 86% of the Nile water finds its origin in Lake Tana, demands its own share of the water. With the vastness of Lake Tana, it is hard to grasp why 2.5 million farmers, in this region of Ethiopia alone, depend on food aid to survive. The Ethiopian government says it is because agriculture, which so much of the population depends on, is at the mercy of the erratic seasonal rains and the 1929 treaty.

The very waters that stream untouched through Ethiopia are used in Egypt in massive irrigation projects. With the construction of the Aswan High Dam, Egypt controls the Nile's flow and farmers have multiplied their harvests. Not satisfied with this, President Hosni Mubarak has launched the construction of the ambitious Toshka Project, which will create a second Nile Valley in the middle of the Western Desert. The Sheikh Zayed Canal will transfer one tenth of the Nile's water from the artificial Lake Nasser to half a million acres of desert. The project –scheduled for completion in 2017– intends to create 2.8 million new jobs and attract over 16 million people to the new towns which are planned.

Given the continent's acute shortage of water, other countries in the region question whether it is justified for Egypt to grow crops in the desert at such a cost. In Ethiopia, where only 3% of the land is irrigated, the government is determined to utilise the Blue Nile to lift itself out of poverty. In 1999, the ten Nile States signed the Nile Basin Initiative and promised to collaborate on water management. Dialogue would be 'the new gift of the Nile,' they promised each other. Up to now, talks are taking place, but an agreement is yet to emerge. The United Nation's World Food Programme says that with nine to fifteen million Ethiopians in need of food aid, the talking should not go on indefinitely. The only solution is to consume less and agree on mutual cooperation.

PREVIOUS Ethiopia, Bahir Dar. A man is fishing in Lake Tana. This is the source of the Blue Nile where 86% of the Nile water originates.

RIGHT Ethiopia, Bahir Dar. The Blue Nile leaving Lake Tana for Sudan and Egypt. Because of the Nile Treaty, which dates back to 1929, Ethiopia is prohibited from using this water for irrigation purposes.

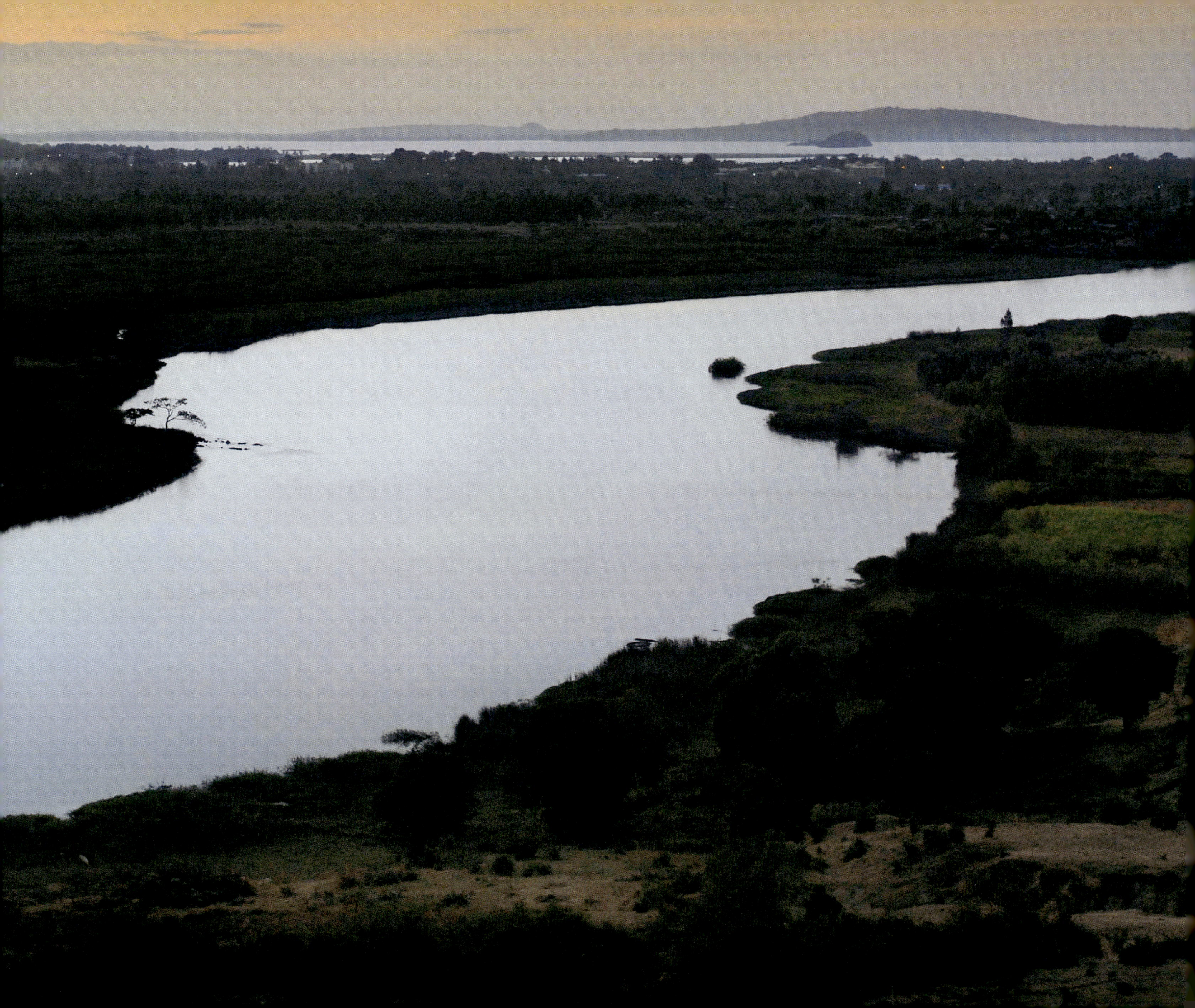

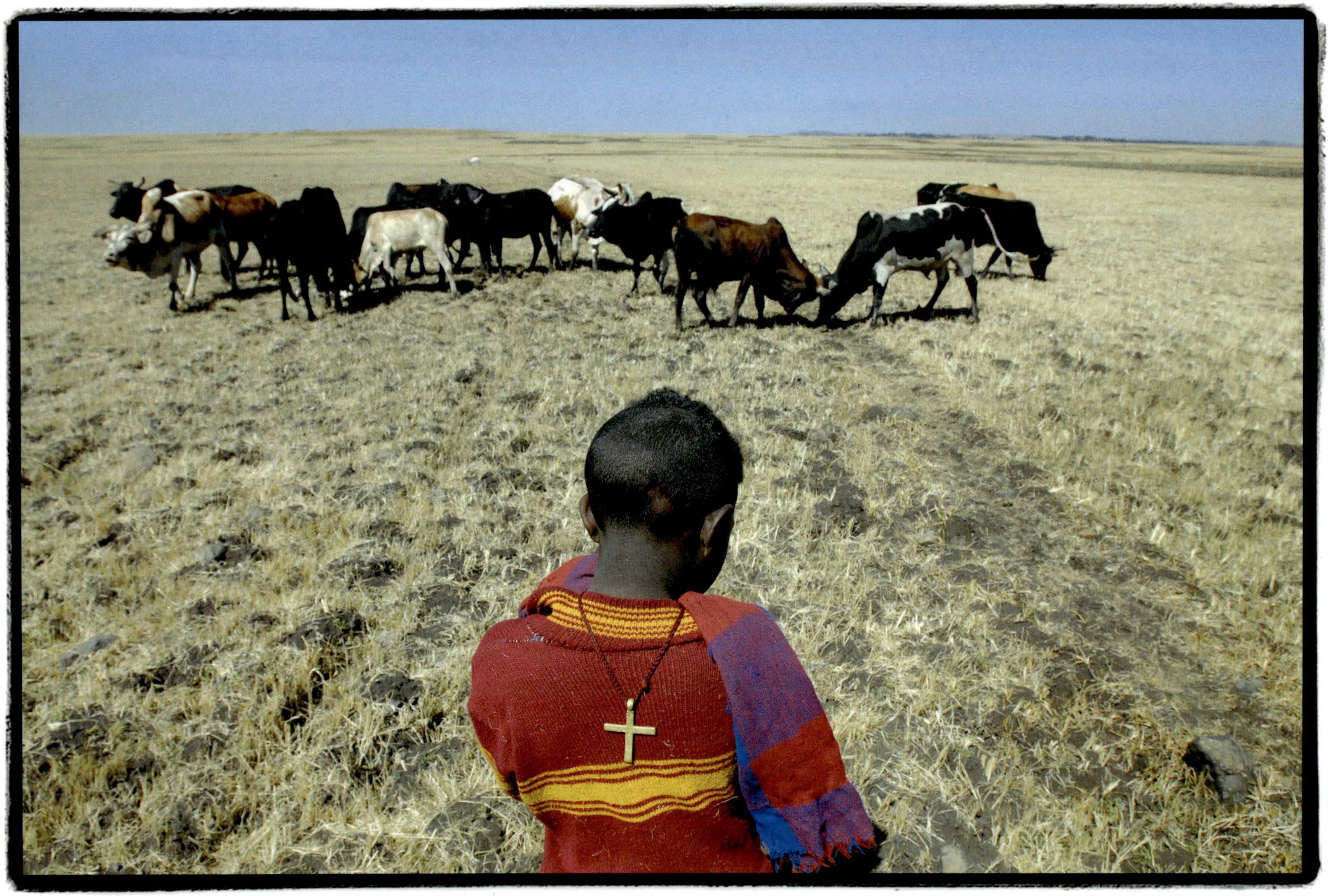

Ethiopia, Bahir Dar. Farmers living next to Lake Tana depend on rain for their harvests. They are too poor to buy water pumps and Ethiopia is prohibited from starting big irrigation projects.

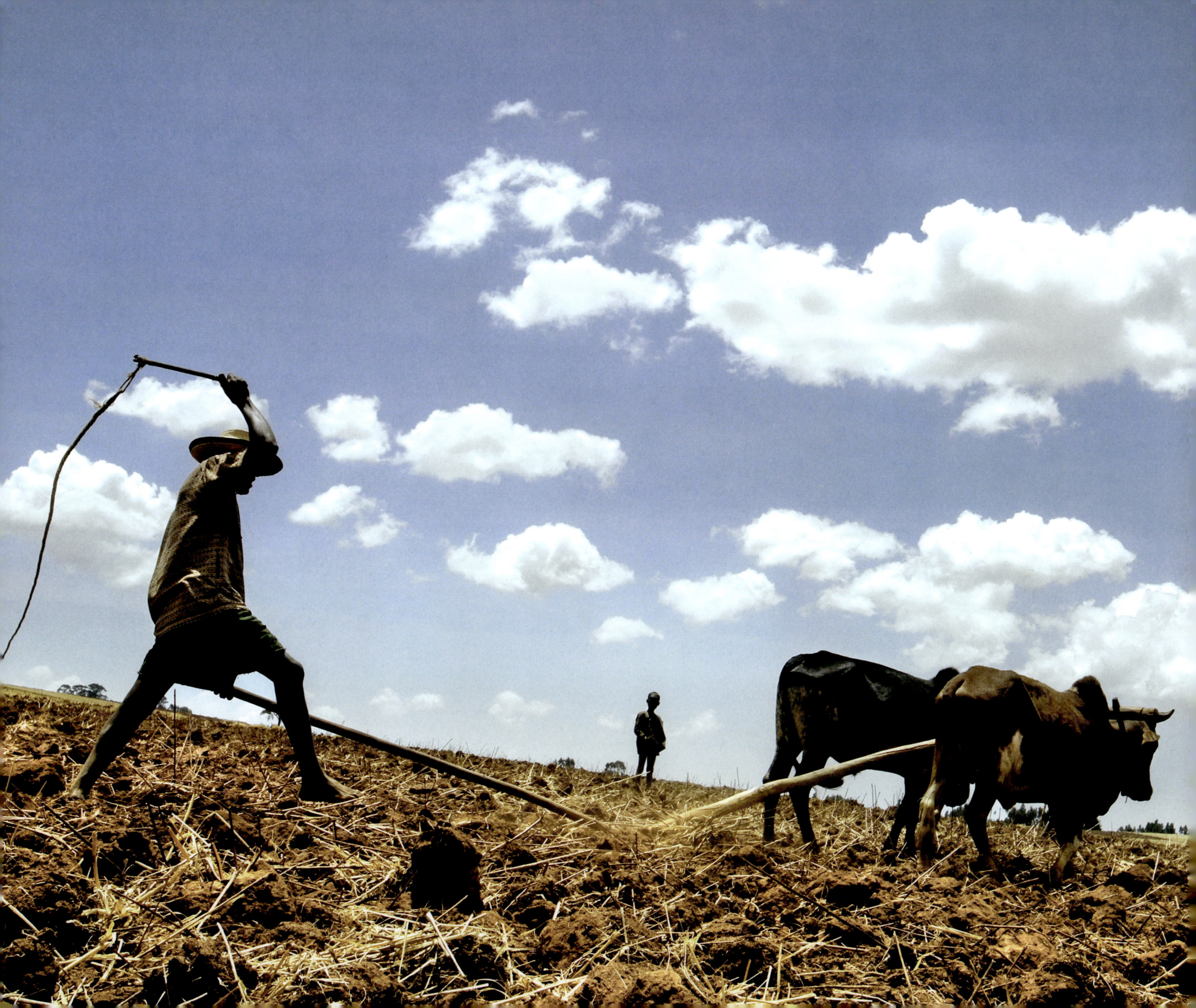

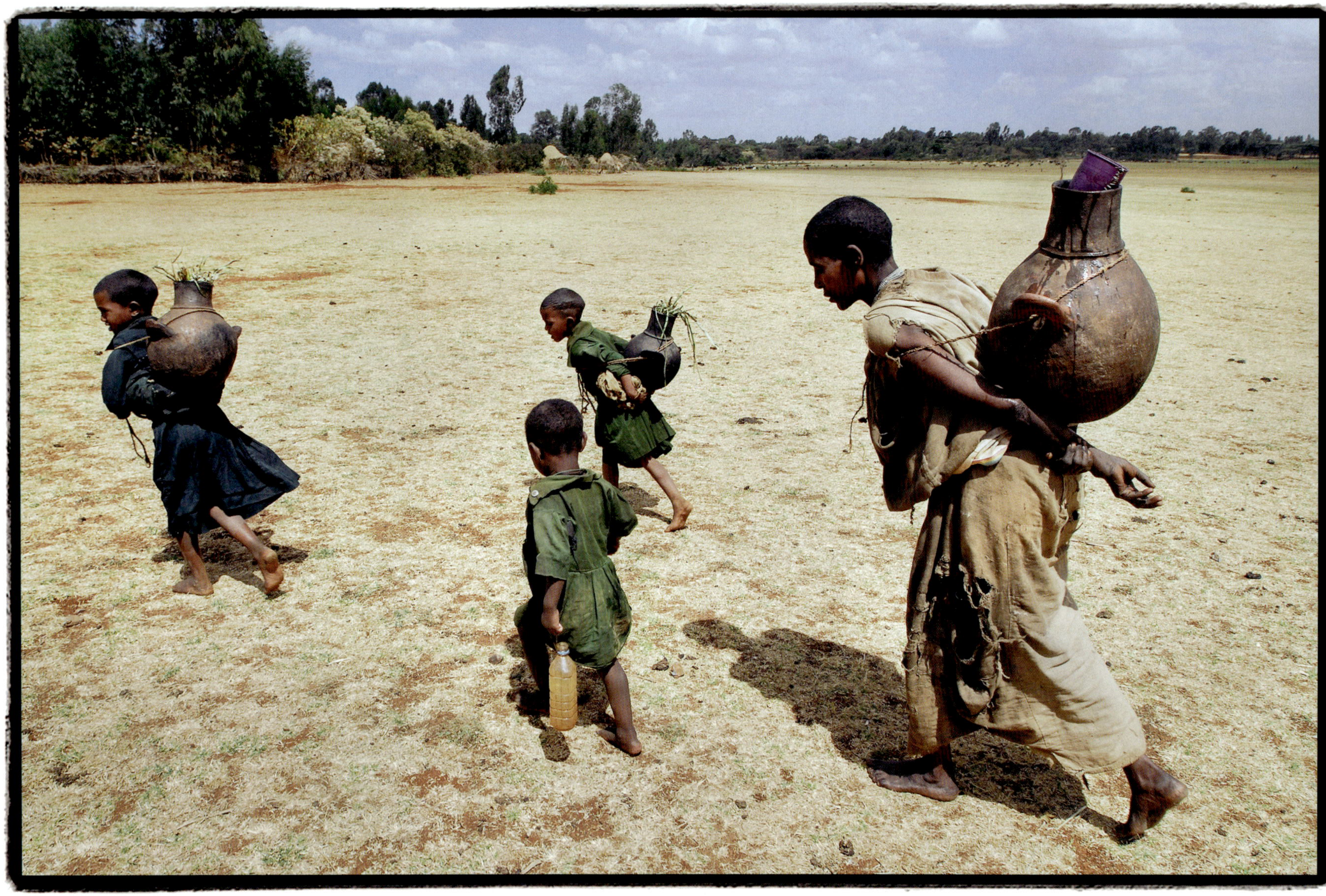

RIGHT Ethiopia, Nazareth. Men are carrying sacks containing lentils into a food aid warehouse of the United Nations World Food Programme (UN WFP).

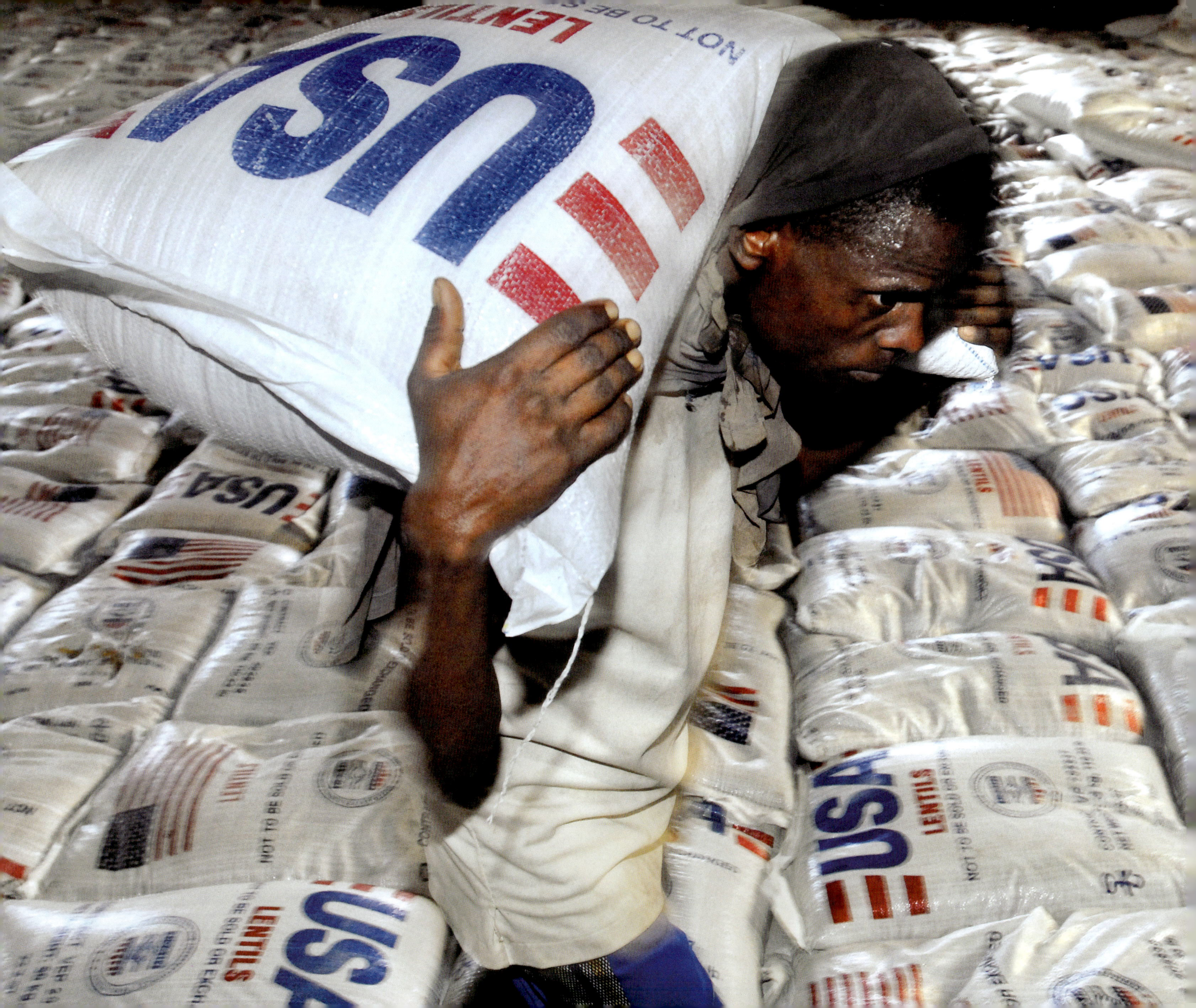
USA
LENTILS
NOT TO BE SOLD OR EXCHANGED

RIGHT Egypt, Aswan. Tourists visit the Aswan High Dam across the River Nile, and Lake Nasser. This is the world's largest fresh water reservoir. It extends up to 550 km from the dam and covers a total surface area of 5,250 km^2. Water experts say this artificial shallow lake in the middle of the desert is an immense source of water evaporation: between 10 to 16 billion m^3 each year. Ethiopians say that the construction of several smaller dams along the deep ravines of the Blue Nile in Ethiopia would have been much more water efficient.

LEFT Egypt, Toshka. The Sheikh Zayed Canal diverts water from Lake Nasser to irrigate half a million acres in the Western Desert. The Toshka Project involves the construction of a 50km-long main transfer canal and four additional 22km side branches. Environmentalists are sceptical about the possible success of this farming project in the middle of the desert. The Toshka Project is nicknamed "Mubarak's pyramid."

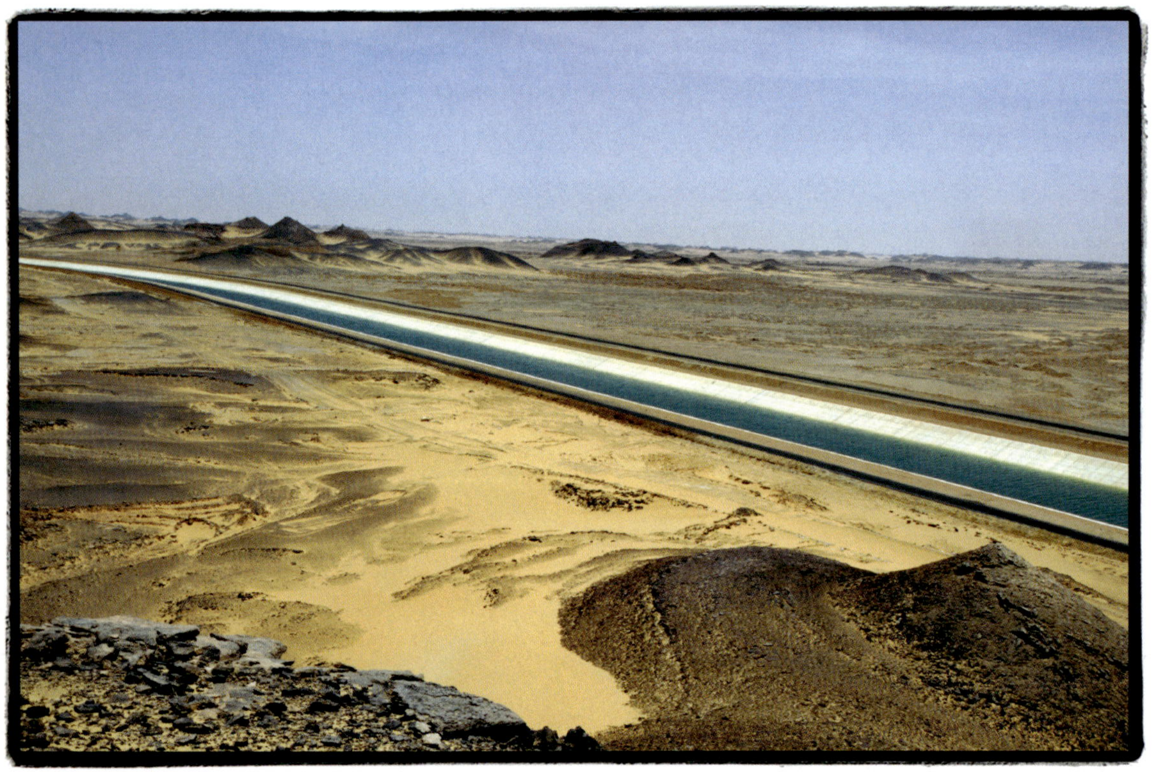

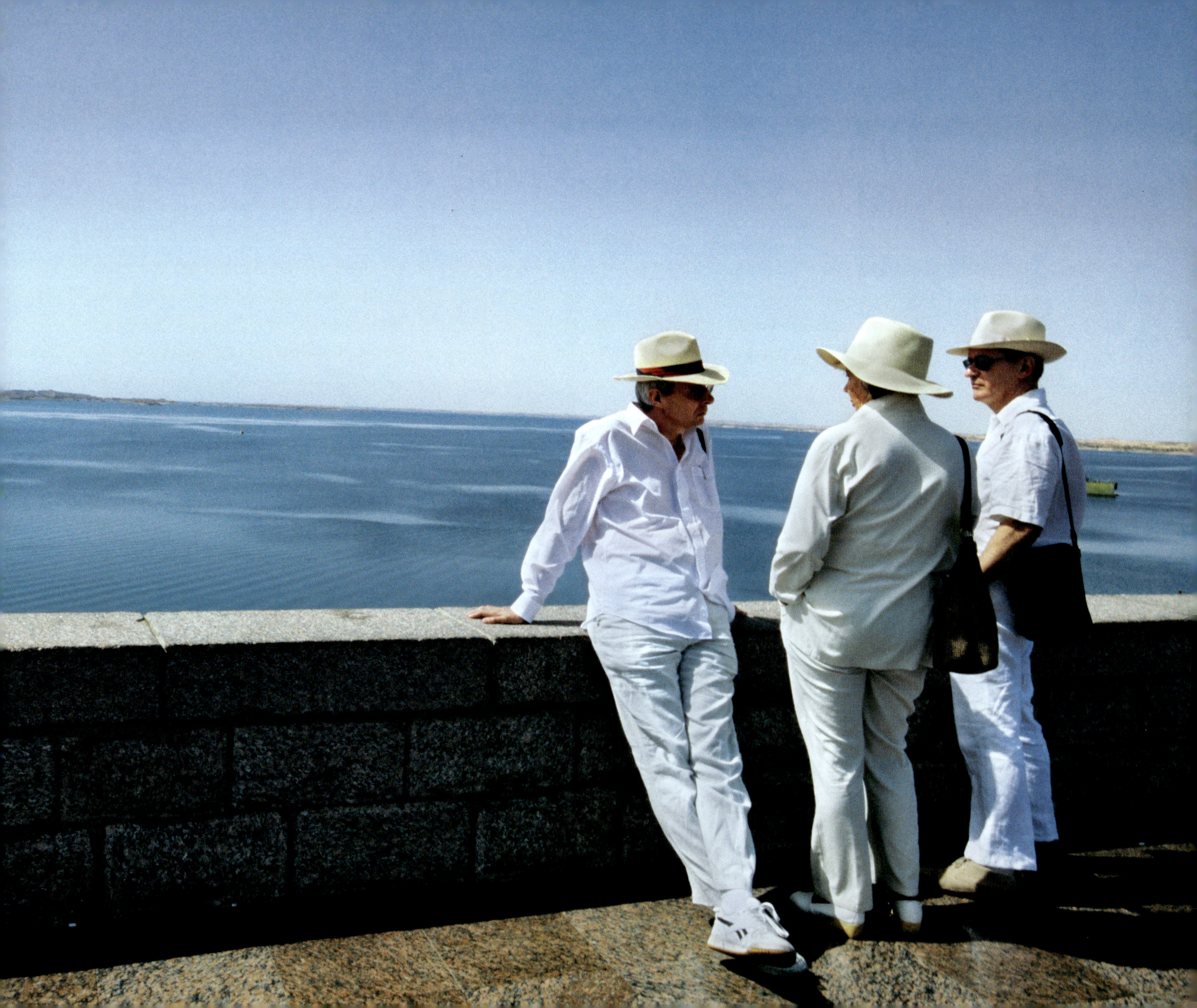

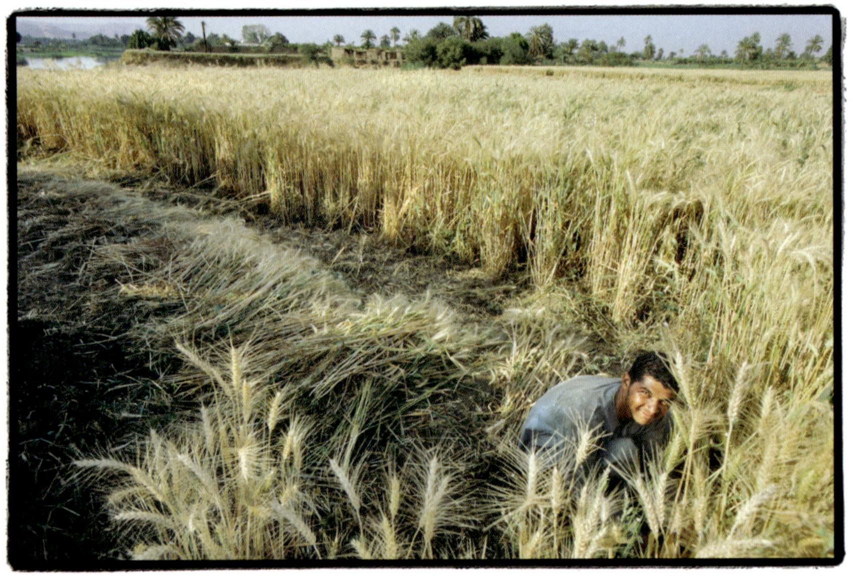

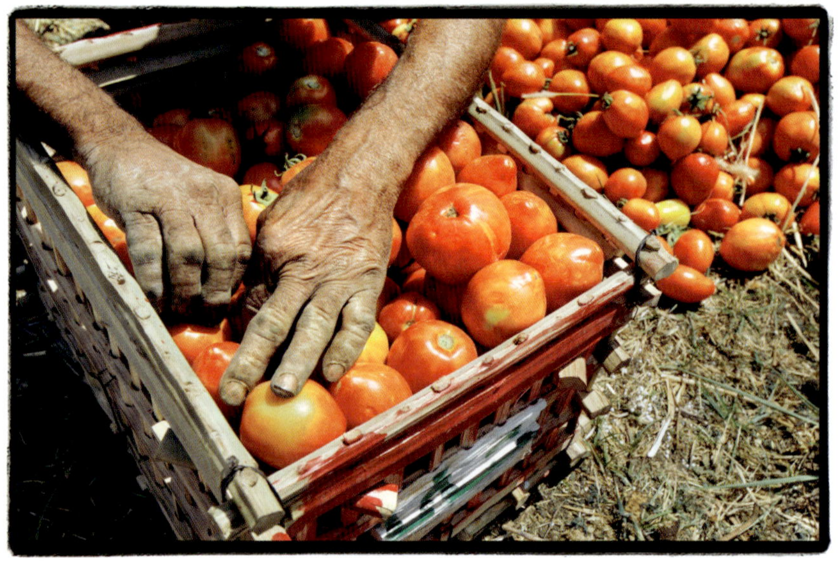

Egypt, Nile Valley, Luxor. Farmers in their fields on the shore of the River Nile. With the construction of the Aswan Dam and the High Dam, the flow of the Nile River is under control year round and farmers have been able to double or even triple the number of harvests.

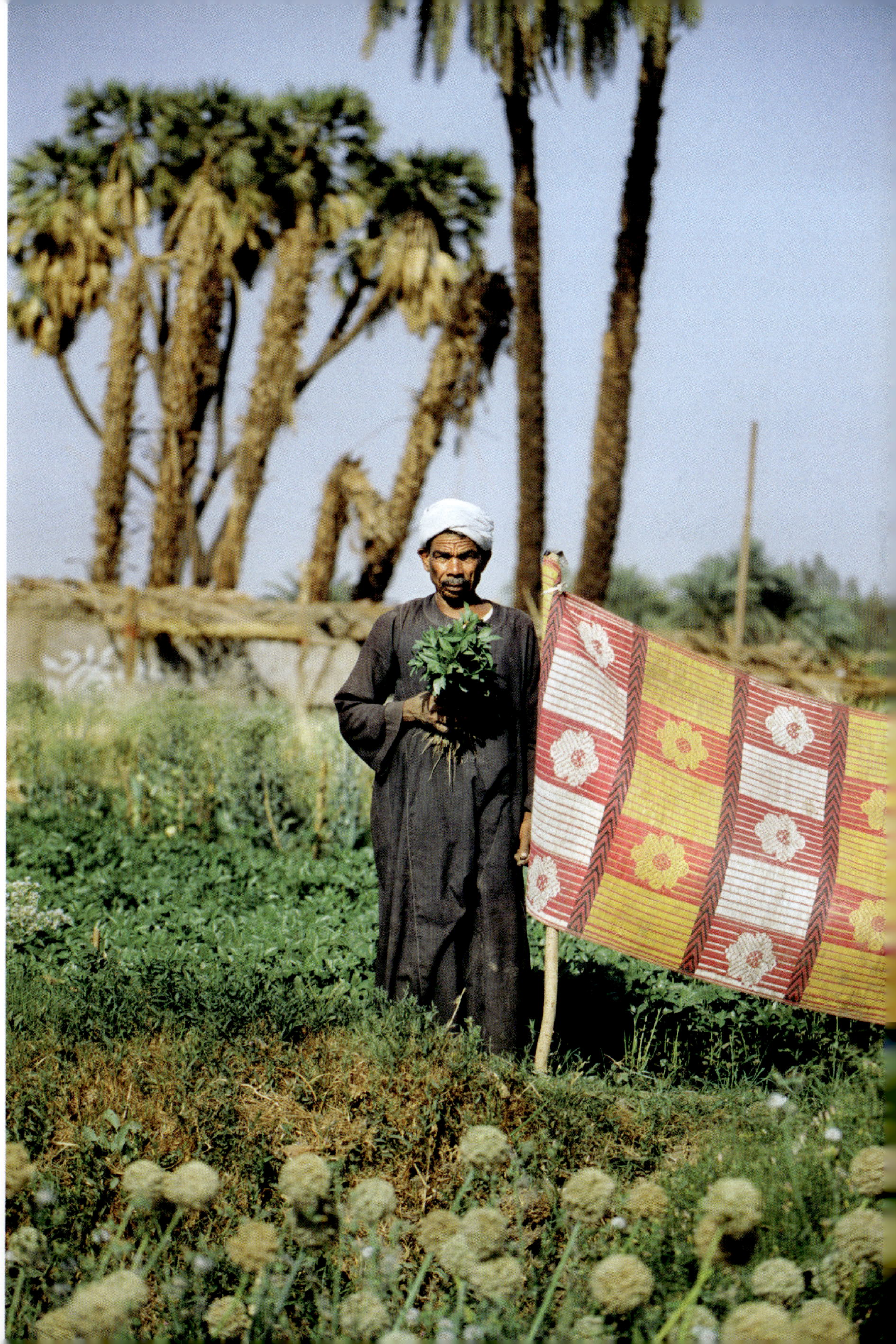

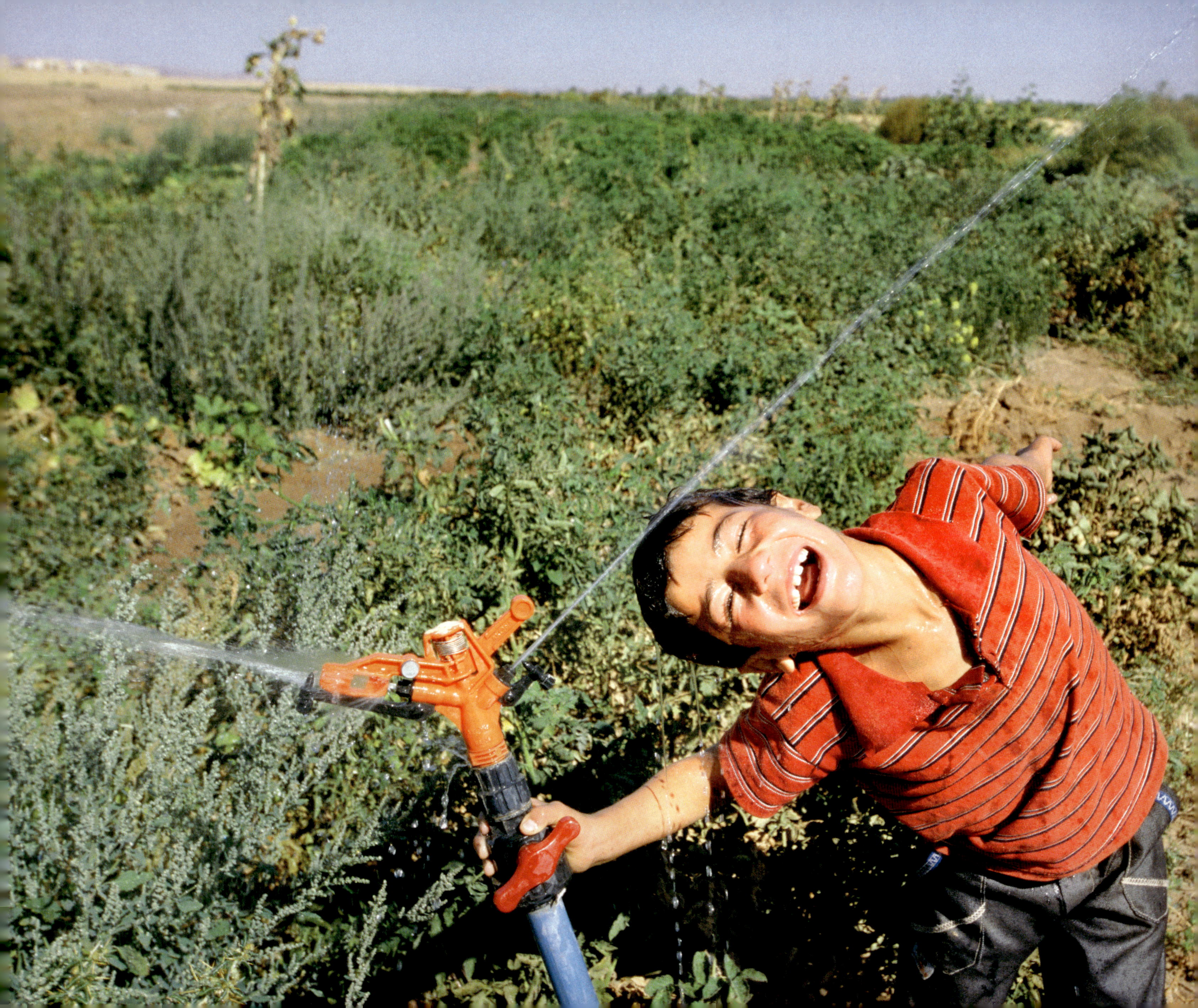

Turkey
The world's biggest irrigation project

Turkey is water-rich by Middle Eastern standards. However, its water supply still amounts to only about one-sixth of the water available per inhabitant in Western Europe. With the rapid population growth, industrialisation and rising standards of living, Turkey's water resources are becoming less than adequate.

Turkey aims at promoting economic growth and social stability by the optimal use of land and water resources. In 1977, the country started one of the world's biggest construction projects in Southeast Anatolia, the most backward region of the country. The project consists in the construction of 22 dams, 19 electrical power stations, and 2 irrigation tunnels.

The US$ 32 billion Southeast Anatolia Development Project (referred to as GAP) is expected to bring life-saving water to a region about the size of Austria and provide almost one-quarter of the country's electric power. The total area of the project covers about 10% of Turkey, and includes 9.5% of the country's total population. With these dams Ankara hopes to create 3.3 million jobs and become self-sufficient in food production. Turkey estimates its national income will grow by 12%, and the gross regional product of Southeast Anatolia will increase more than fourfold.

But not everybody is happy. The project is developed in the area of the Euphrates and the Tigris. These watercourses flow through Iraq and Syria before reaching the Arabian Gulf. Syria and Iraq are worried that the project, of which the Ataturk Dam and the Sanliurfa Tunnel System are already in operation, will in the end limit their own access to water. Turkey agreed to supply 500 cubic meters per second to Syria. But the latter country wants more because it already has a shortage and cannot maintain its economy at its current level without a bigger supply. In the past, Ankara threatened Syria to reduce its water supply if Damascus continued supporting the Kurdish PKK. And Iraq has threatened to bring Turkey before an international tribunal for not recognising the international agreements of downstream countries on dam projects.

The project is expected to give Turkey an enormous strategic advantage in a part of the world chronically plagued by water scarcity. Many experts predict that during the 21st century, whoever controls the water supply in the Near East will dominate economic and political events to a great extent. Former UN Secretary General Boutros Boutros-Ghali once warned an American think tank that "the next war in the Near East will not be about politics, but over water."

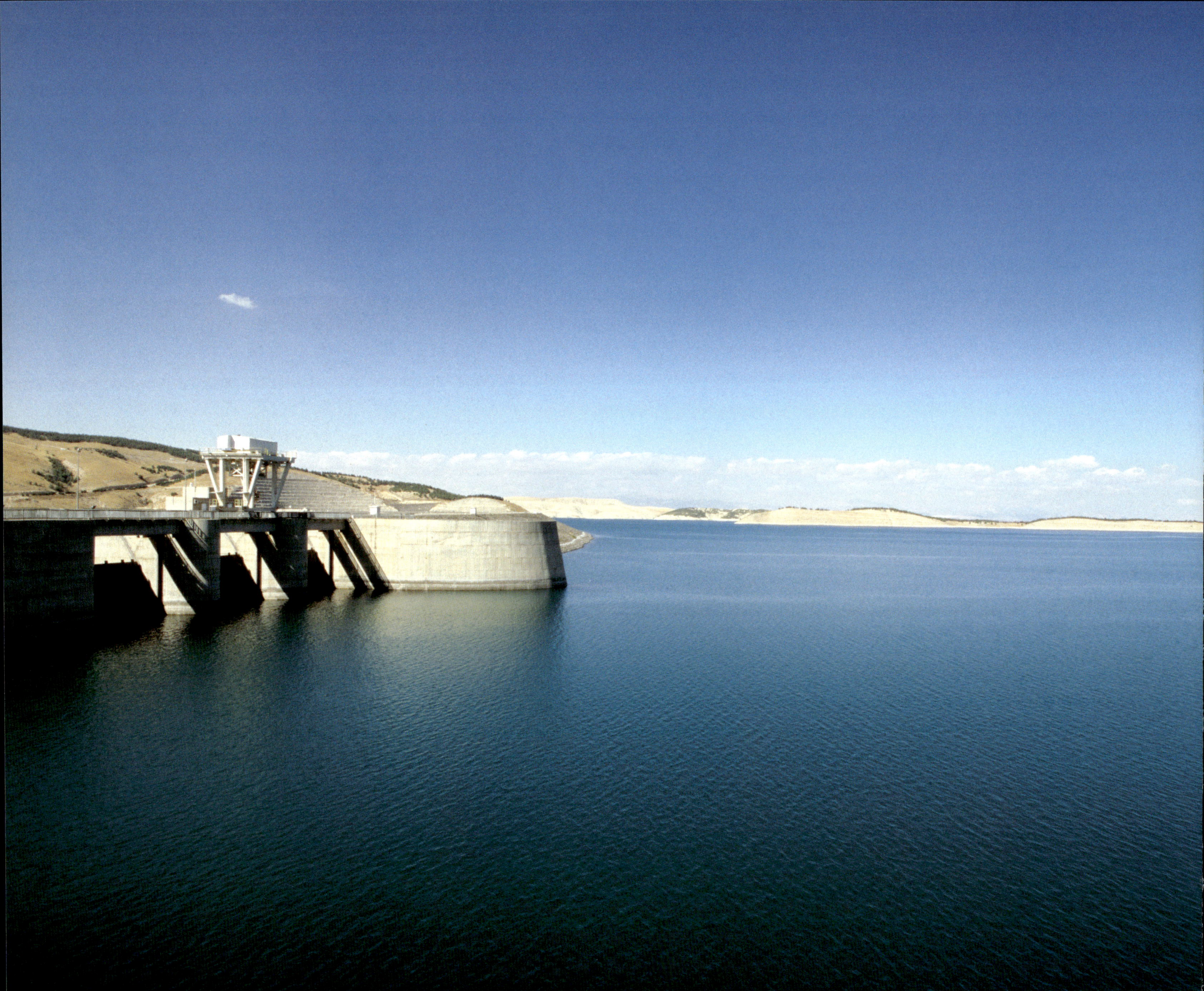

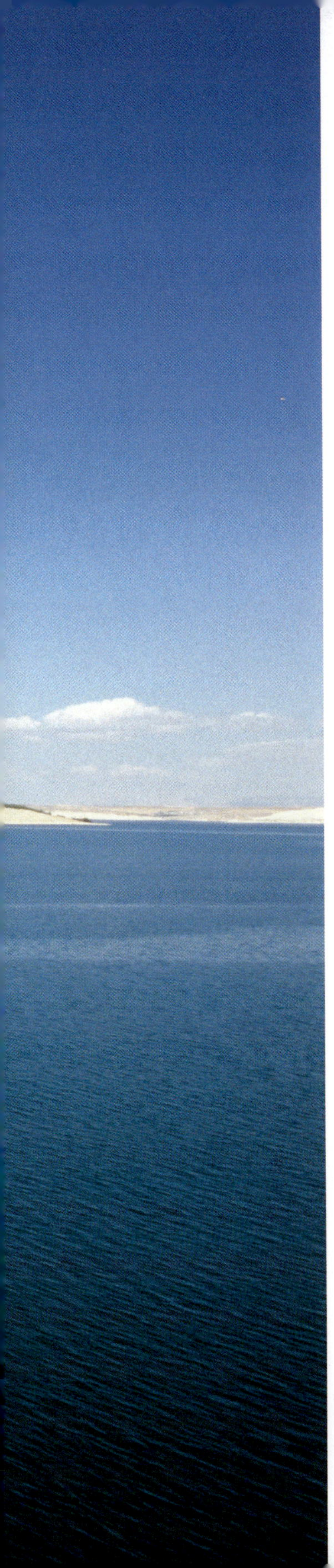

Southern Anatolia. The Ataturk dam on the Euphrates River is the centrepiece of the Southeastern Anatolia or GAP Project. Completed in 1990, the dam is one of the world's largest earth-and-rock fill dams. Water impounded behind the dam is fed to power-generating units at Sanliurfa that have a capacity of 2,400 megawatts. From there the water is gravity-fed to vast irrigation networks in the Harran Plain and elsewhere in the region.

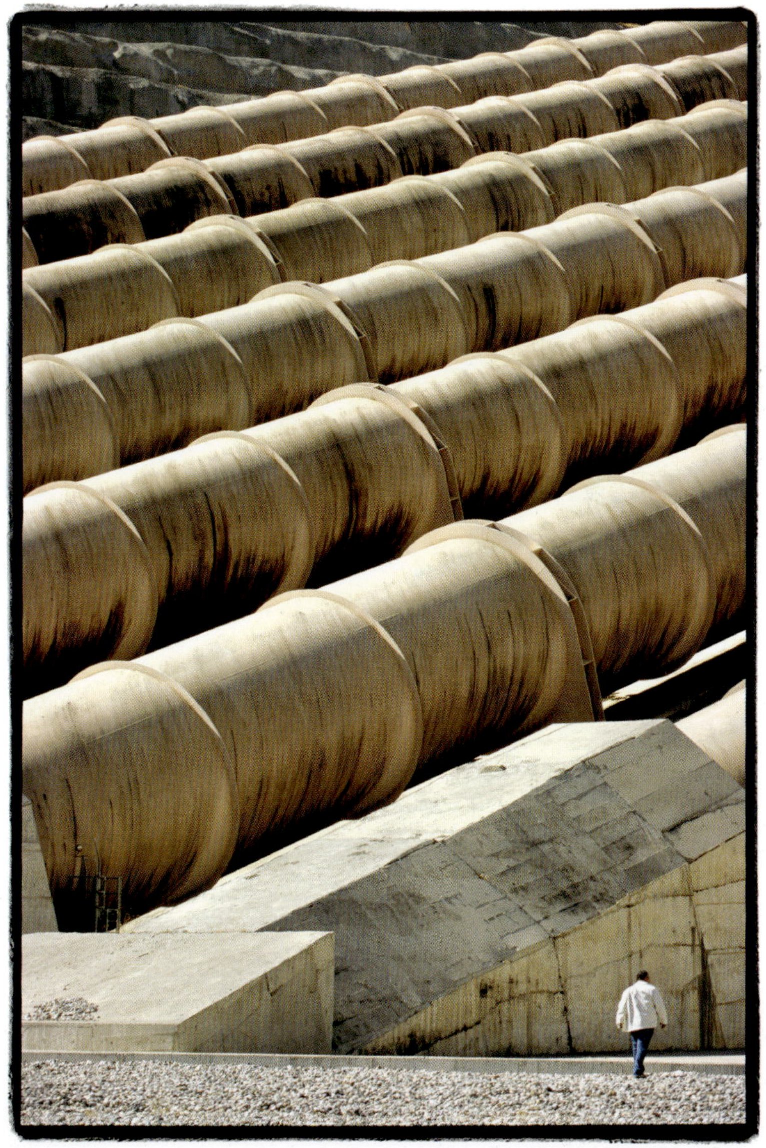

LEFT Southern Anatolia. The canal between Sanliurfa and Mardin. The canal is transferring Euphrates River water from the artificial lake of the Ataturk Dam towards desert land in the great plain of High Mesopotamia.

RIGHT Southern Anatolia, Harran Plain. This plain used to be desert land. Irrigation water from the Euphrates River has transformed the landscape into green thirsty cotton fields. In a region where water is scarce one wonders where all this water comes from and what the consequences are for people living downriver. Since the 1980's, Syria and Iraq, the southern neighbours of Turkey, have put forward claims regarding the waters of the Euphrates and Tigris originating in Turkey. They have tried various methods to hinder the works started by Turkey on these rivers.

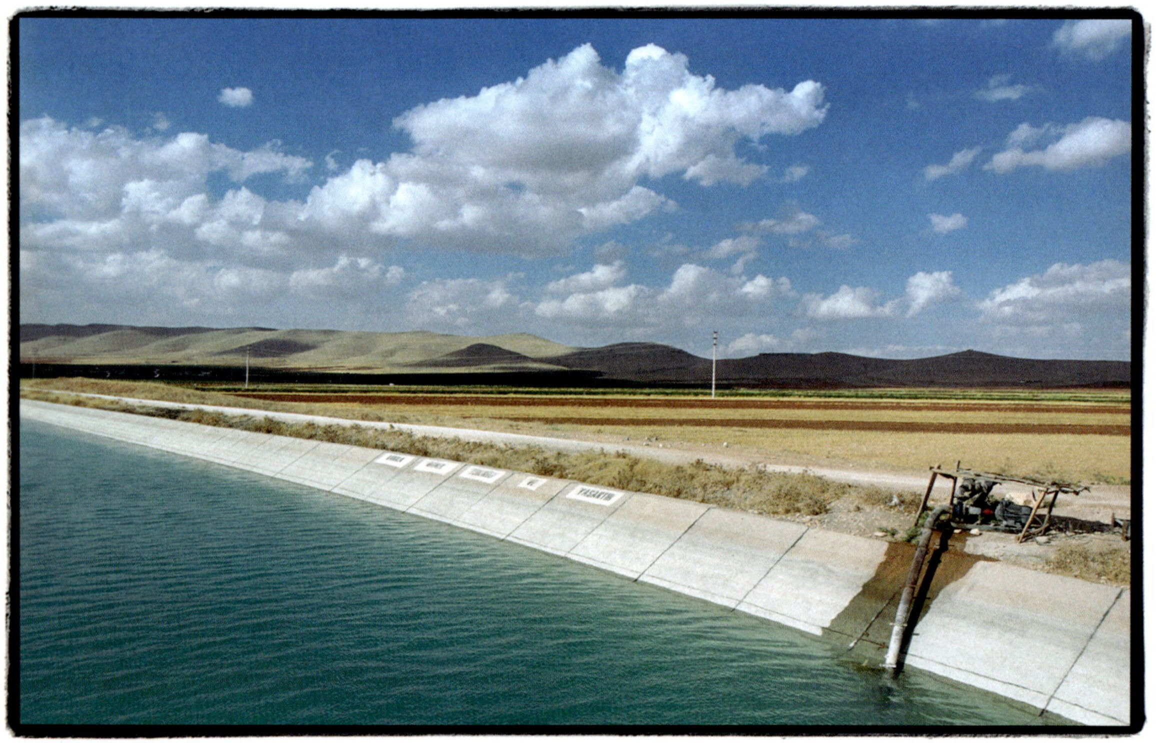

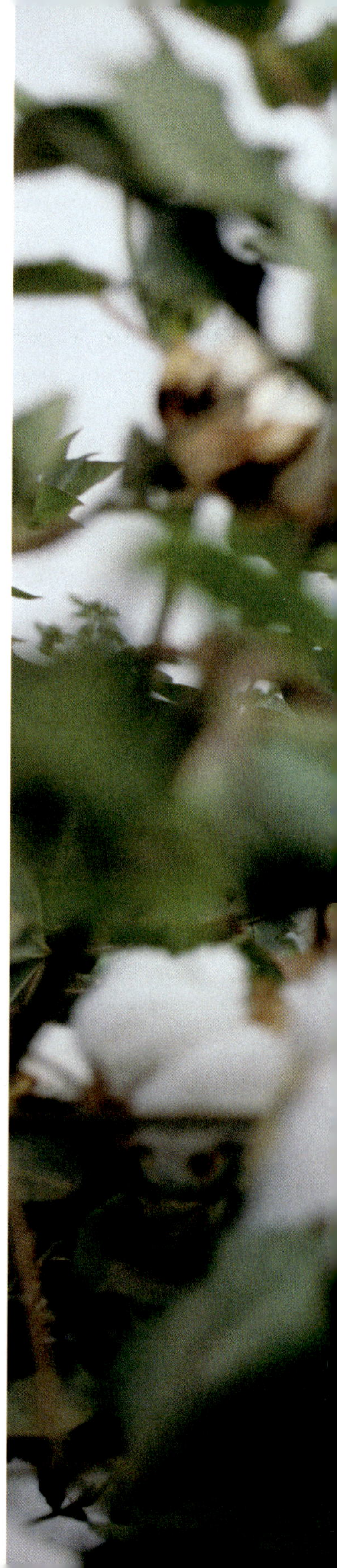

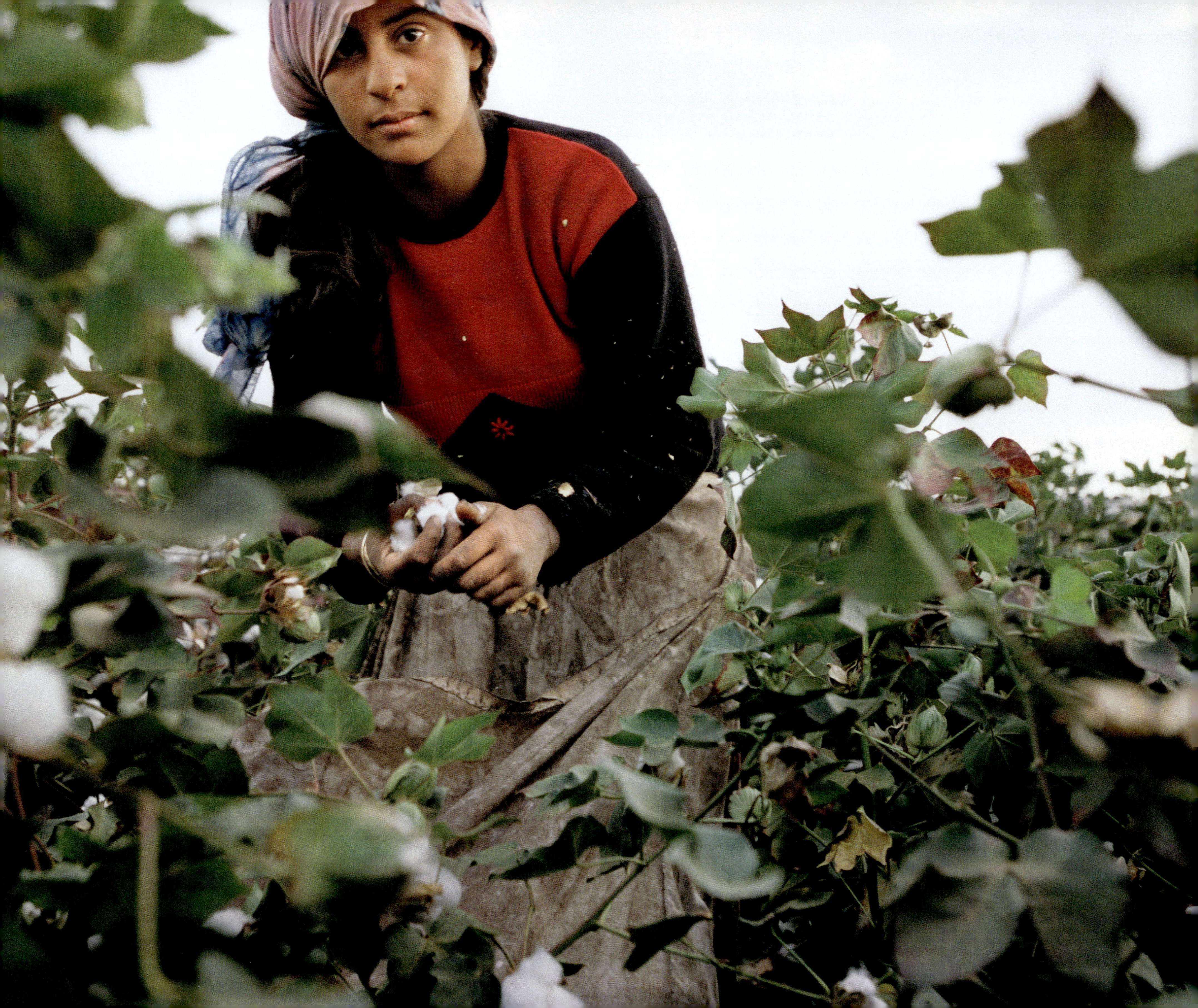

LEFT Southern Anatolia, Hasankeyf. Hasankeyf is one of mankind's oldest settlements. As the construction of the Ilisu Dam on the Tigris River has started, 52 villages and 15 small towns, including Hasankeyf, will be flooded in the near future. The dam will be the second biggest in Turkey after the Ataturk Dam on the Euphrates River. Hasankeyf is a rich treasure of Assyrian, Christian, Abassidian-Islamic and Osmanian history in Turkey. Numerous cultural experts and activists in Turkey have appealed to the government to save Hasankeyf by changing the design of the Ilisu Dam.

RIGHT Southern Anatolia, Halfeti. This village near Halfeti was flooded by the construction of the Birecik Dam on the Euphrates river. This dam is part of the Southeast Anatolia Development Project or GAP.

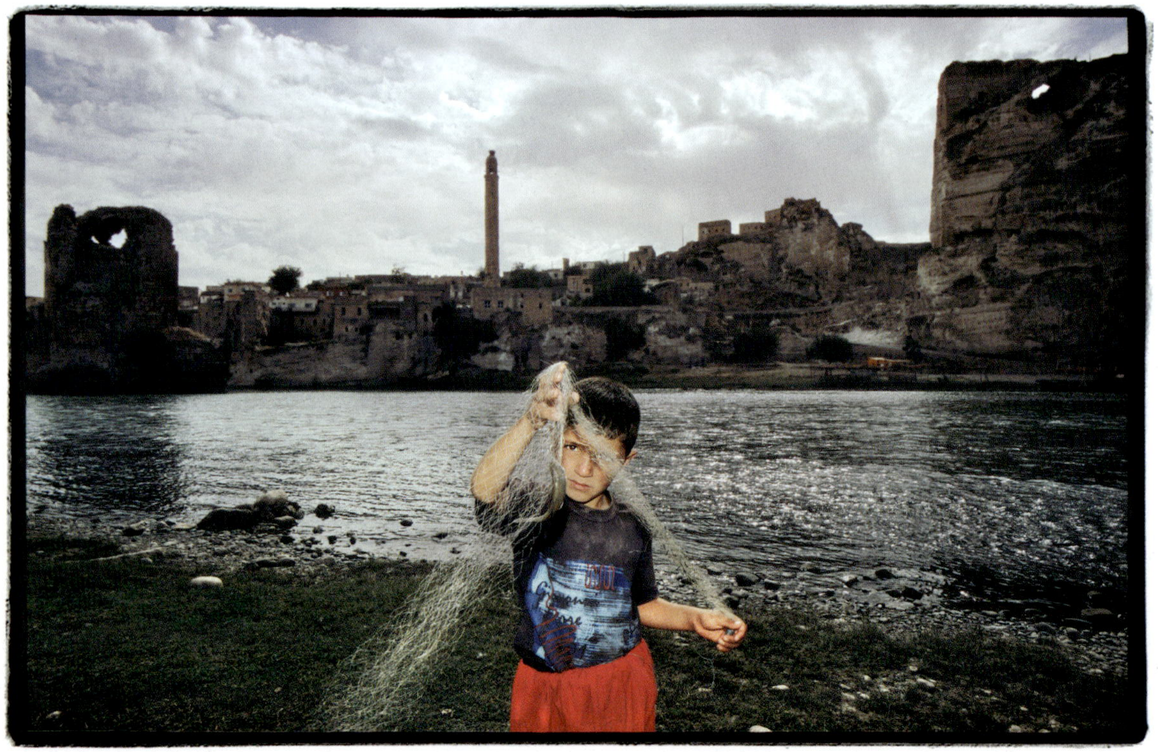

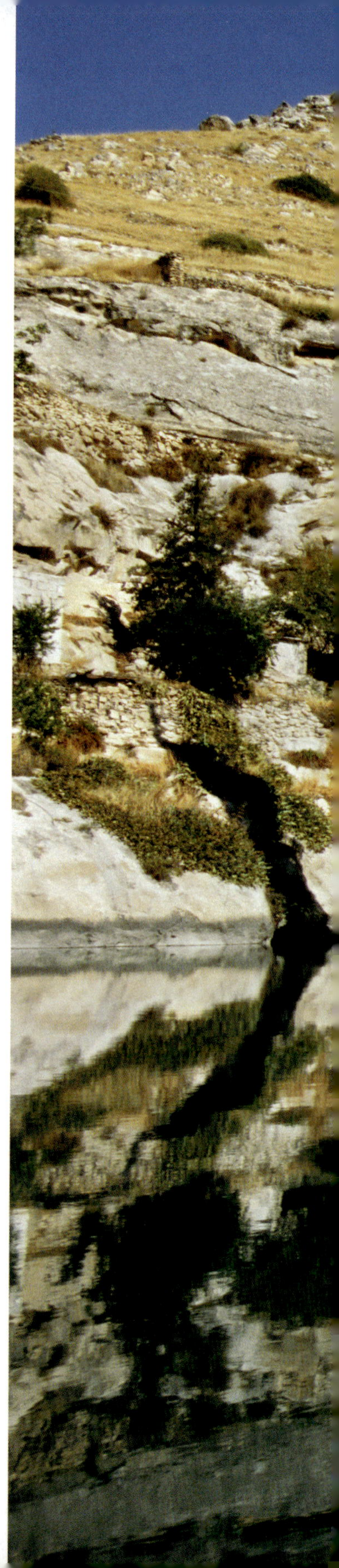

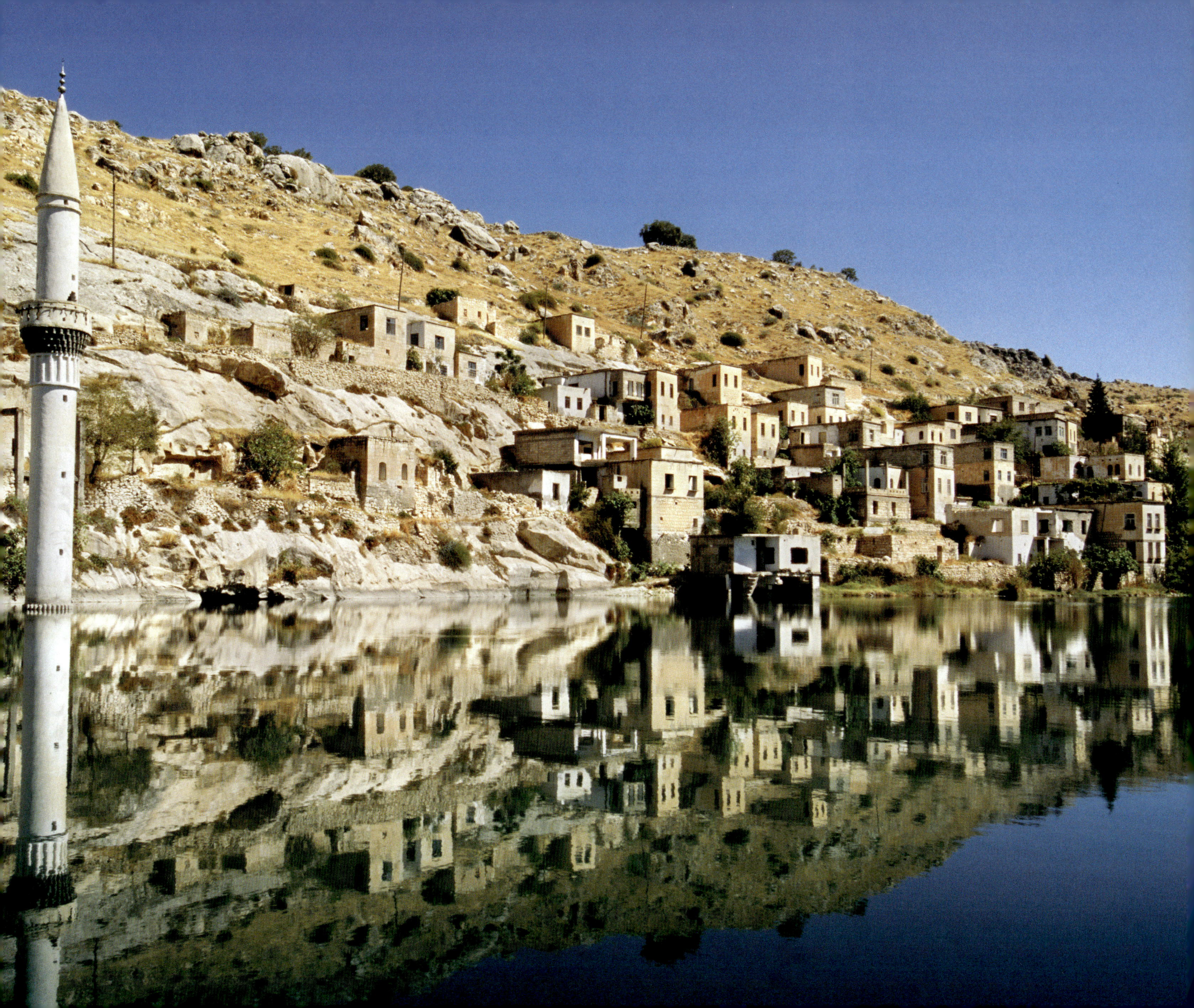

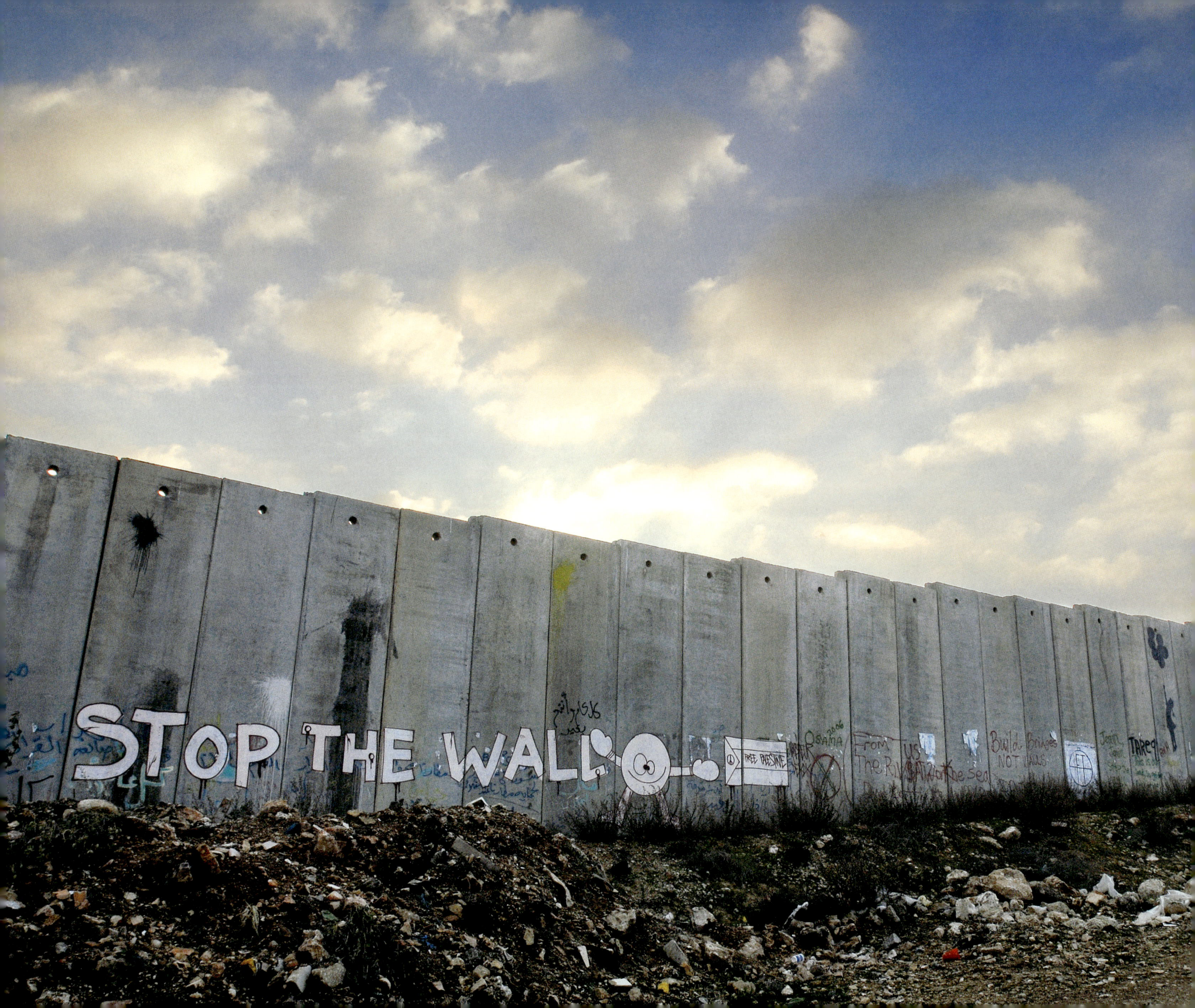
STOP THE WALL
Build Bridges
NOT WALLS

Israel
Formula for conflict

In the Middle East, 5% of the world's population is trying to survive on 1% of the world's water. Since the creation of the state of Israel, water has always been a priority for its government. The water reserves are hardly enough to provide for the needs of 6 million Israelis, let alone the 13 million inhabitants of the rest of the region. The threat of drought and desertification is real.

This is the perfect formula for conflict. The late King Hussein of Jordan said he would never start a war against Israel, except for water. Yitzhak Rabin said, to Israel water is a matter of life or death. In 2002, Ariel Sharon threatened to declare war when the Lebanese government planned to tap water from one of the tributaries of the Jordan River in order to provide water to villages in south Lebanon.

In 1967, Israel took control of Jerusalem and the West Bank and declared all the water sources military zones. Since the construction of the wall in 2002, thousands of Palestinians have been cut off from their farmland and water sources. Under the West Bank lies an important underground aquifer but the Palestinians are prohibited from pumping water. In the meantime settlers are allowed to use the most advanced pumping equipment. Palestinians claim the Israeli colonies use 80% of the aquifer although they make up only 10% of the population.

Israel also diverts up to a third of its water supply from the Jordan River. Where every single Israeli has 300 litres of water per day, the Jordanians have only 80 litres. Friends of the Earth Middle East say more than 90% of the Jordan is being diverted and the river is now in danger of disappearing altogether. The Dead Sea has also shrunk by 30% in the last 50 years. And the level of the Sea of Galilee near the Golan Heights has dropped in recent years, sparking fears that Israel's main reservoir will salinate.

In Gaza, the continuing conflict prevents the Palestinians from building sewerages and water-cleaning systems. As a result, the groundwater is being polluted by agricultural chemicals and by waste from landfills and the burning of refuse. Israel claims the Palestinians' water shortage should be resolved by a more efficient use of water, without considering redistribution. The Palestinians, however, want their rights to water to be recognised.

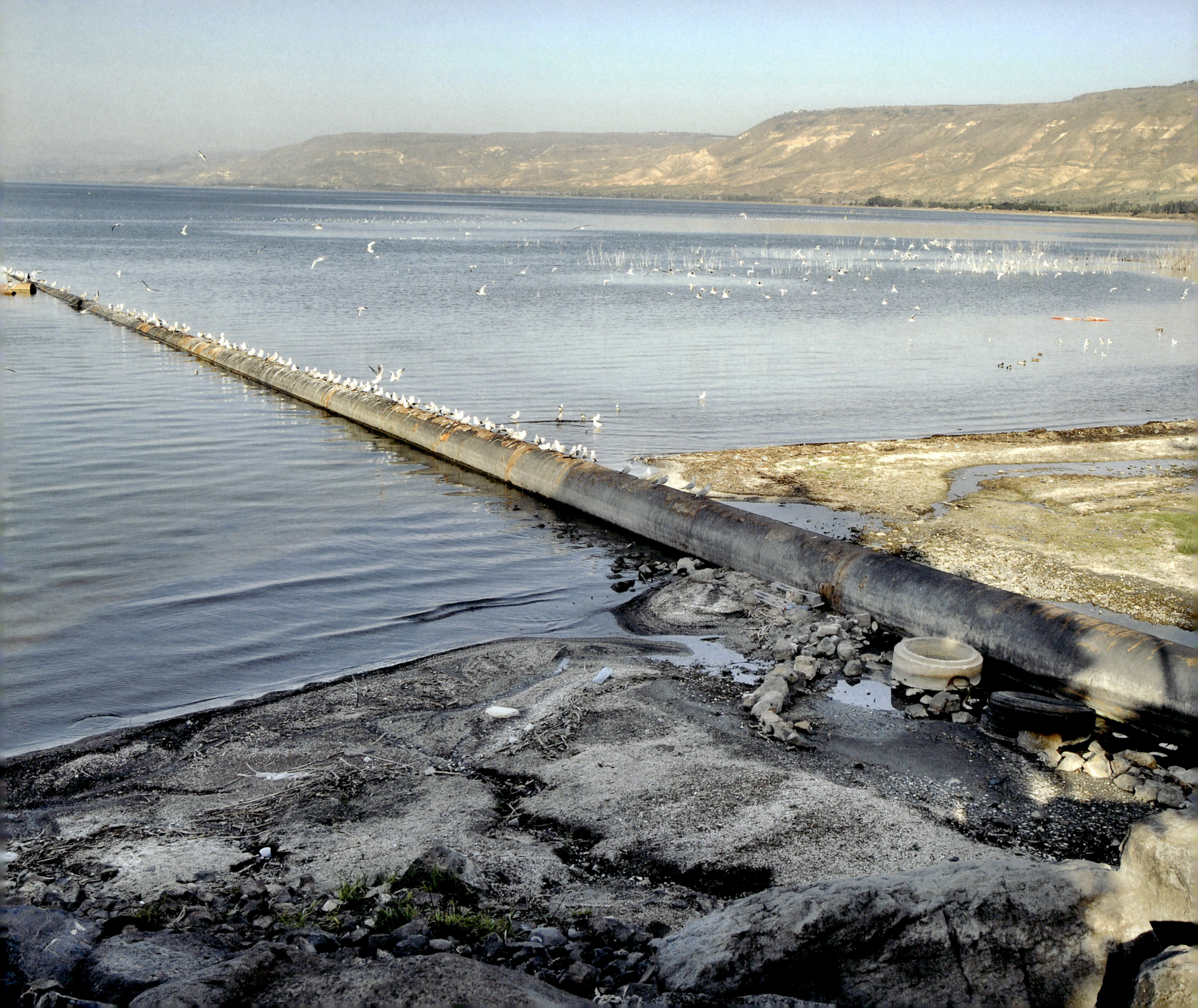

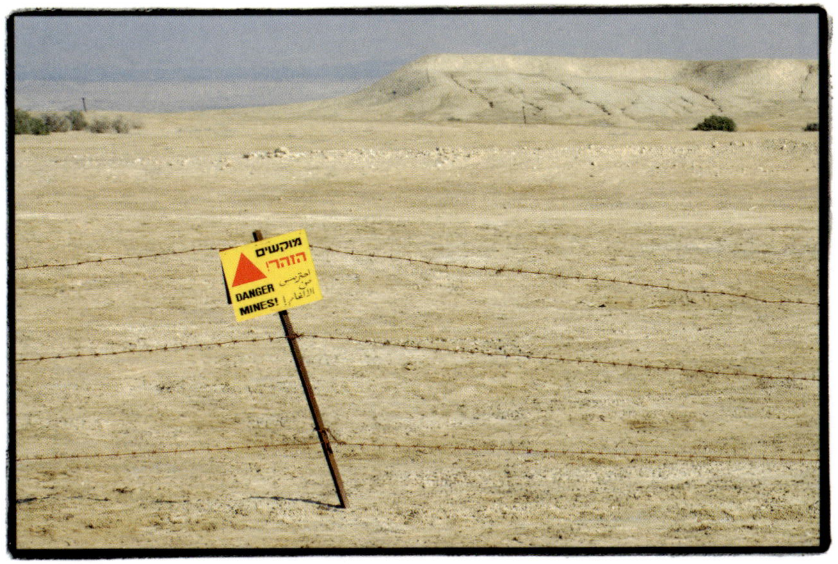

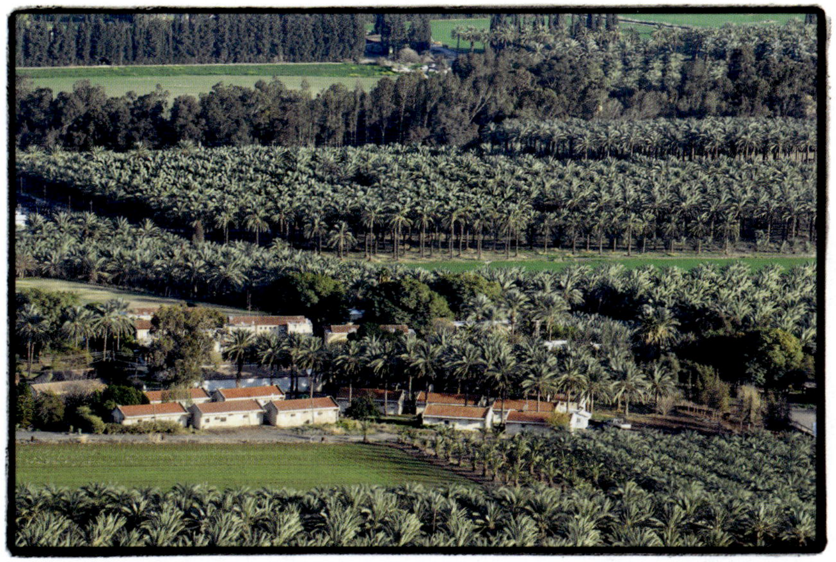

LEFT Sea of Galilee. The upper Jordan River and its tributaries –the Sea of Galilee, the Yarmuh, and the lower Jordan River– supply approximately one-third of Israel's water needs. The Palestinians in the West Bank no longer receive water from this source since the Israeli occupation of the Golan Heights and the Jordanian West Bank in 1967. Over the past two decades, Israel has over-pumped the water of the Sea of Galilee and the levels have dropped dramatically. During the summer of 2001, the level sank to an unprecedented low and alarmed scientists worried that the lake would soon reach the level where salinity seepage into the fresh water would cause irreversible damage.

MIDDLE and RIGHT The Jordan Valley just below the Sea of Galilee and in the West Bank just above the Dead Sea. In the West Bank, the Jordan River is a highly militarised area that is protected by mine fields. The Palestinians have no access to the river or its water.

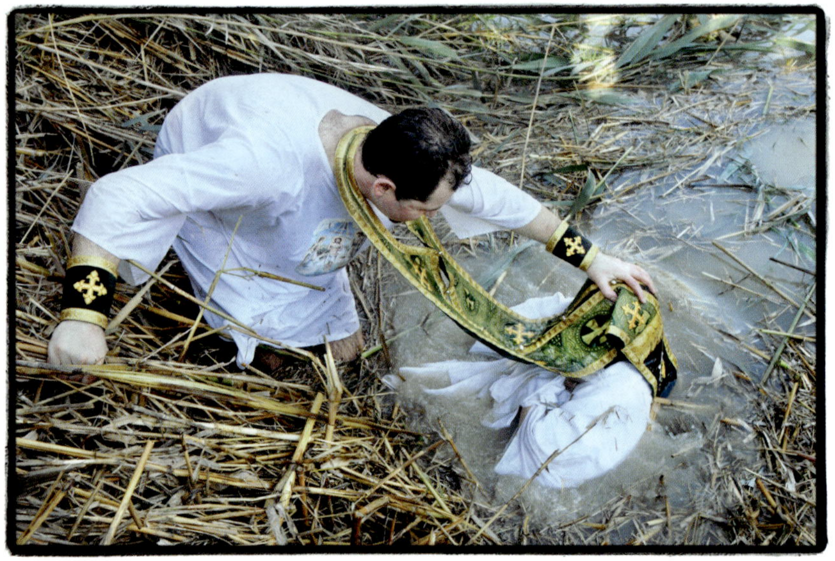

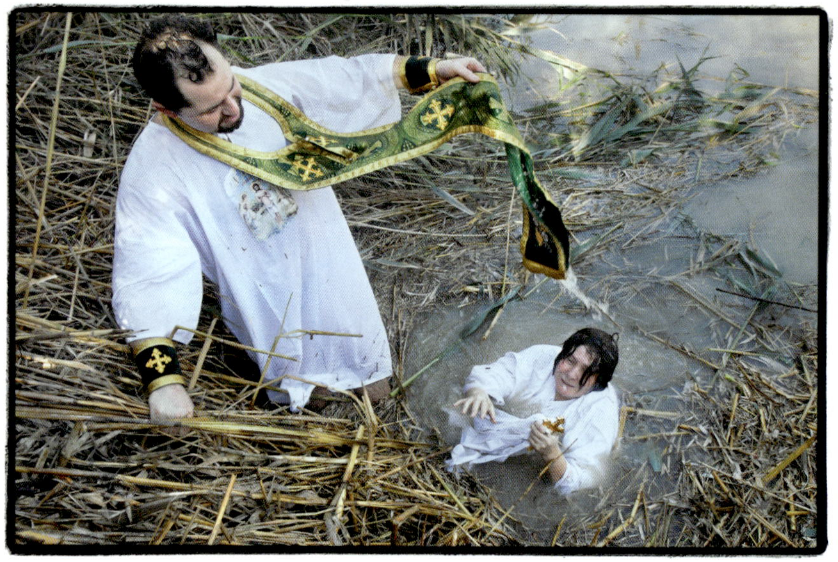

River Jordan, near Jericho. Orthodox ritual at the site where Jesus Christ is said to have been baptised.

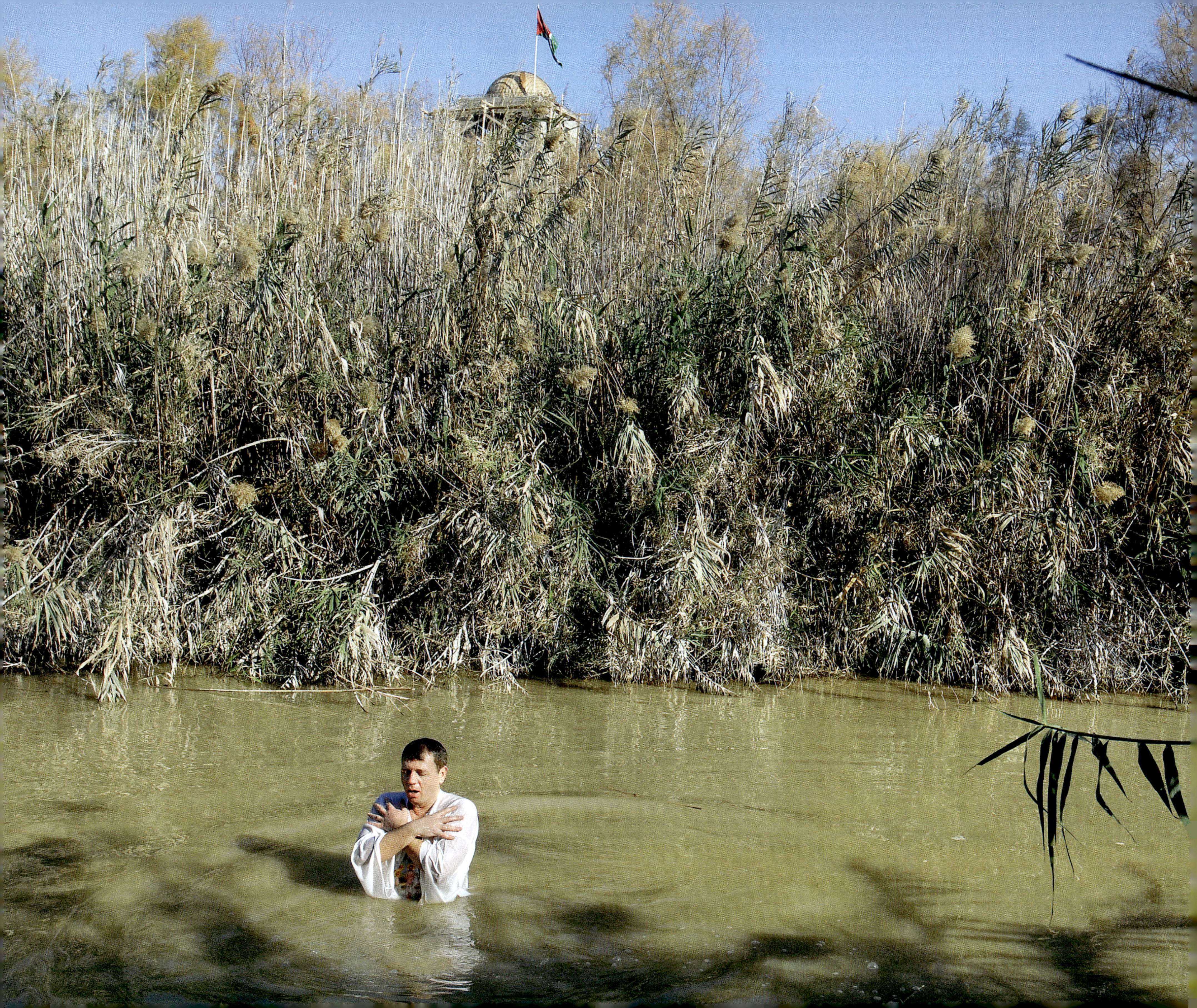

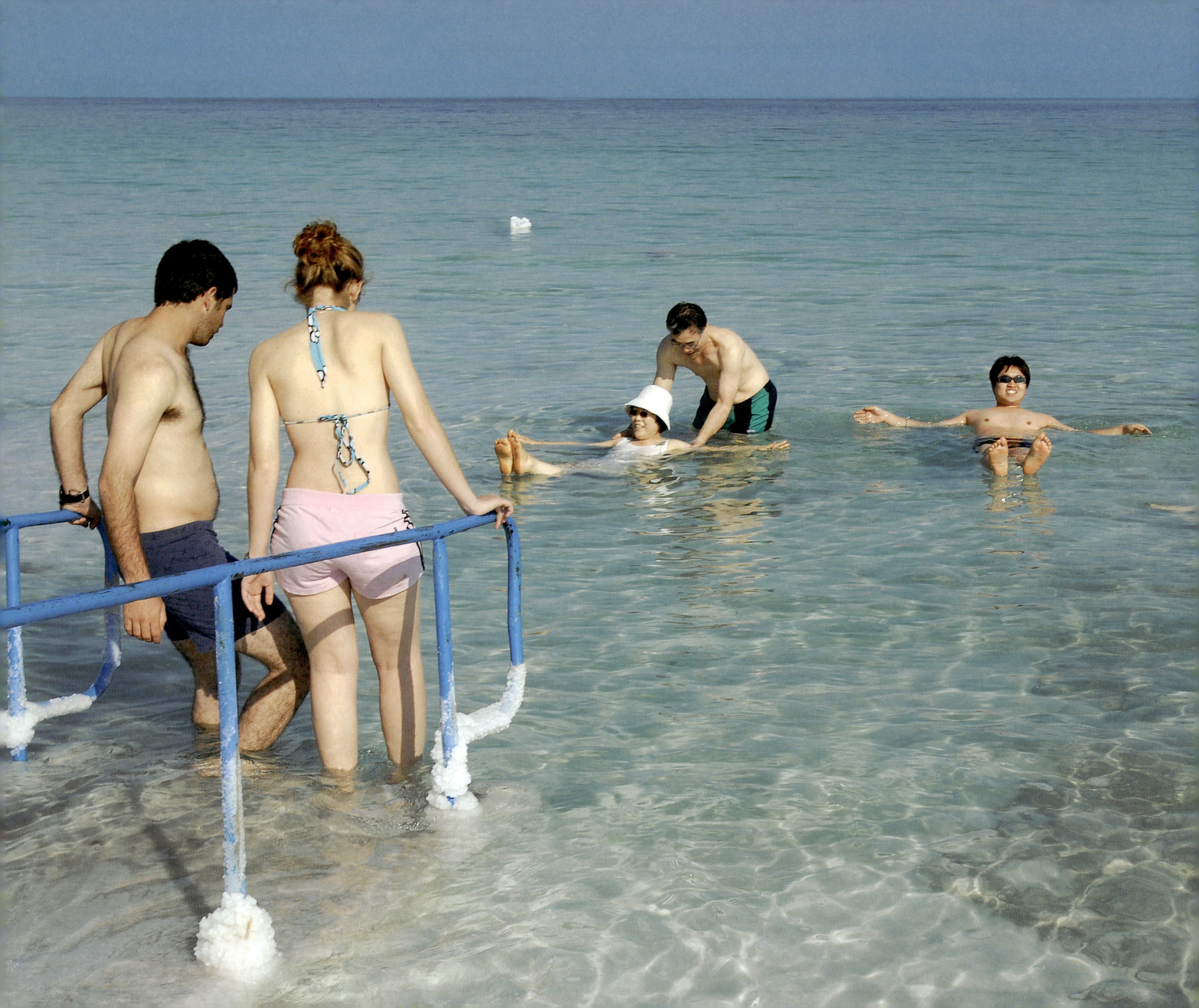

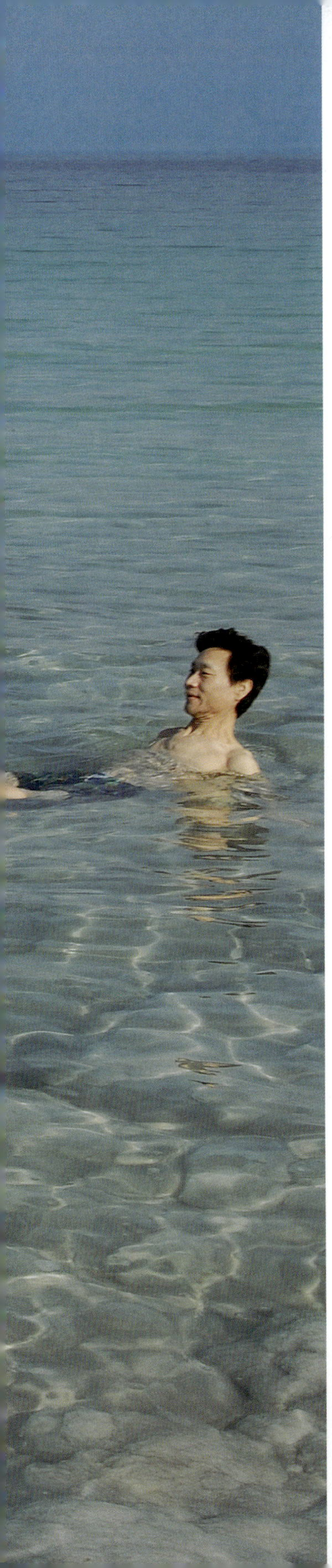

The Dead Sea. In less than half a century the water level of the Dead Sea has fallen by 25 meters and its surface area has shrunk by a third. In the past few years, the water level fell at a rate of 80-100 cm per year, and it is getting worse. The water that once flowed into the Dead Sea from the River Jordan has been diverted by Israel, Syria and Jordan for agricultural and hydroelectric projects. Environmentalists are now warning that drastic action has to be taken to avert an ecological disaster.

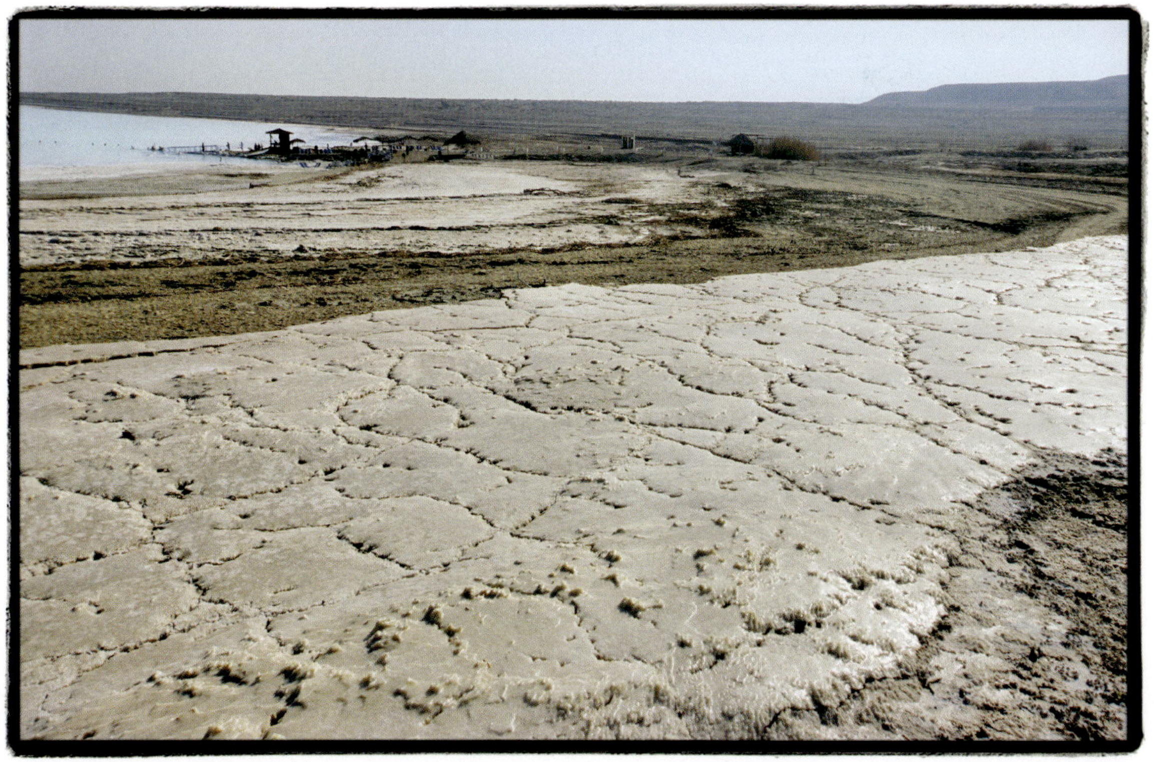

In June 2002, the government of Israel decided to erect a barrier along the entire West Bank in order to prevent the uncontrolled entry of Palestinians into Israel. Most of the barrier is located on the West Bank, far away from the Green Line. It infringes the human rights of more than 200,000 Palestinians. Thousands of people are enclaved between the barrier and the Green Line or between the barrier and another eastern depth barrier. More than thirty communities are separated from their main farmland which lies on the Israeli side of the barrier.

LEFT The seperation barrier at Abu Dis, and RIGHT at Tulkarm. The separation barrier on the Palestinian side is either an eight meter high concrete wall - twice the height of the Berlin Wall - with armed watchtowers and cameras, or a 30-100 meter "buffer zone" with electric fences, trenches and military patrols. On the Israeli side the separation barrier is hardly visible, as along the Yitzhak Rabin Highway near Tulkarm.

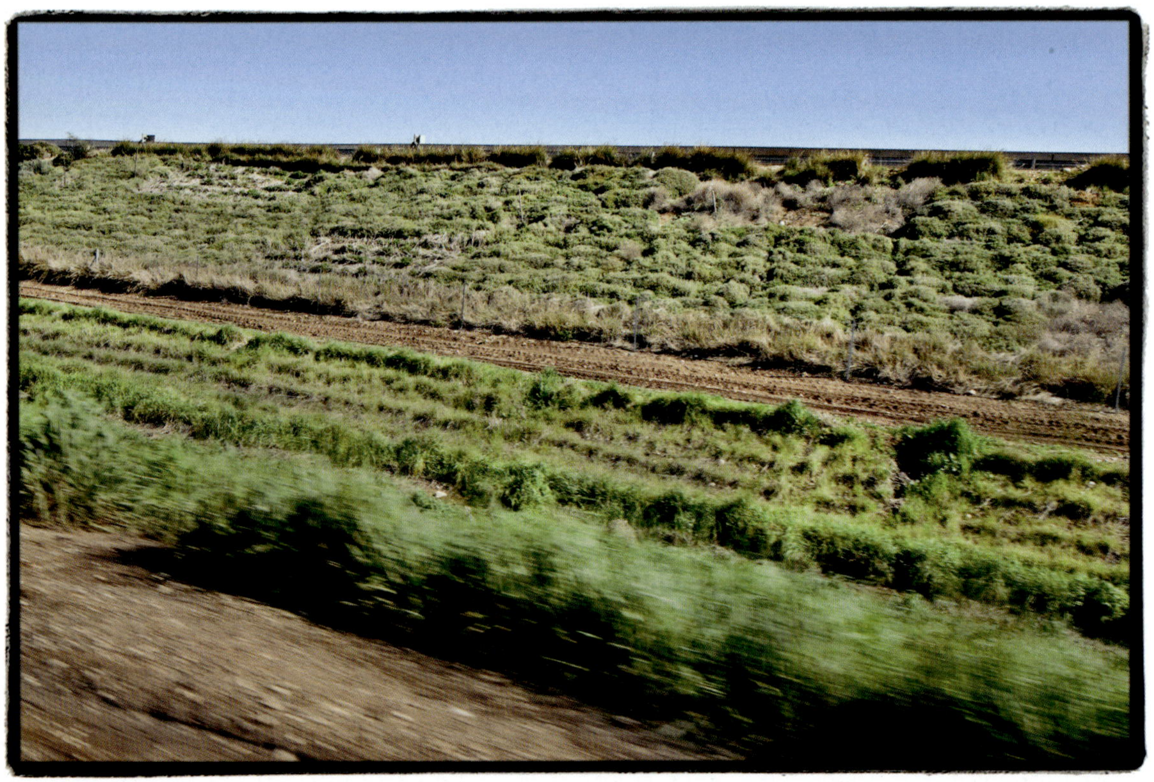

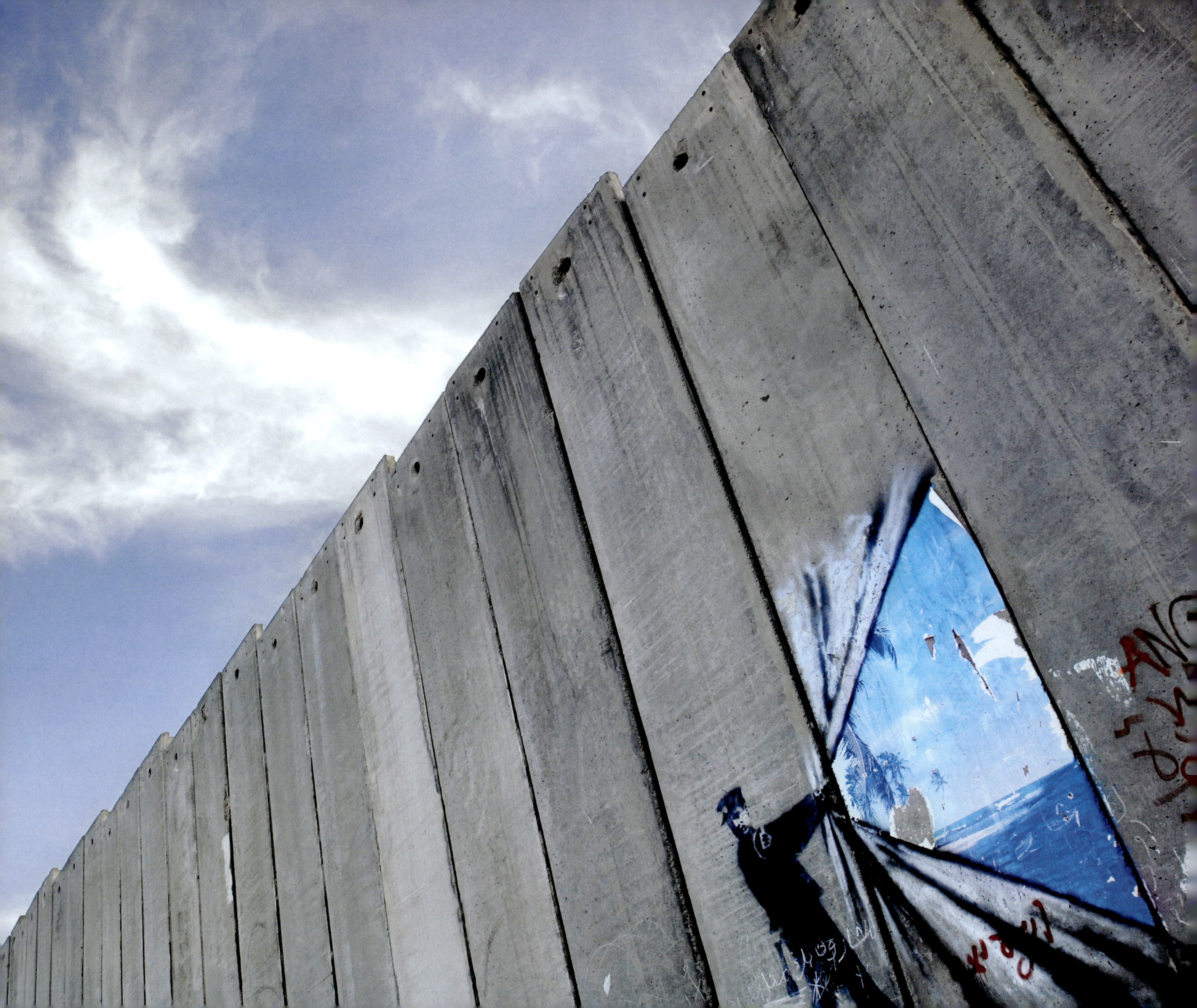

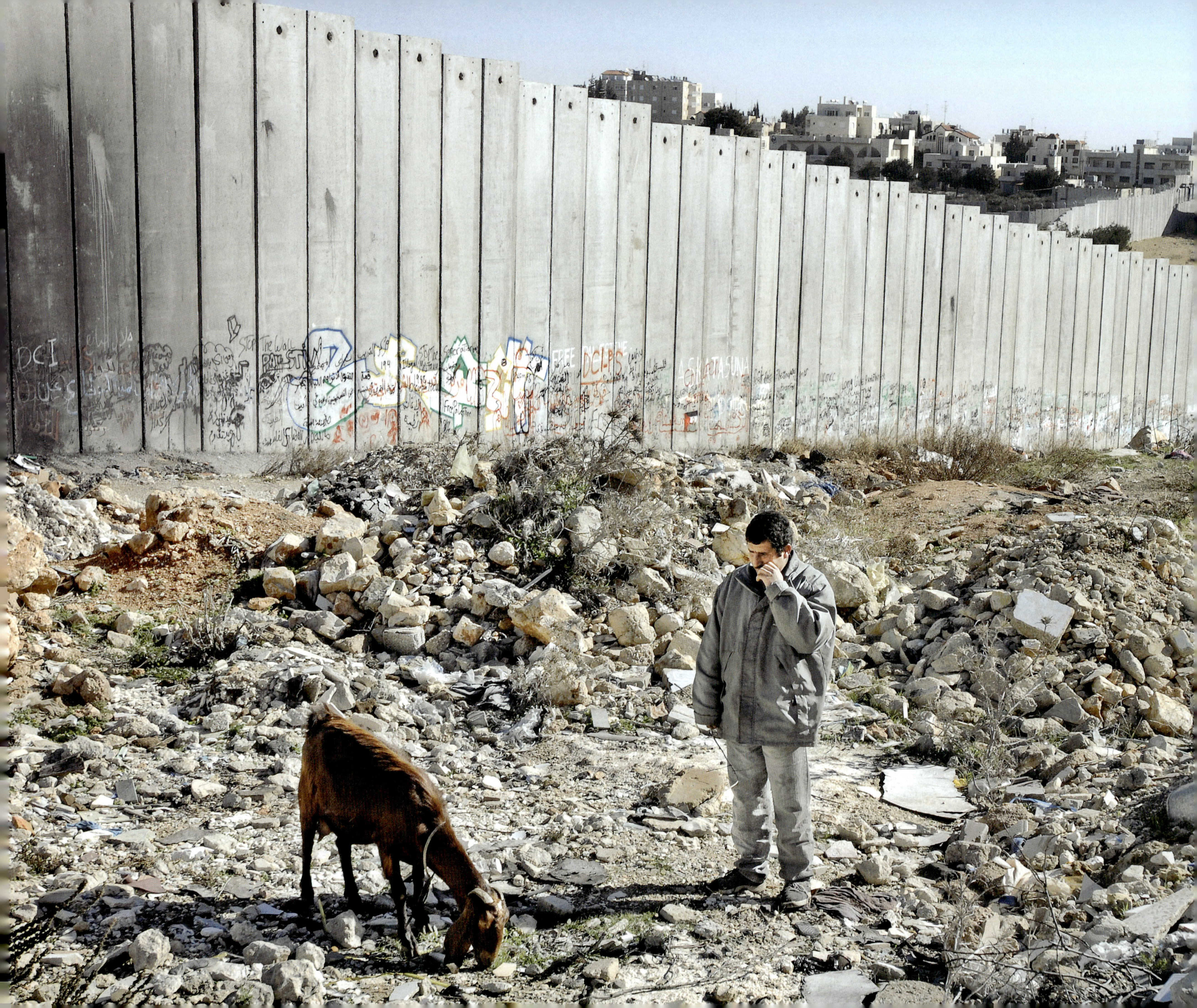
DCI
Stop The Wall
FREE PALESTINE
DCI-PS
ASKATASUNA

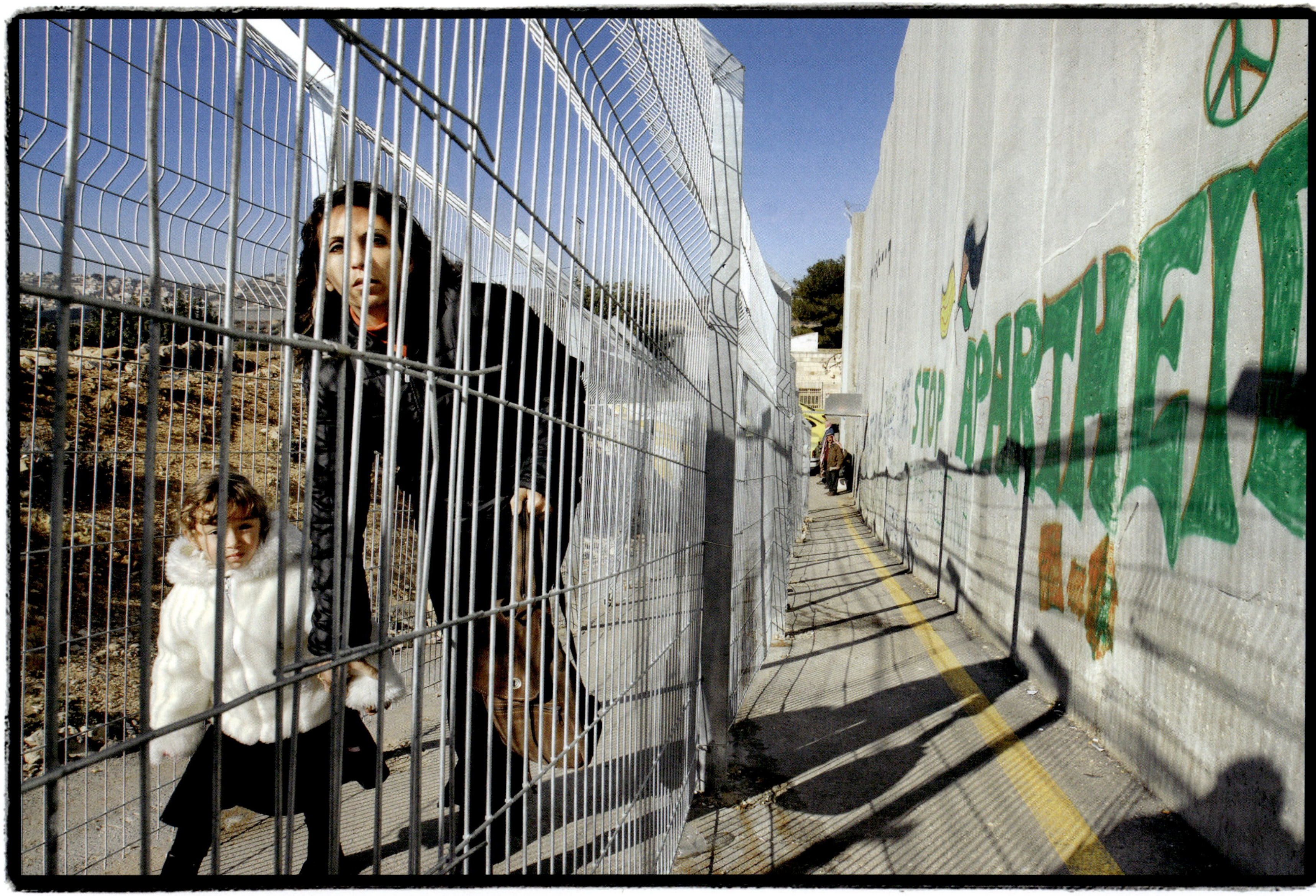

Israel has set up two separate law systems in the Occupied Territories. The rights of individuals are based on their nationality. This regime is the only one of its kind in the world, and is reminiscent of regimes from the past, such as the Apartheid regime in South Africa.

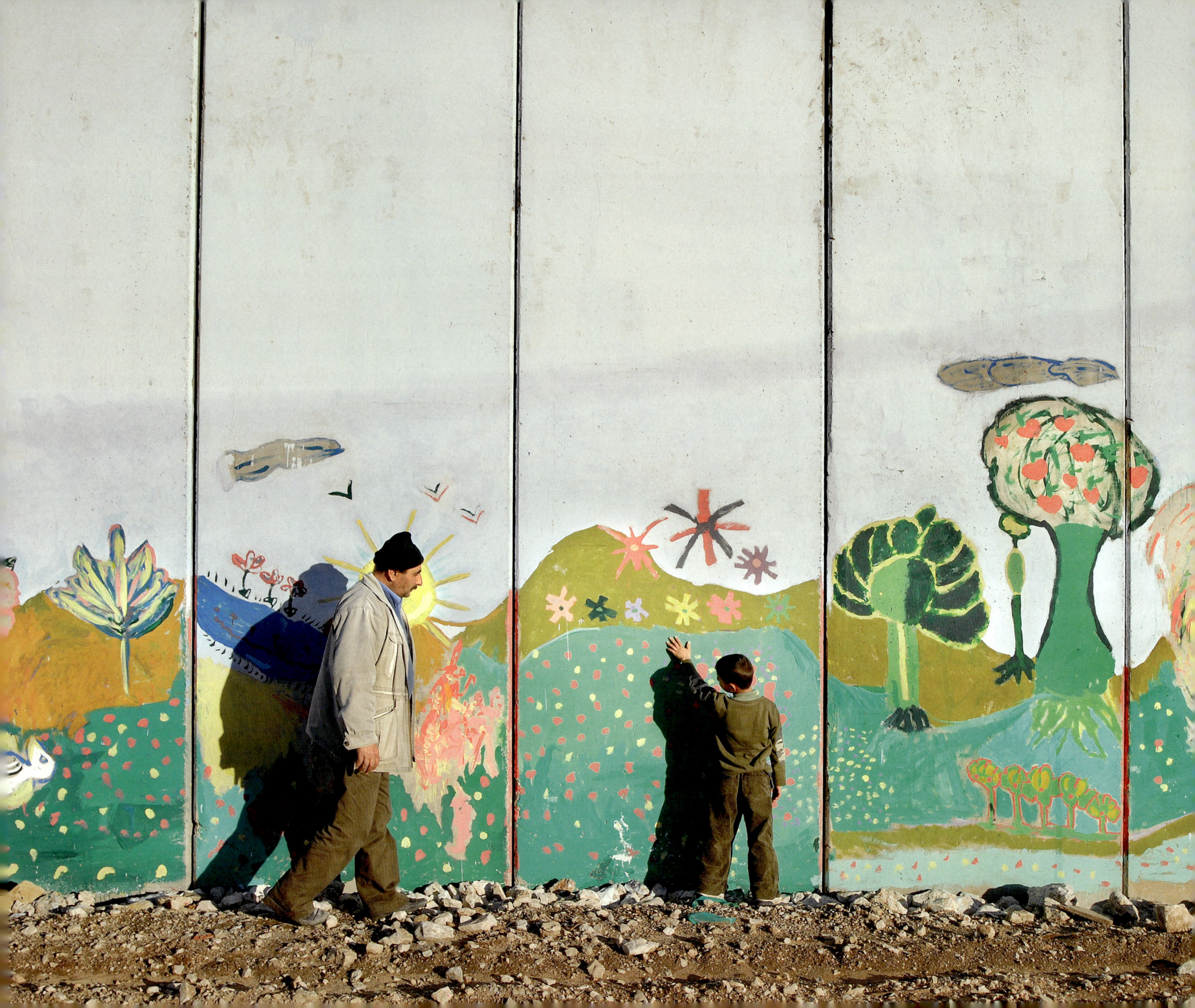

Israel has implemented a system of permits, through which it restricts Palestinians' access to their farmland situated west of the barrier. About 25% of the applications to obtain entry permits are rejected. Israel has also limited the number of gates through which Palestinian farmers can reach their land, and the gates are only opened two or three times a day for very short periods. Israel refuses to compensate Palestinians who are denied access to their lands for their loss of income. Many roadblocks hinder Palestinian farmers from transporting their produce inside and outside the West Bank. The harm to the farming sector is liable to have drastic economic effects and drive many families into poverty.

LEFT West Bank, Masha. This farmer has been isolated from his village by an eight-meter high separation wall which was constructed on his greenhouses. No compensation was given for this illegal act and for his US$ 25,000 loss. On the other side of his house is the industrial settlement of El Kana. The settlers live just a stone's throw from his house and have already smashed his solar water heating panel. Between his house and the wall is a military road where heavy army vehicles drive up and down. Every night the military traffic wakes up and traumatises his children.

RIGHT West Bank, Jayous. Shareef Omar Abu Azam (64) used to be a wealthy farmer. Although the Green Line is six kilometres away from his village, all his fields - a total of 175,000 m^2 - are now on the other side of the separation barrier. "Our water wells are also on the other side of the enclosure," Shareef says. "Now we are forced to buy water from another village. Access to our farms is only possible three short times a day. To cross the gate you need a special permit. I got one for myself and for my wife, but I'm 64 and my wife 60. We can't manage 3,200 olive trees on our own. I have four daughters and two sons but the Israelis refuse to give them a permit. And now the Israelis are talking of closing the gate. That means that we'll have to make a detour of 15 kilometres. Many farmers have no transportation so they won't be able to reach their land. The Israelis have their reasons to make life impossible for us: they are planning to expand the existing settlement and build 1,500 new houses."

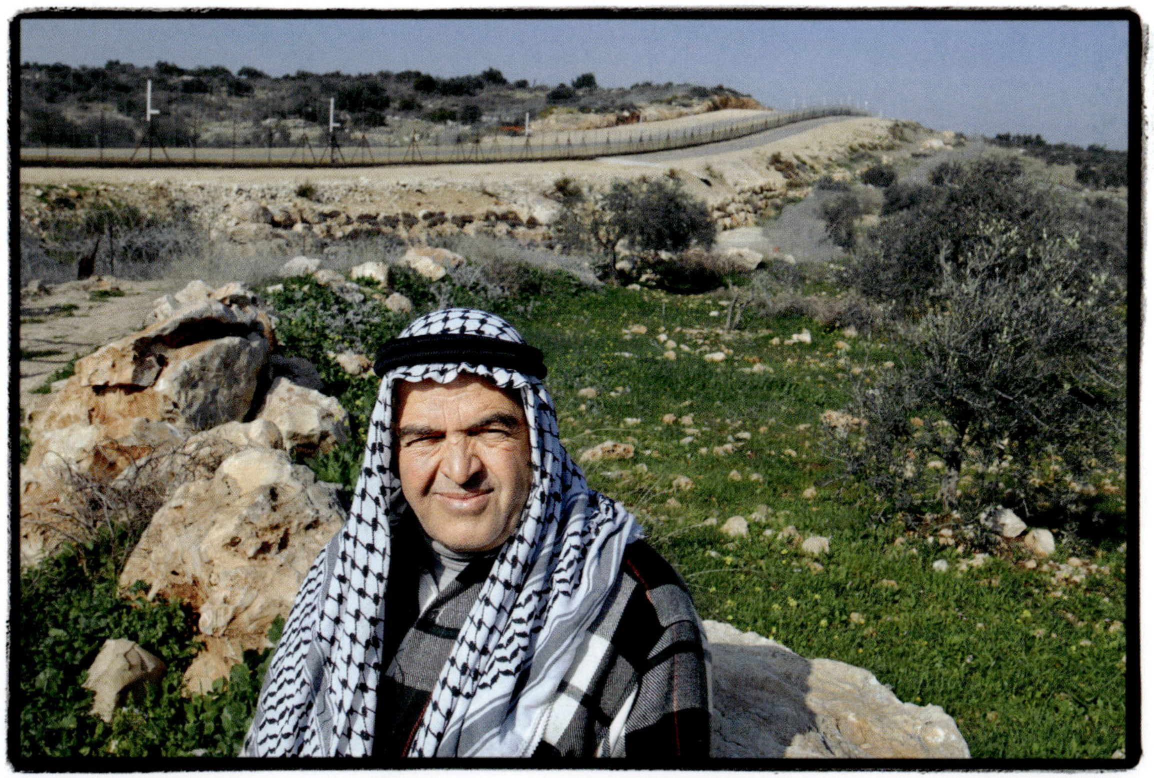

LEFT West Bank, Sulfeet, Marda. Hani Amer Abu Nedal (38) stands at an invisible line that he may not cross. "Although I have land with olive trees on top of this hill, I'm not allowed to go there," he says. His village is situated below Ariel, one of the biggest settlements in the West Bank. "If we come too close to the settlement the Israeli soldiers will threaten us. They destroyed our harvests before. They even made holes in olive trees and filled them with acid. On the other side of the village is a military control post where they also prevent us from going to our fields. In total I have lost 10,000 m2 of agricultural land. And then there is the pollution: in some areas the waste water from the settlement is running down the mountain into our land and rivers."

RIGHT West Bank. Hussein Mohammed Suida was a banana farmer untill last year. "Look at these banana trees, they look miserable," he says, while walking on the farm. "Banana trees need a lot of water during the summer. With good winter rains, the summer is fine, but the past rainy seasons have been dreadful. And then there is the settlement pumping water from the El Oja spring, our only water source. While we have water shortages during the summer, they are spraying their gardens." Hussein quit farming after he lost US$ 7,000. "Even with enough water it would be difficult. Before we used to sell our bananas to Jordan. Now our goods have difficulties passing the many road blocks. Before I could reach Jericho while smoking a single cigarette, now you can smoke a whole pack and still be on the road. Israel is also dumping their subsidised bananas on the West Bank. They are destroying us." Hussein has now opened a tiny shop where he sells soft drinks and small food items. "I hope this will be a supermarket next time I see you," he jokes. With the poverty rates growing in the West Bank, there is little chance for that to happen.

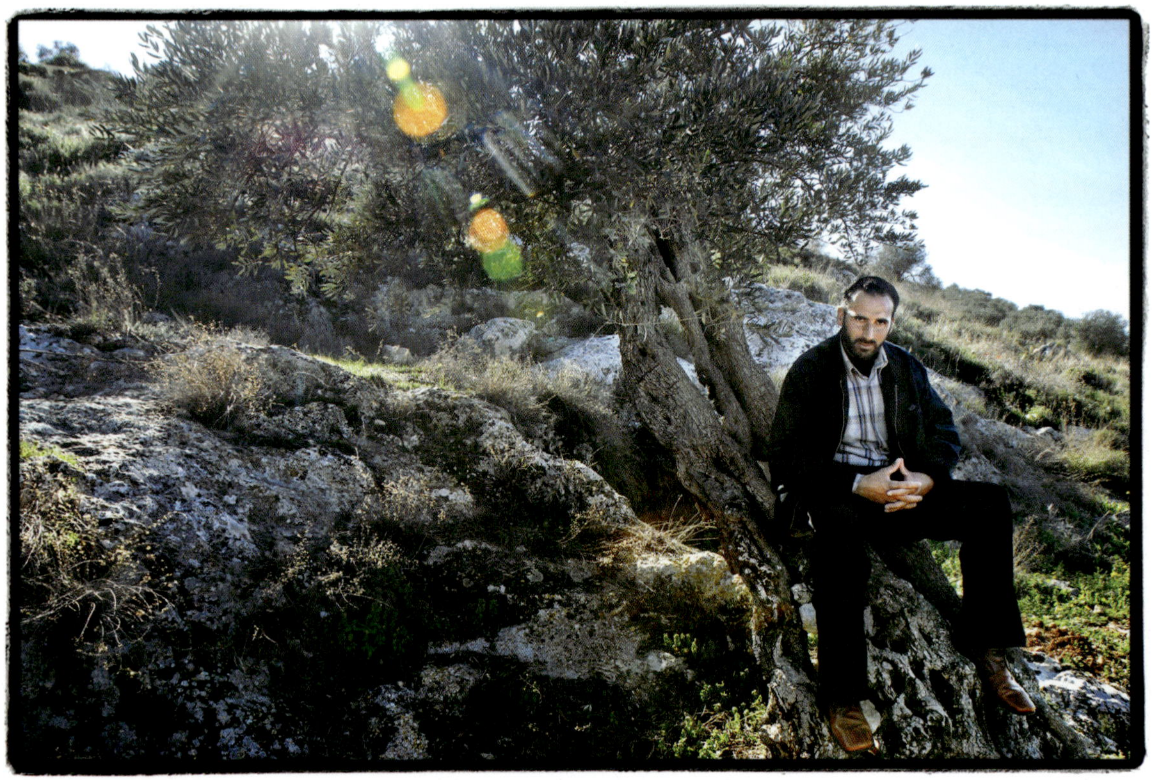

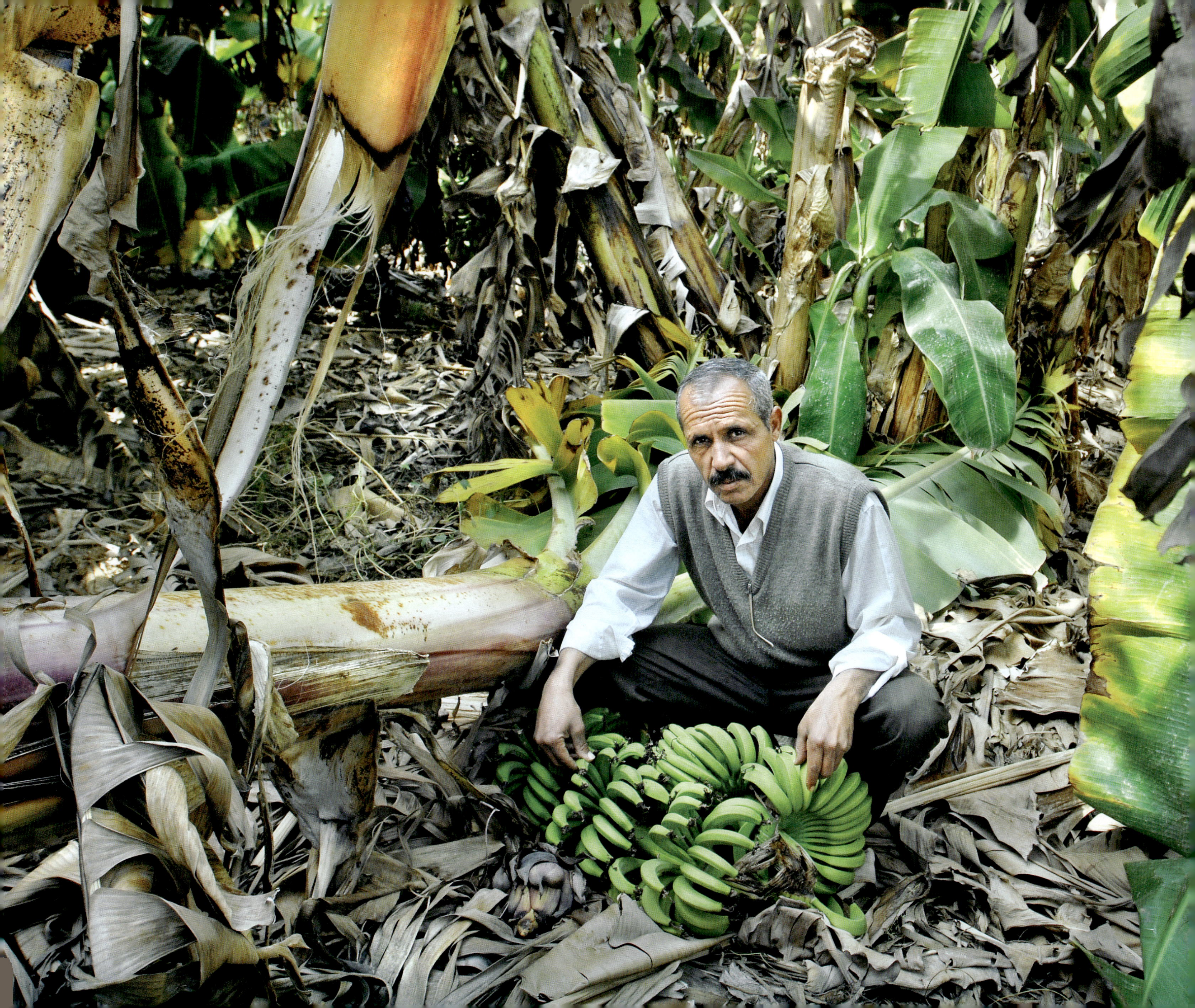

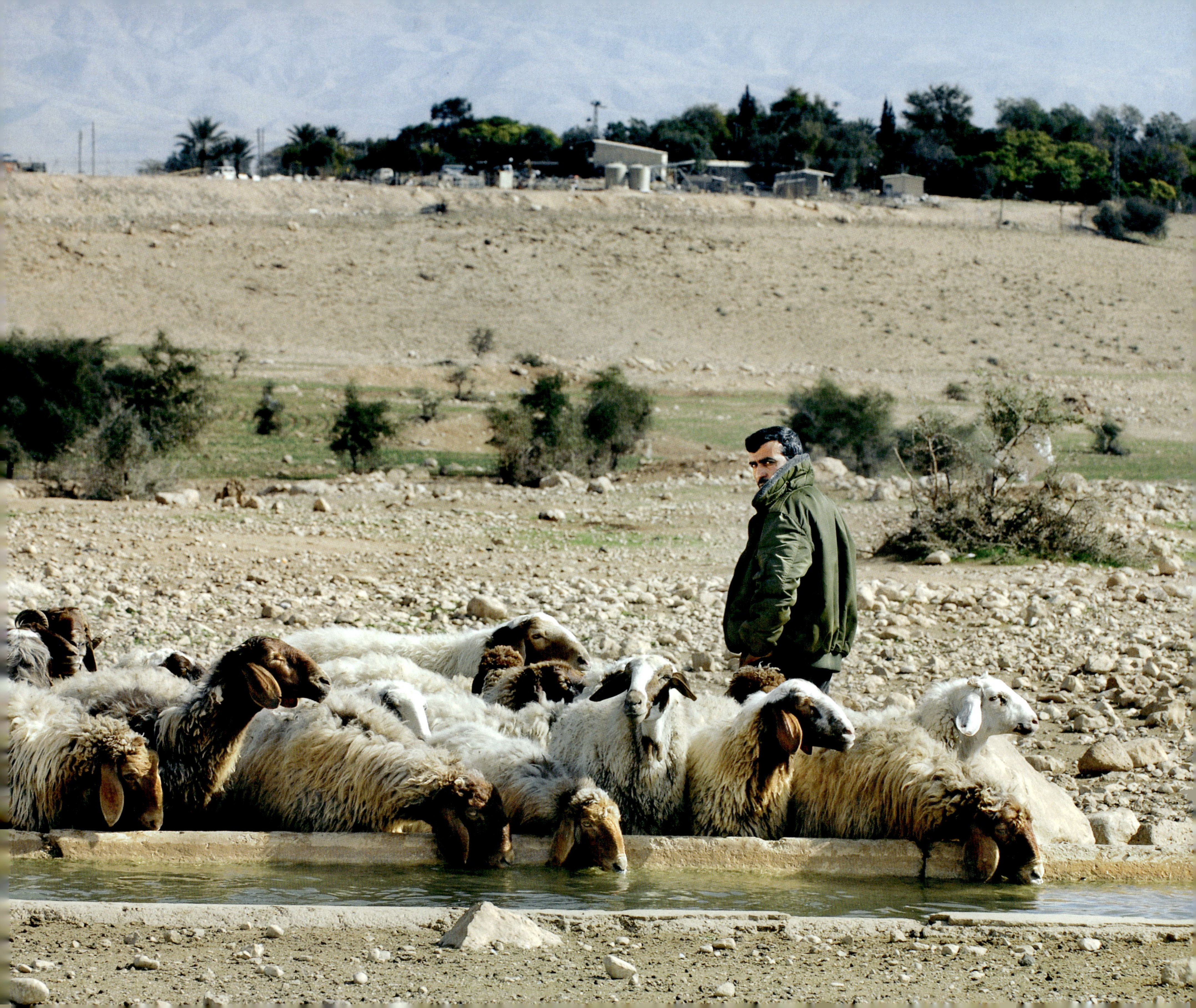

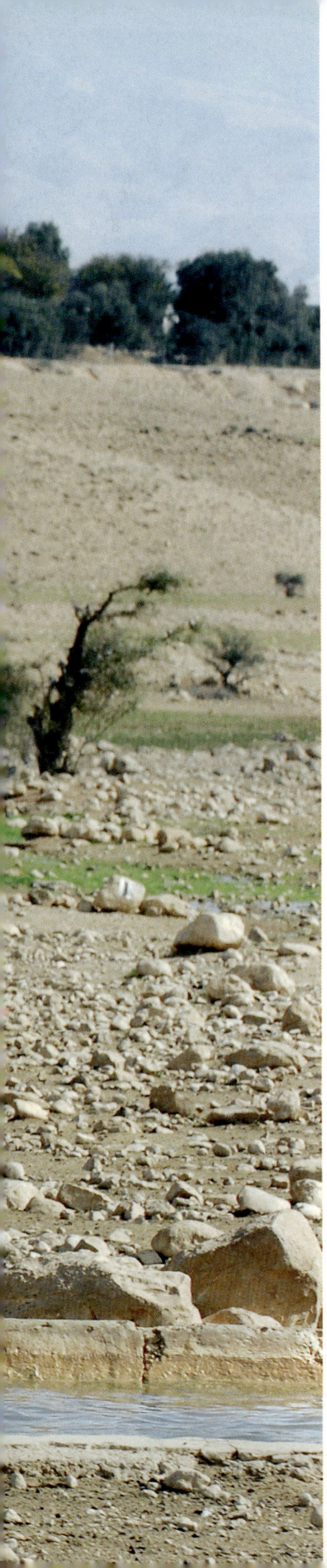

LEFT West Bank, El Oja. Sheep drinking from the El Oja spring. This used to be a spring with a high capacity all year round, but now it runs out of water during the summer. On the hilltop is a settlement which pumps water from the same spring. The quantity and regularity of water supply to the settlements in the West Bank is much greater than that to the Palestinians. This discrimination is especially blatant during the summer, when the water supply to Palestinians is reduced to meet the increased demand for water in the settlements.

RIGHT Ariel settlement, West Bank. In the past, Israel used 'imperative military needs' to justify expropriation of land to establish settlements and argued that the action was temporary. The settlements have for some time been 'facts on the ground' and Israel now demands that some of them be annexed to Israel. But all settlements on the West Bank violate international law, which prohibits the occupying power to transfer citizens from its own territory to the occupied territory (Fourth Geneva Convention, Article 49).

The Israeli governments have implemented a policy intended to encourage Israeli citizens to migrate to the West Bank. Accordingly, the settlers are entitled to significant financial benefits from six different government ministries. The Ministry of Construction and Housing for example gives generous loans for the purchase of apartments, while the Israel Lands Administration gives significant price reductions in leasing land, and the Ministry of Education provides free transportation to school.

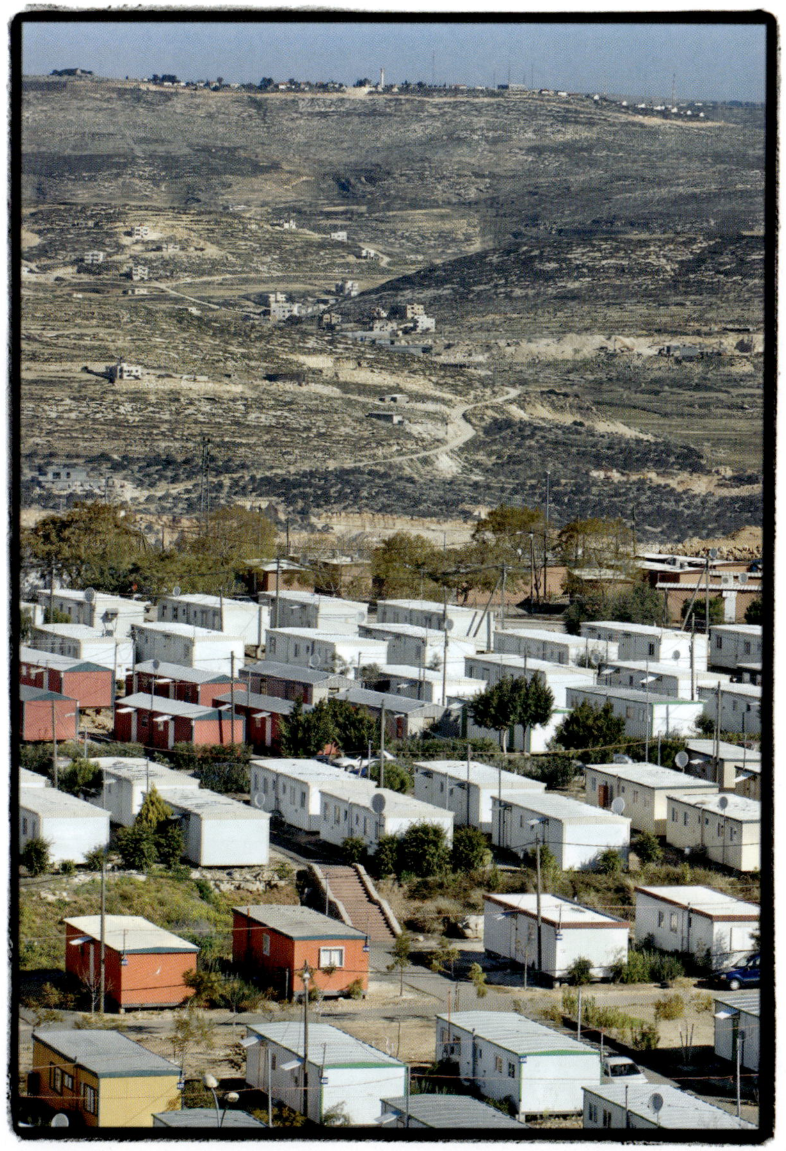

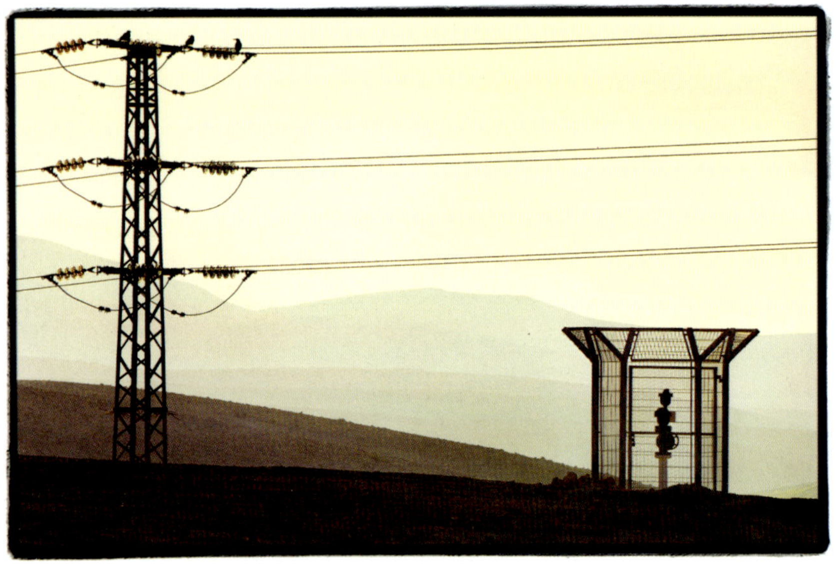

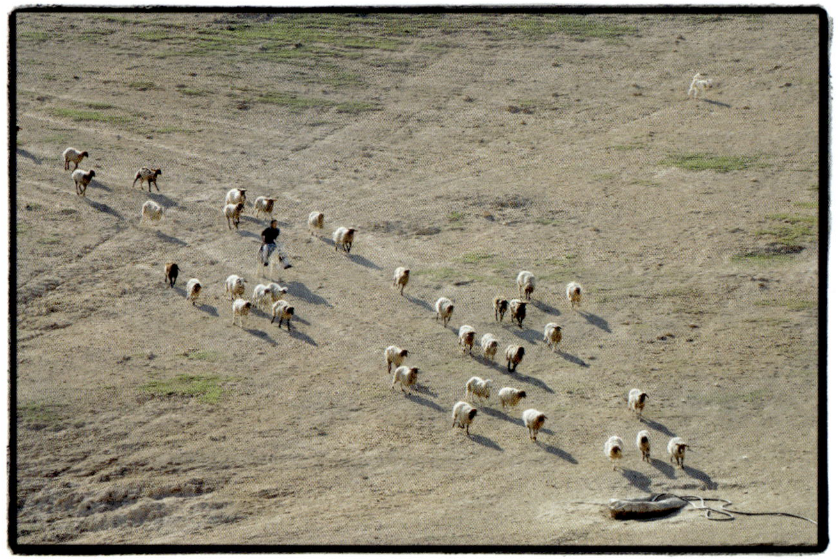

Negev Desert, Bedouin villages.   Although the Bedouins living in the Wadi Nam area are surrounded by electricity pylons and two water pipes going to a military camp and an industrial zone, they have no electricity or tap water. Article 3 of the 1959 Water Law stipulates that every person in Israel has the right to receive and use water, subject to the instructions of the law. The Bedouins from Wadi Nam went to court and were given access to water, but at a point 15km away. The Bedouin Arab citizens living in unrecognised villages in the Negev face water discrimination. They suffer from water shortage, poor water quality, high prices and waste of precious time in the process of accessing water. The government wants the Bedouins to leave their villages and move to recognised Bedouin cities.

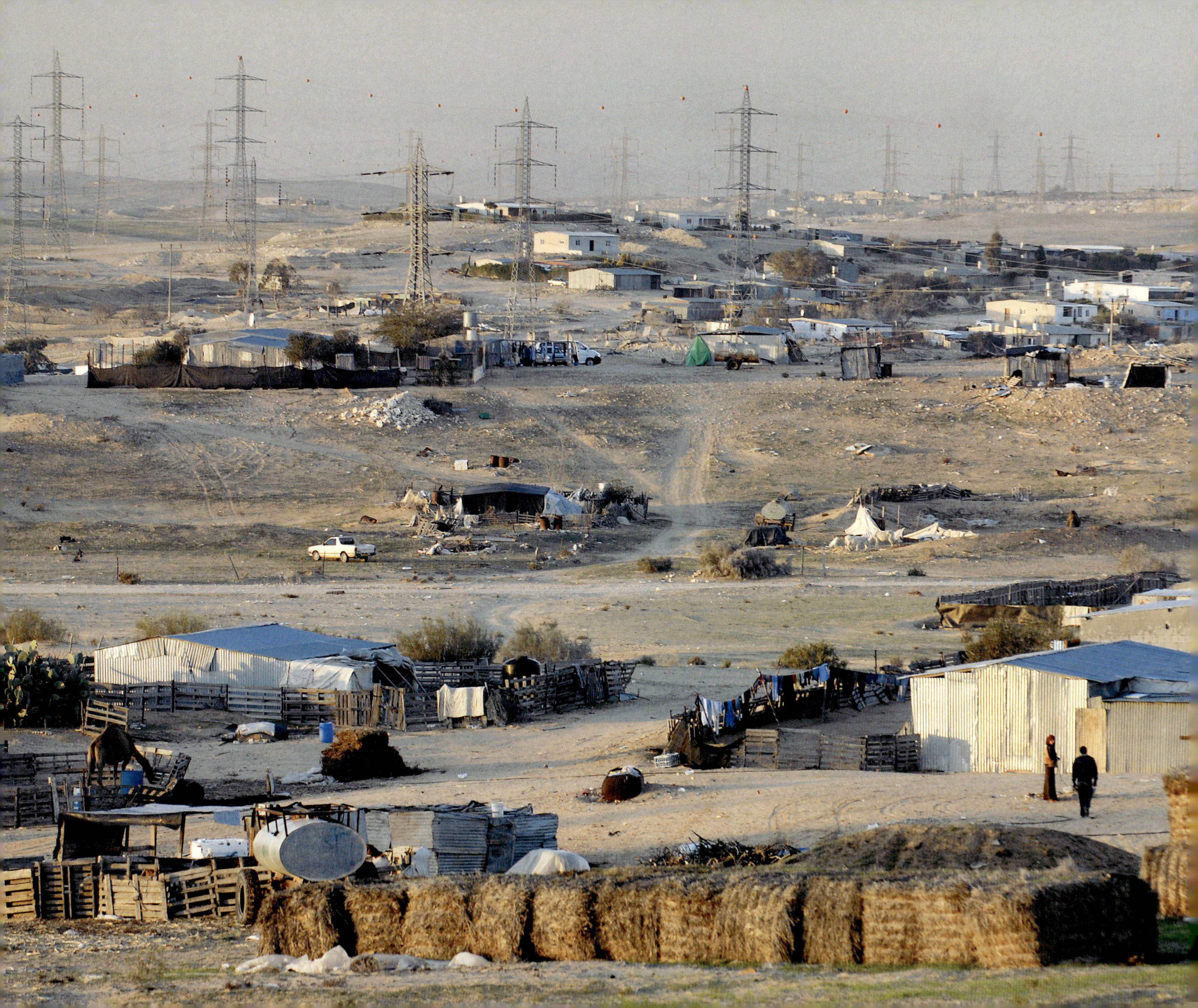

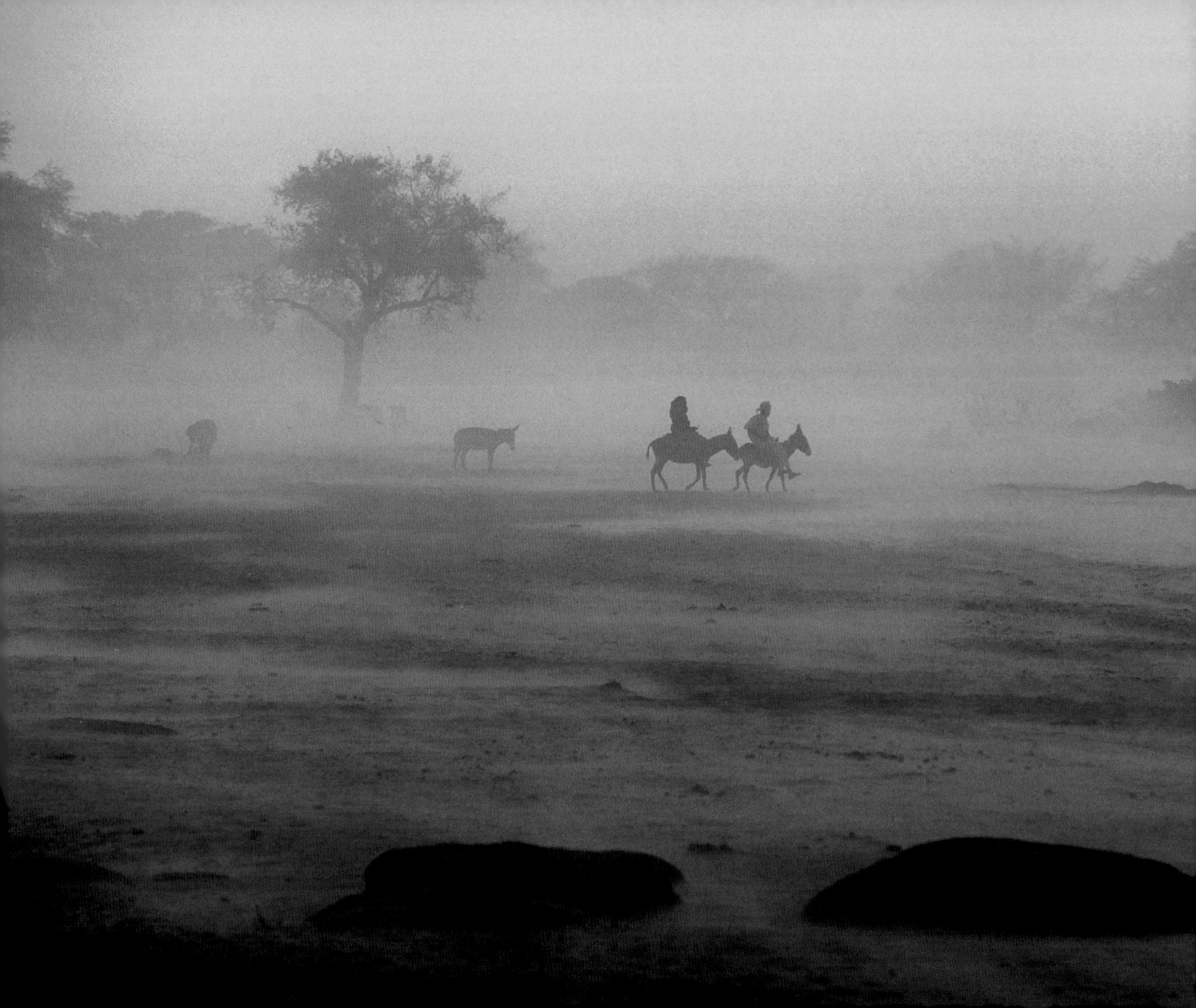

Chad
War refugees

Sudan has one of the largest concentrations of livestock in Africa. As much as 50% of the country's rural dwellers are herdsmen. In recent years, increasing drought cycles and the Sahara's southward expansion have degraded the land and reduced its capacity for livestock-raising. In addition, rural residents are depleting the forests for fuel wood, further harming the land.

Shortages of water and land have created conflicts between nomadic and sedentary tribes. Forsaken, desperate and hungry, groups of Darfurian farmers attacked government outposts in 2003. They claim that the Darfur region has been grossly neglected by the Khartoum Government - a trend that goes back to colonial times. The response to the attacks was the Janjaweed, a militia group recruited from the camel-herding Arabs. The Sudanese government denies that it supports the Janjaweed, but has provided arms and assistance and has participated in joint attacks.

According to the Coalition for International Justice the estimated number of deaths in the conflict is now around 400,000. As many as 2.5 million are thought to have been displaced. Eastern Chad has been hosting more than 234,000 Sudanese refugees. The world media have described the conflict as both "ethnic cleansing" and "genocide." Despite a ceasefire signed in April 2004, a large number of the refugees still have no access to humanitarian relief, and major security and human rights issues remain cause for grave concern.

Darfur may well be the first war triggered by the global climate change. According to a report of the UN Framework Convention on Climate Change, between 25 and 40% of Africa's natural habitat could be lost by 2085. While it is risky to reduce the cause of any conflict to a single factor, a growing number of aid workers, officials, and experts agree that climate change is likely to stretch the tense relations between nomads and sedentary farmers in many regions to the breaking point. Whenever there is less land, and less water to irrigate the land or to water the thirsty herds, competition for that land or water may turn violent.

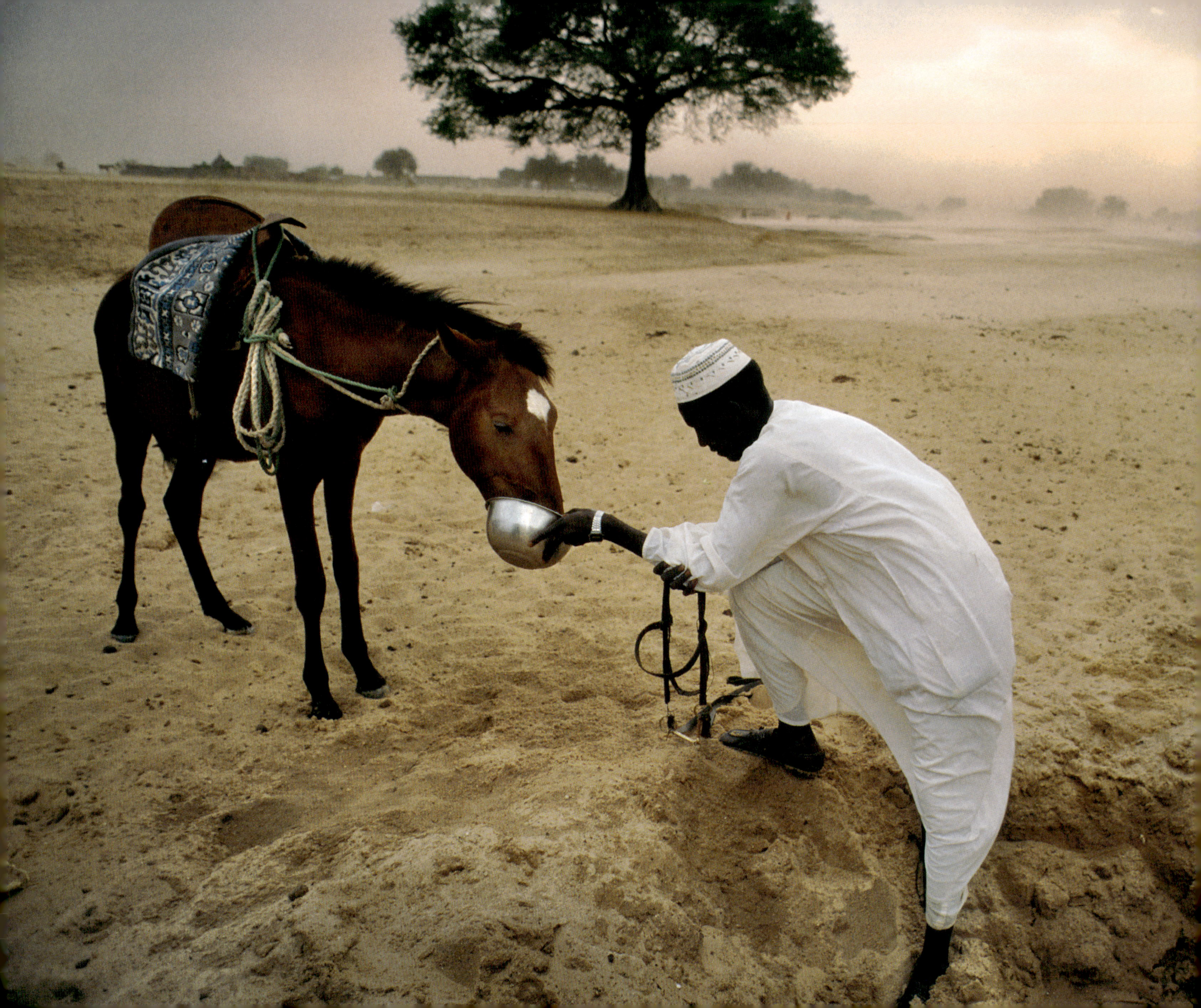

LEFT Birak, Sudan border. A refugee collects water for himself and his horse by digging a hole in the dry river bed. Unless they have money to buy piped water, this is the only water source for the refugees. As livestock drink from the same water sources, many refugees are suffering from water-borne diseases.

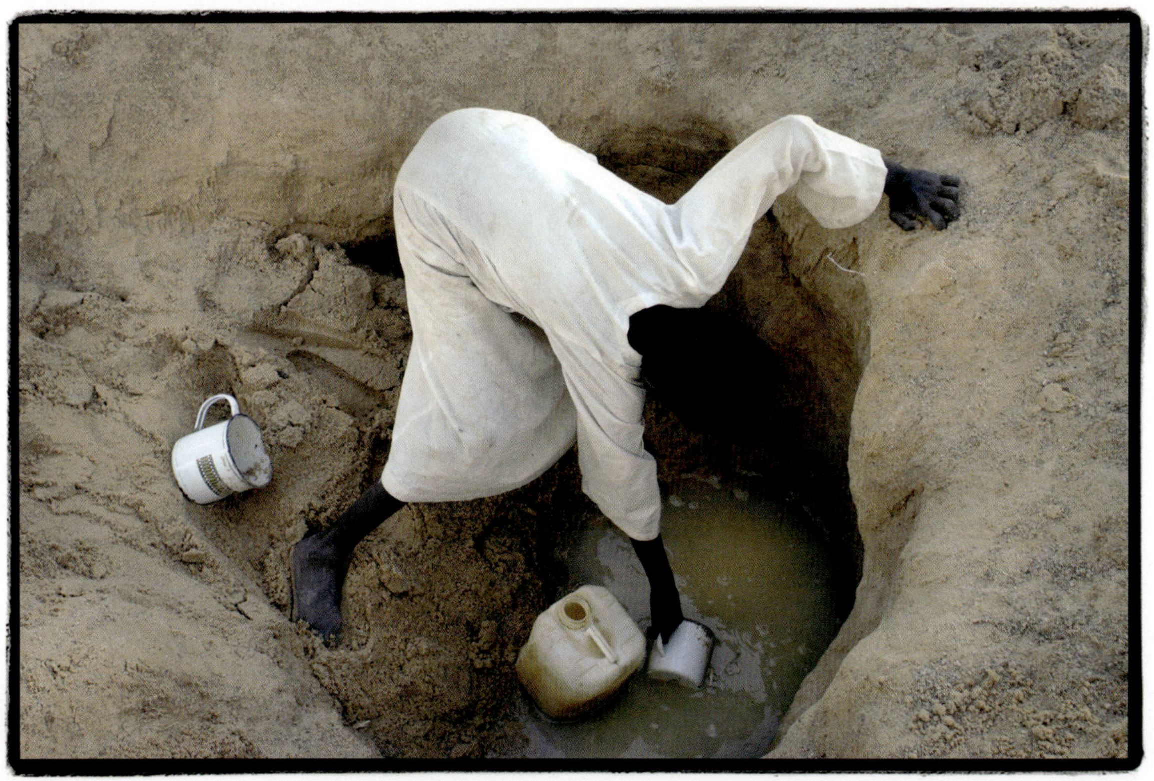

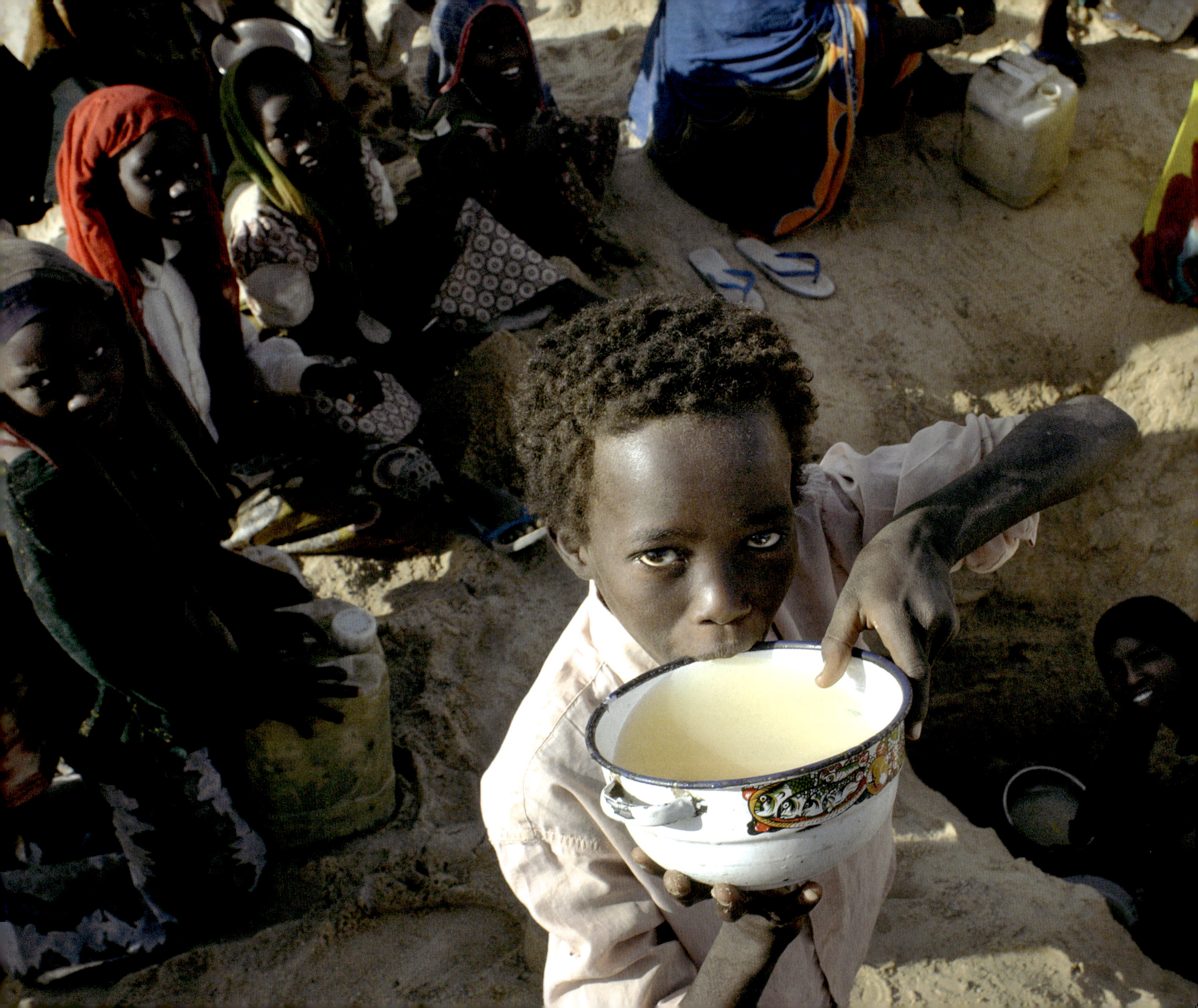

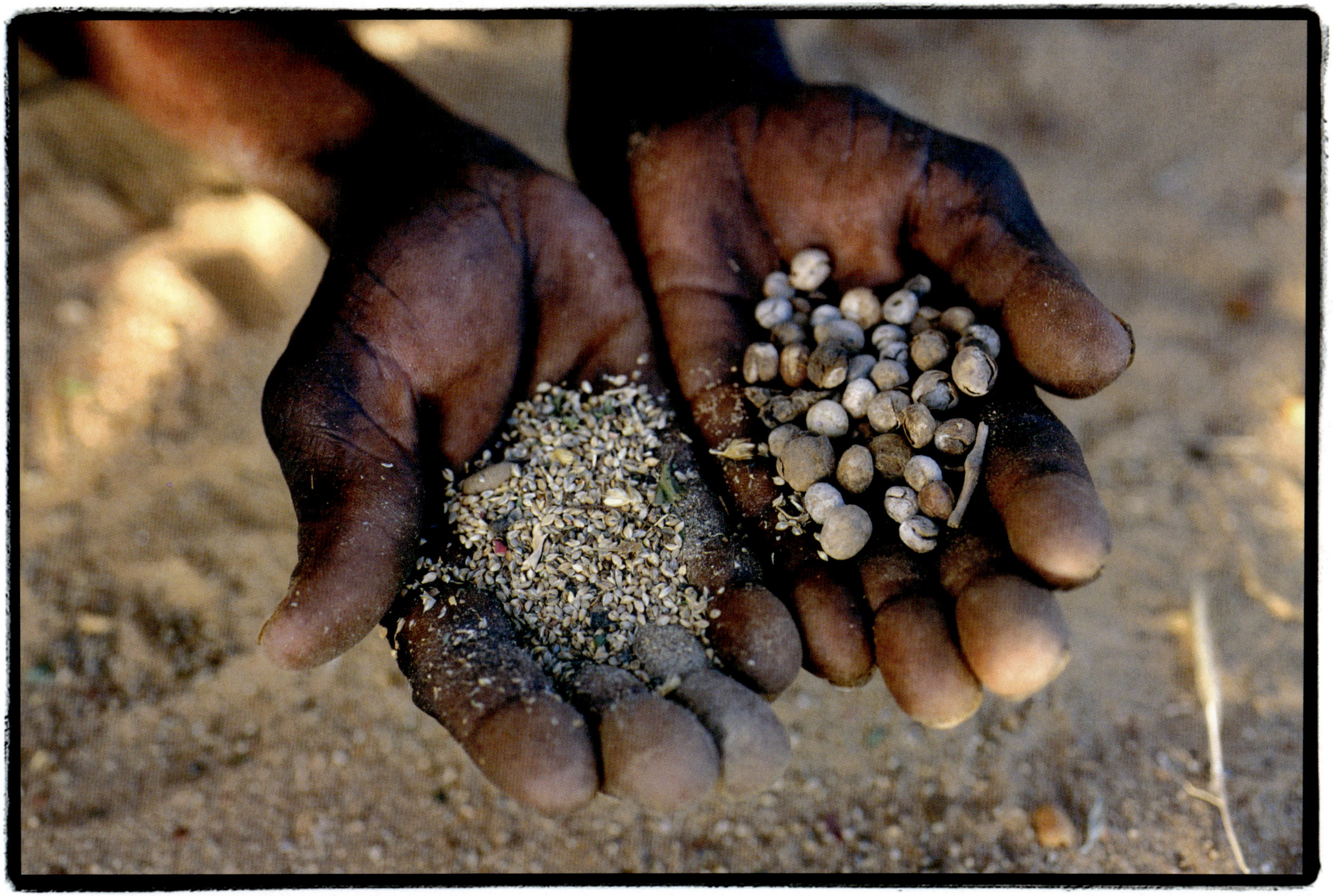

RIGHT Tine, Sudan border. A woman is showing seeds she gathered in the refugee camp in the border town of Tine. With the beginning of the dry season and food resources scarce, most of the people in the camp are surviving on millet alone.

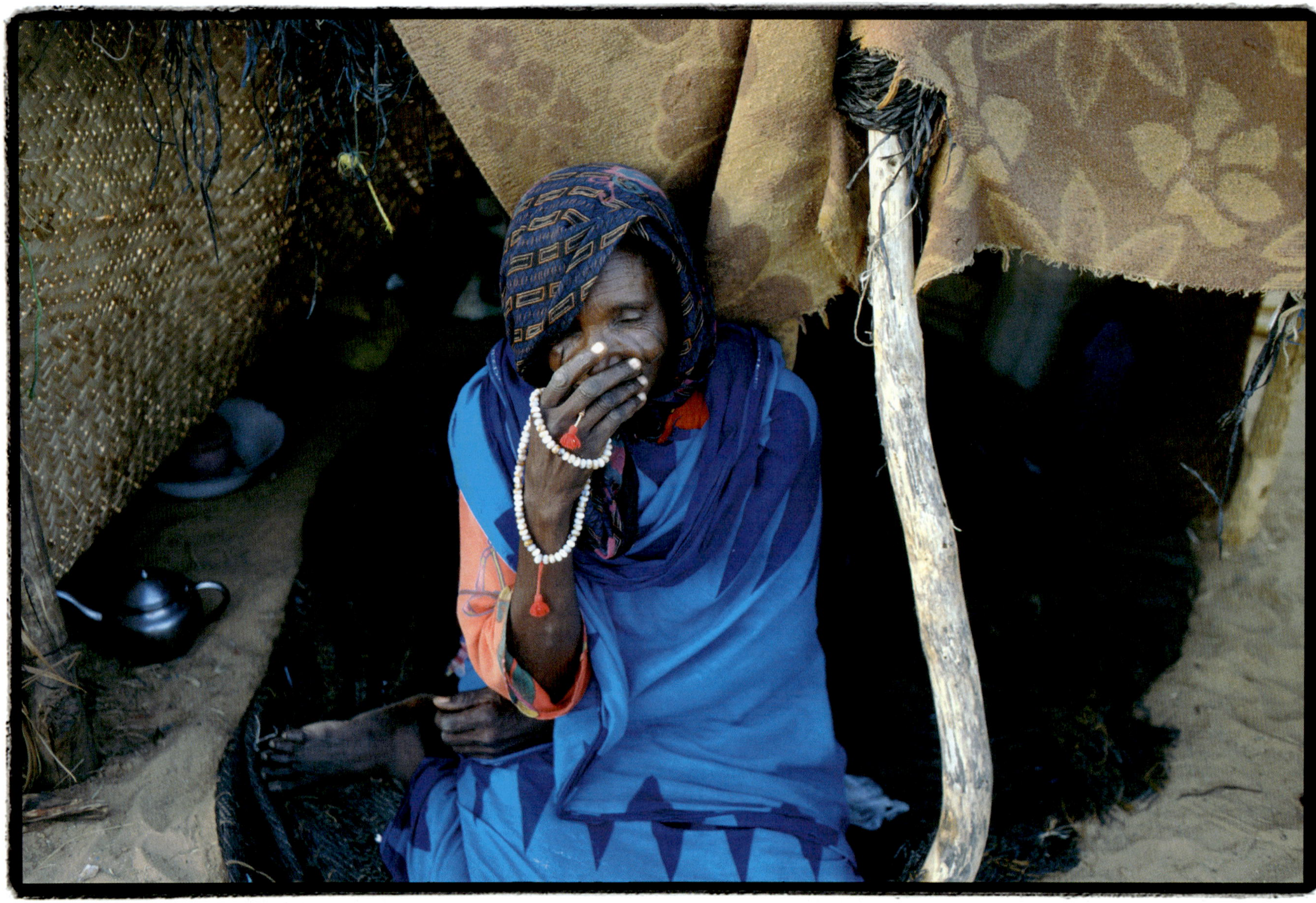

LEFT Sudan border. Patient waiting in the temporary Médecins Sans Frontières (MSF) dispensary at the refugee camp in the border town of Tine.

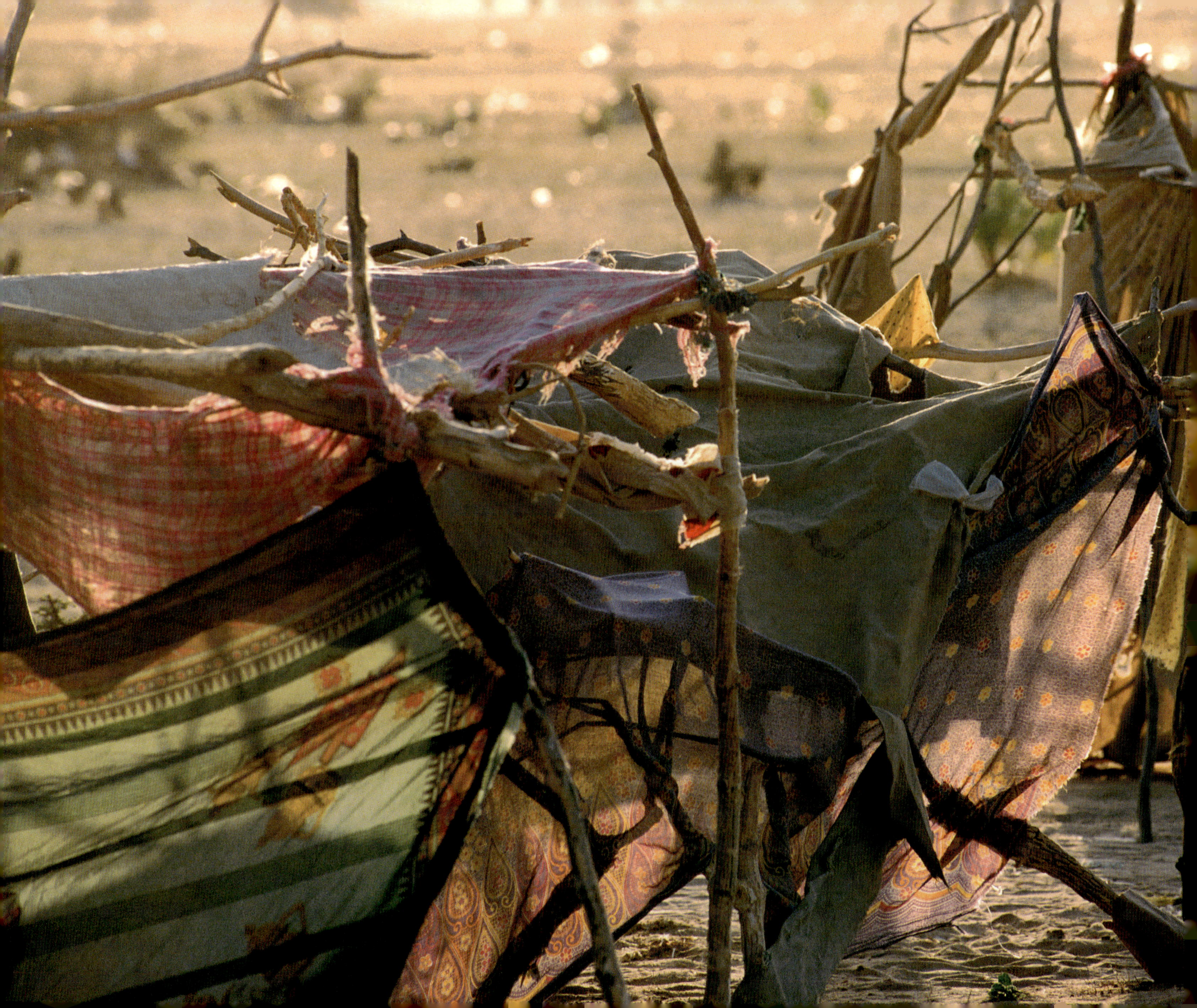

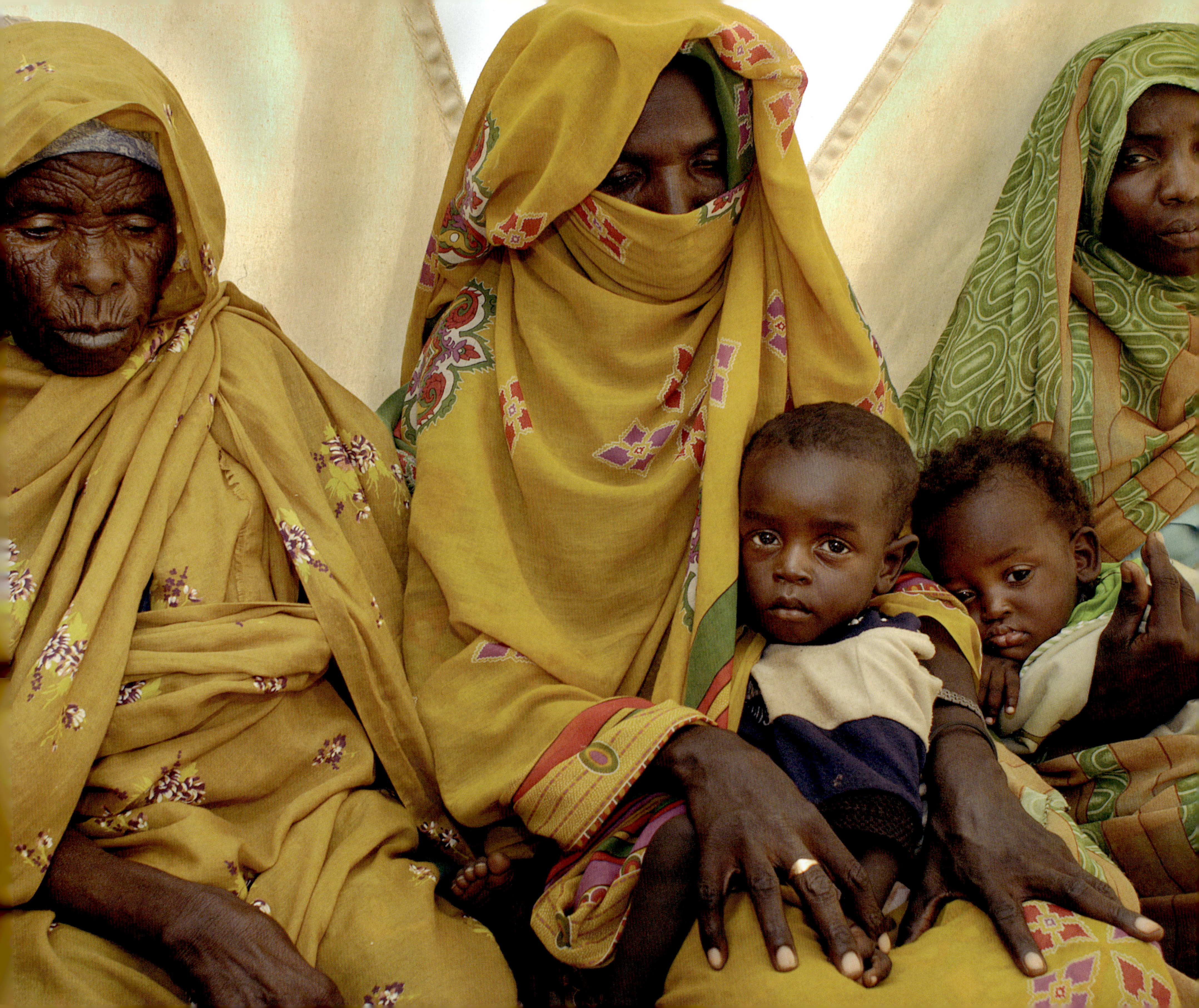

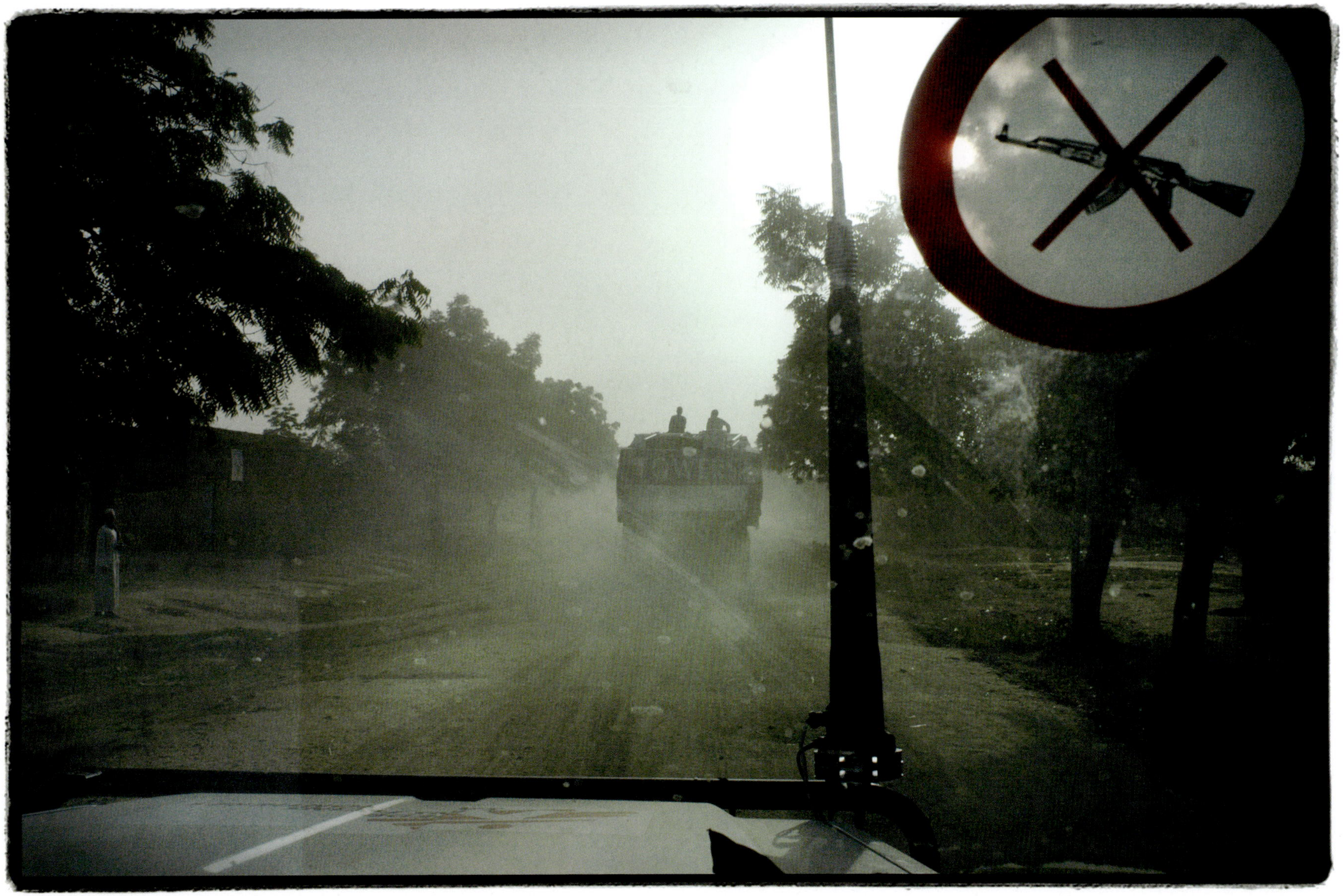

\> the world's poor often pay higher prices for water than people in Europe or the US

\> many countries spend less than 1% of their national income on water

\> almost everywhere, water is wasted, and as long as people are not facing water scarcity, they believe access to water is obvious and natural

\> a third of Spain could be desert within 50 years

\> in the world's worst slums, people often pay 5 to 10 times more than wealthy people in the same cities

\> visitors to Spain's beaches may face water rationing

\> Spain's construction boom has led to the building of thousands of illegal wells

\> modern engineering has caused irrigation disasters on a massive scale

6 WATER MANAGEMENT

\> water losses in municipal distribution systems in the Middle East often exceed 50% of the water supplied for urban use

\> different farm management practices can prevent the degradation of water quality

\> excessive irrigation brings salt to the surface and can create salt marshes

\> while the world population tripled in the last century, the use of renewable water resources grew sixfold

\> a government report gives warning that a third of Spain may turn into desert because of climate change, poor farming practices and tourism

\> many of the world's aquifers are being withdrawn at a rate that may empty them within 50-100 years or less

\> in Third World countries people can spend up to 20% of their income on water

\> lack of political interest prevents many people from enjoying their basic water rights

> 1,000 villages in Rajasthan escape the effects of droughts thanks to the revival of ancient tanks and underground reservoirs

> once desertification has started, it is extremely expensive or impossible to reverse it

> Australians are getting used to timing their showers

> the battle to provide water for Spain's southern coast has generated major controversy

> the Spanish government is building several desalination plants to provide water for the near-desert region

> Australia faces major water scarcity in the Murray-Darling Basin as a result of diverting large quantities of water for use in agriculture

> the world cannot increase its supply of fresh water: all it can do is change the way it uses it

> there is room for better management of water in agriculture, which currently takes up 70% of the water we use

> in South Africa many people now receive 6,000 litres a month for free, then pay for whatever they use beyond that

> over-exploitation of water resources is causing water tables to recede at an alarming rate

> in many groundwater tables saline seawater is flowing in to replace the fresh water that has been pumped out

> Spain is about to die of thirst, according to the newspaper El Mundo in the summer of 2005

> the number of golf courses in Spain has doubled in the past five years, to 276: another 150 are planned

> the cost of climate change has been calculated at US$9 trillion

> investors estimate that the expected shortages will cause the price of water to skyrocket, yielding profit to companies active in the field

> Spain loses millions of gallons in leaks from antiquated pipes

> in rural India hundreds of millions of inefficient pumps are being used to deplete the already scarce groundwater

> Saudi Arabia is depleting fossil groundwater and desalinating sea water to grow wheat at a cost 100 times the world price

> water is overused and misused with seemingly little care for the future consequences and future generations

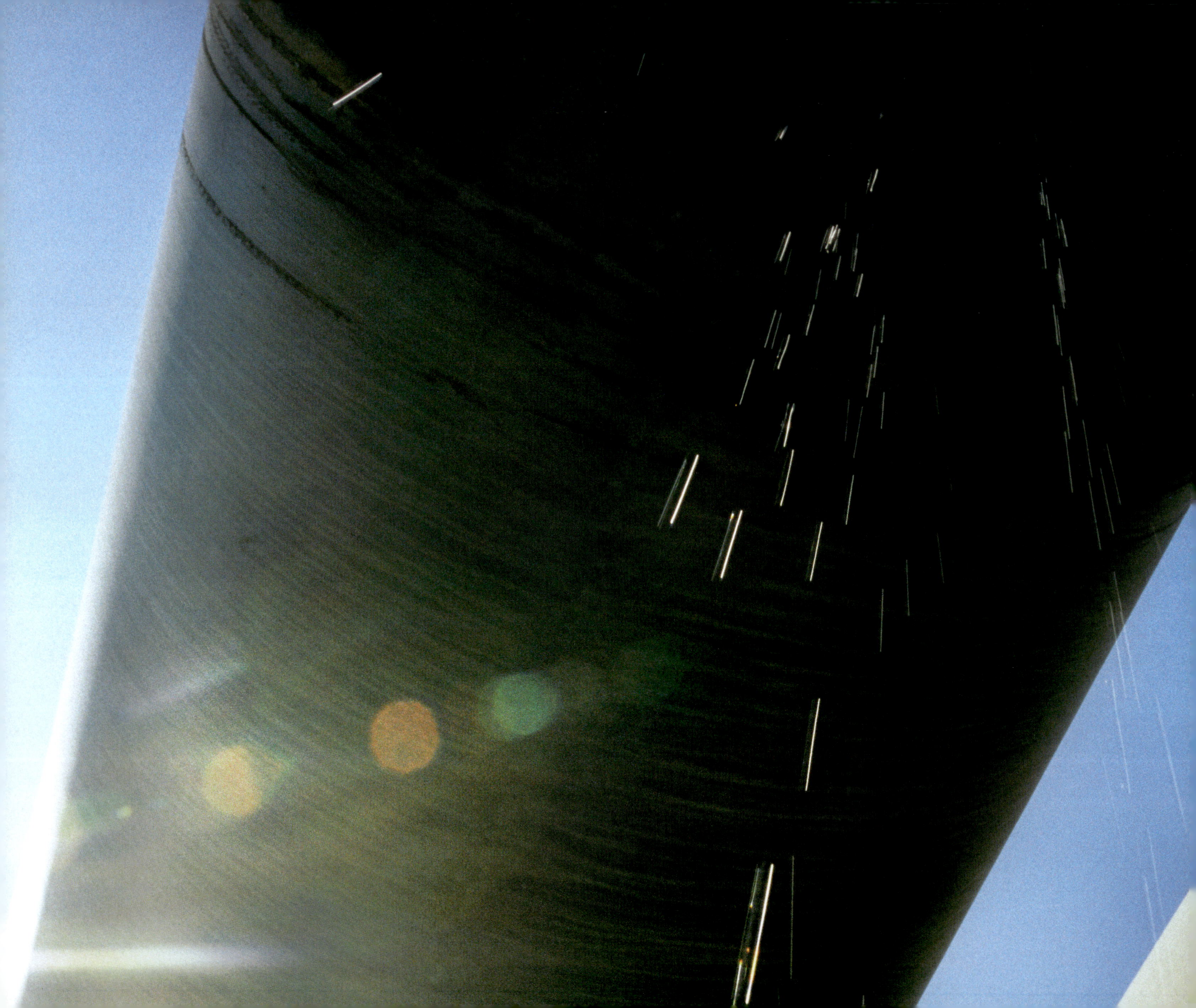

Spain
The cost of golf

In the Spanish province of Murcia massive building zones are being developed. Last year 350,000 holiday homes were built with plans for another 1.65 million. Alongside these new homes well over a hundred new golf courses have been created, each consuming as much water as a town of 16,000 inhabitants. Already two million foreigners have bought a holiday home along the coast and estimates are that by 2020 there could be six million pensioners –mainly from Britain and Germany– living in the area.

At the same time, Spain has experienced its worst droughts since 1947, river and reservoir levels have been at record lows, and, in some places, it has not rained in 15 years. The water boards of two coastal regions have said that they are unable to guarantee water supplies to the 450,000 new homes planned for 2006.

In 2005 a few Spanish farmers became so desperate that they used untreated sewage to water their export crops, leading to outbreaks of salmonella in the UK and Finland.

The warning signs are there, but with tourism accounting for a tenth of Spain's GDP, politicians have shown little inclination to rein in the developers, instead offering flexible interpretations of zoning laws and huge tax breaks to construction firms. The golf business alone was good for 2 billion euros in 1997; a number that has tripled since then.

In order to provide more water to the area around Valencia, Almeria and Murcia, a 4.2 billion euro plan to divert water from the northern river Ebro was introduced, but abandoned after tens of thousands had protested against it. The concerns were that it would encourage misuse of water and that the Ebro's fragile delta would suffer. Instead, the government has now commissioned 17 desalination plants which, when operational, will provide enough water for 800,000 people.

This may well be a case of too little too late. Spain's economic success has already come at a huge environmental cost. According to the UN a third of Spain could be desert within 50 years if the building boom and the excessive water use in agriculture are not halted. This is bad news not just for local people and wildlife. It may also spell the end of tourism and destroy the industry that supplies European supermarkets with cheap fruit and vegetables.

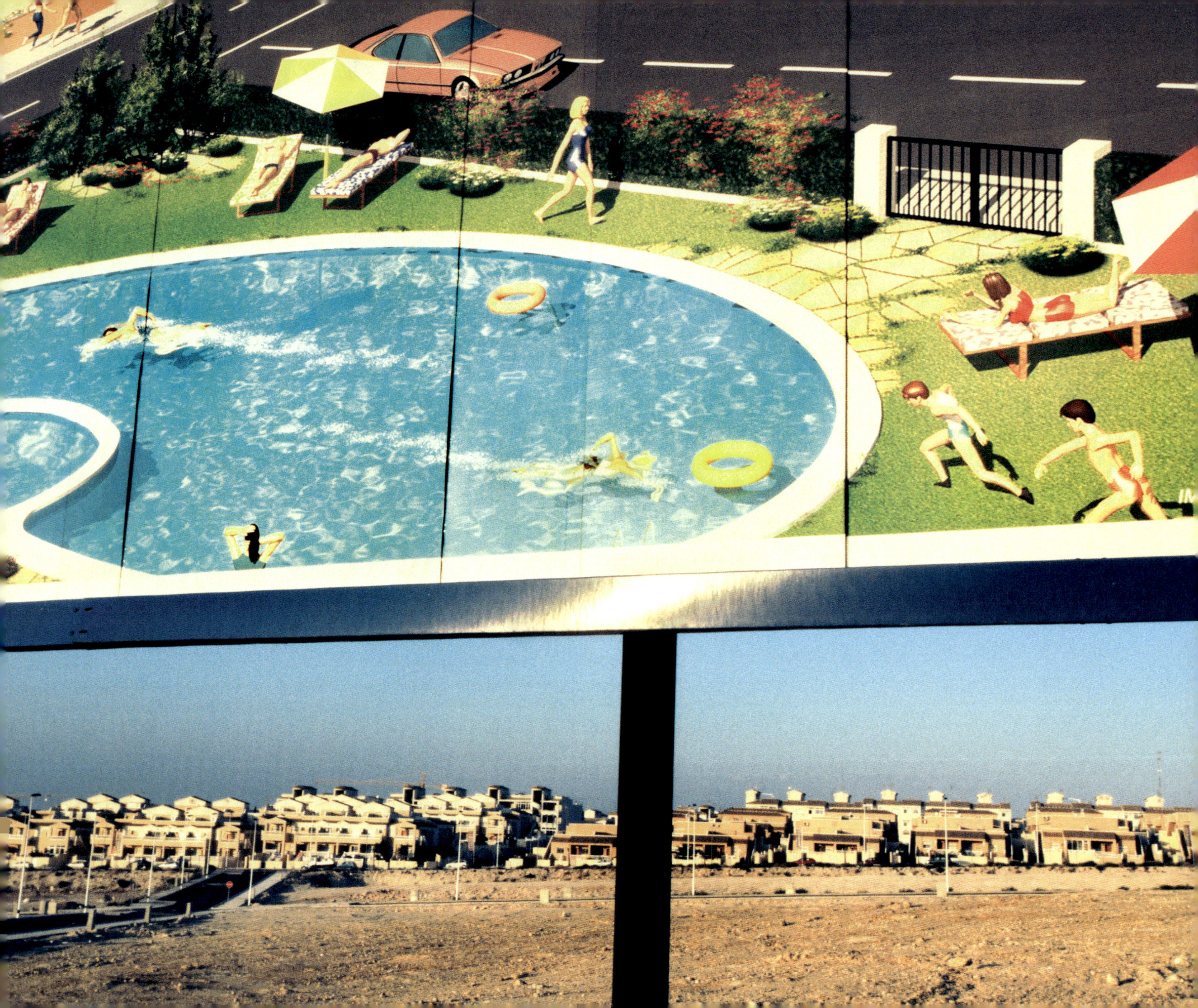

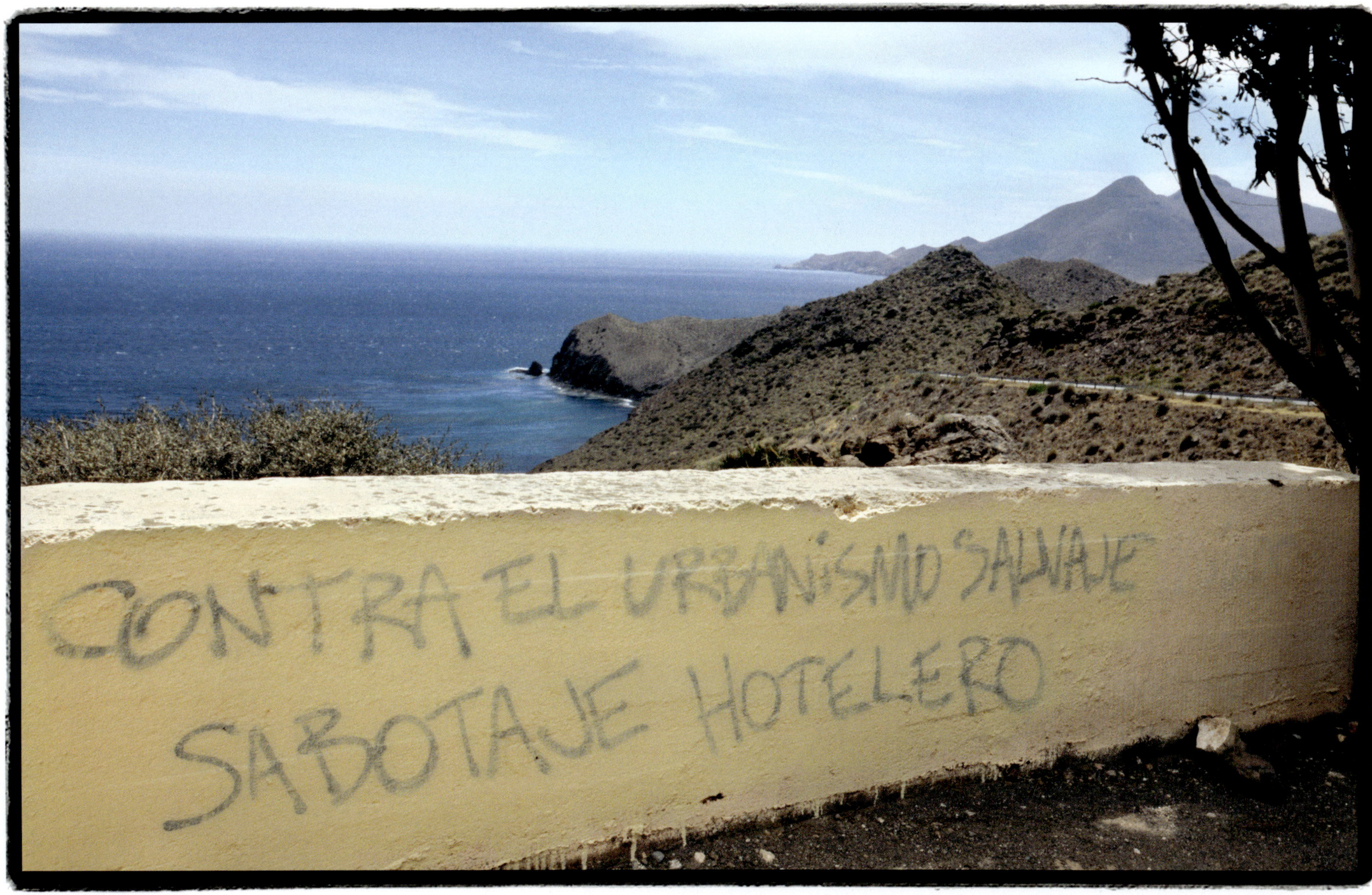

RIGHT Costa Blanca. Protest against the construction of hotels along the coast. By law, in Spain the first 50 meters of land along the coast is common property, but in practice this rule is often ignored. The UN has warned that Spain's desertification will go on unabated if the excessive water use by the construction and tourism industries and agriculture is not halted.

PREVIOUS PAGE Alicante Province, Orihuela. Pipes diverting water of the Tajo River to the southeastern region of Murcia.

RIGHT Costa Blanca, Garrucha. New house with swimming pool in one of the last unspoiled areas along the Garrucha coast.

LEFT Costa Blanca, Cabo de Gata. Construction of the massive hotel "Legal" along the coastal road, just after a sign saying "You are now entering the Natural Reserve of Cabo de Gata".

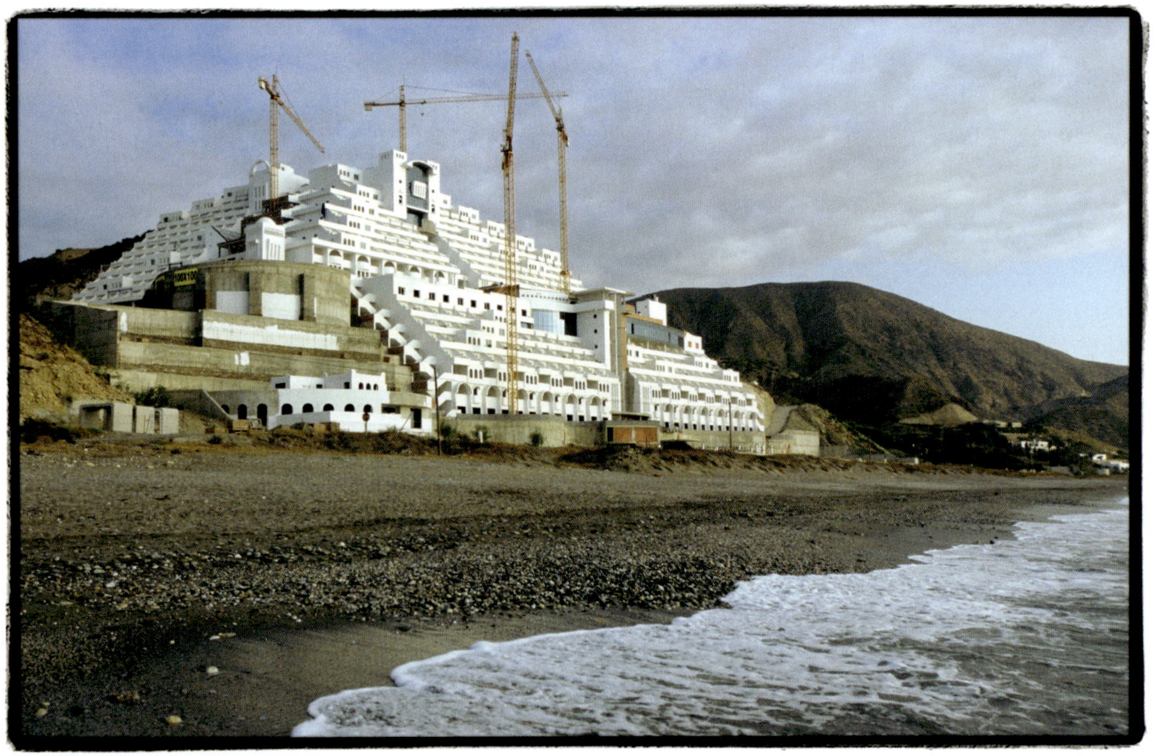

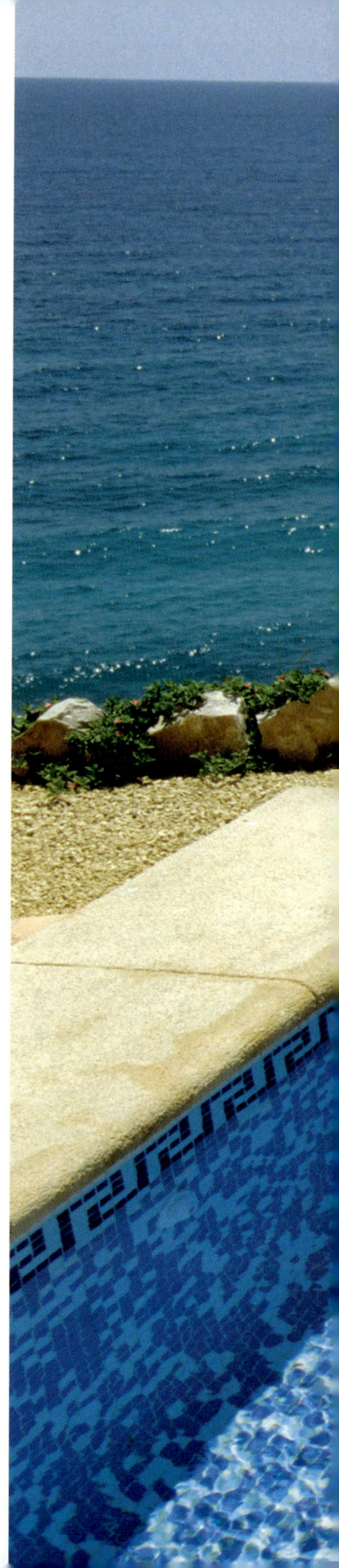

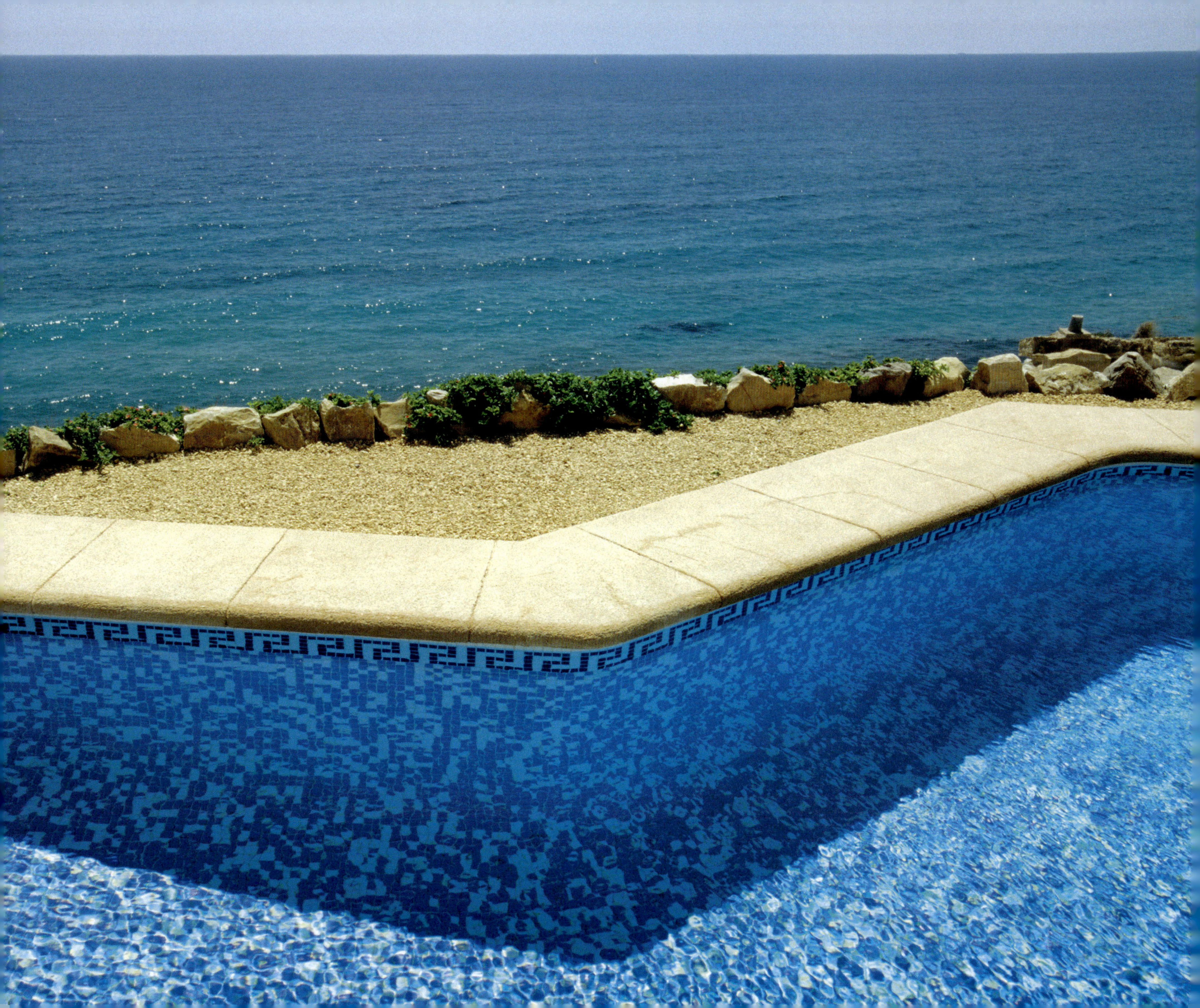

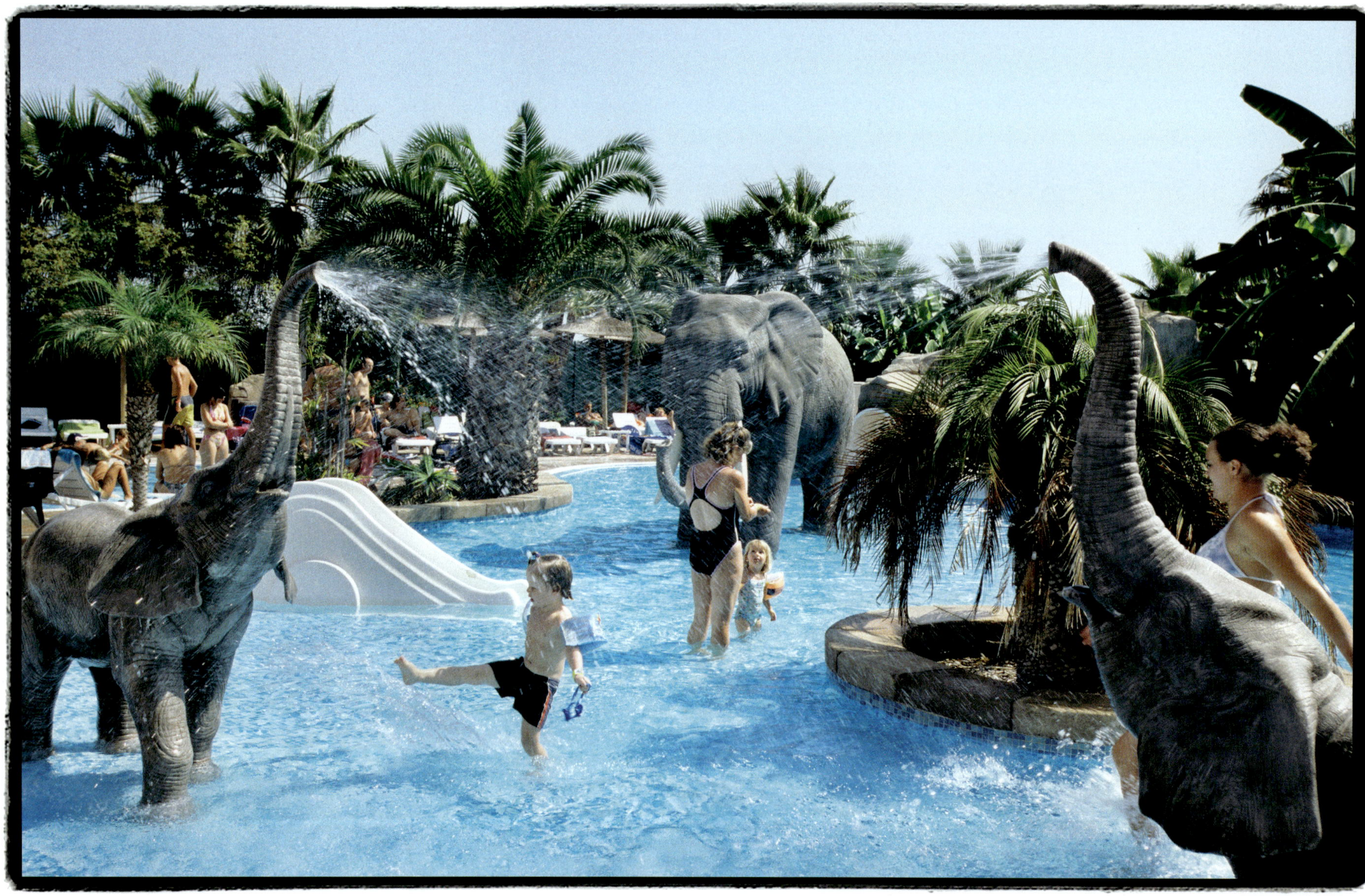

RIGHT Alicante Province. Dry water reservoir in the mountains near Benidorm. In 2006, the water reservoirs were at their lowest level in ten years' time.

LEFT Alicante Province, La Marina. Swimming pool in a campsite along the Mediterranean.

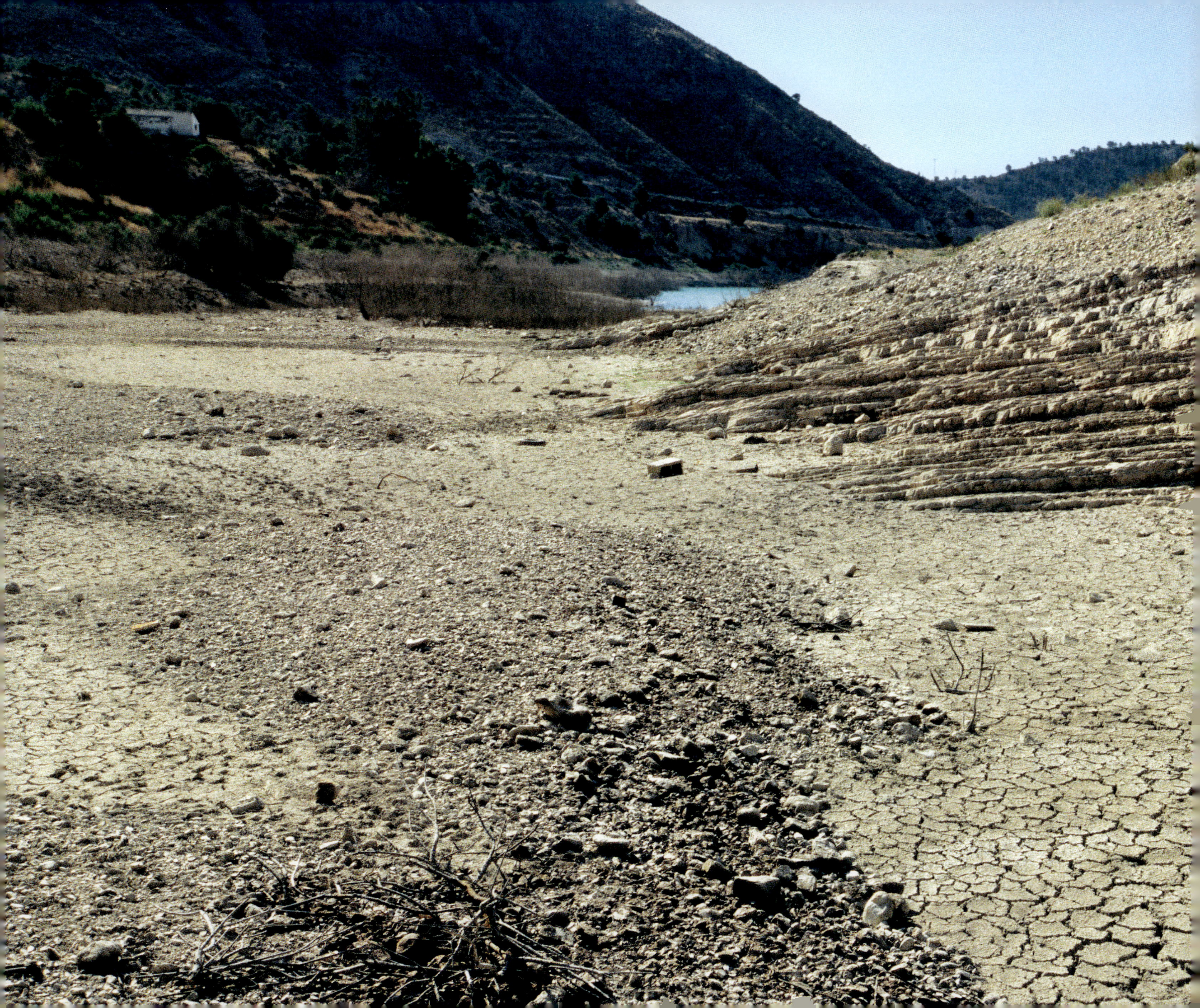

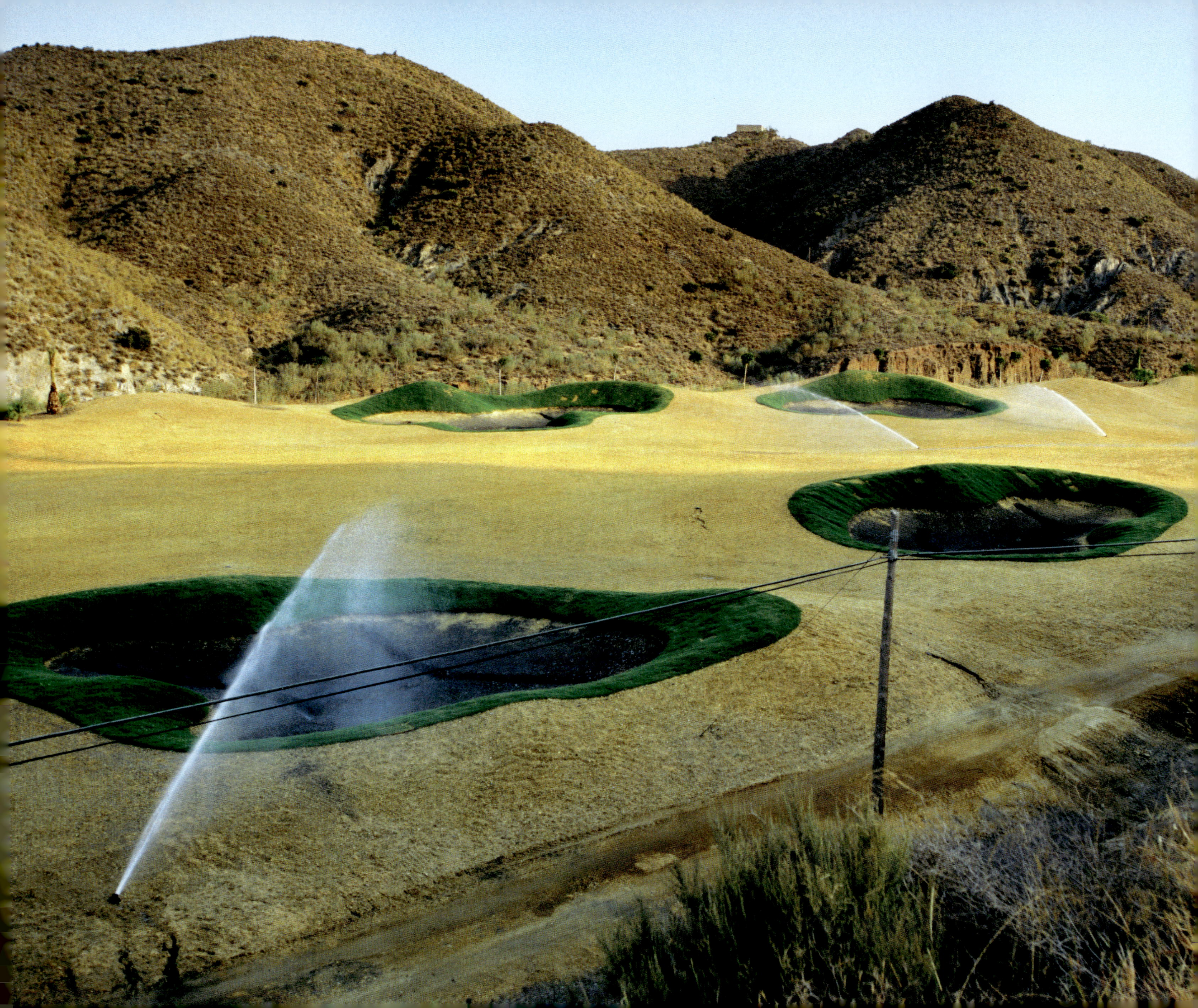

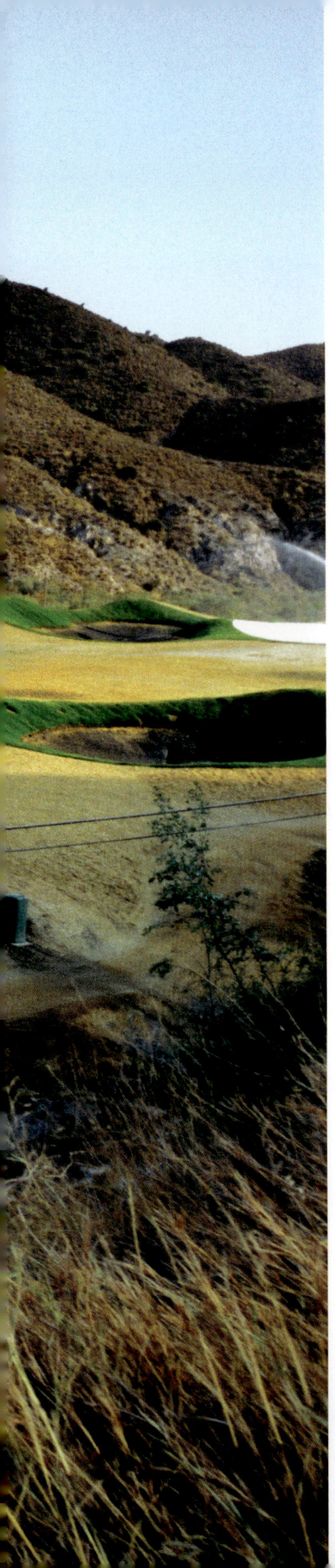

LEFT Costa Blanca, Carboneras. The "Desert Springs" golf course under construction near Cabo de Gata. It rains only 10-12 days each year, so the sprinklers are running 24 hours a day to turn the desert into grass, from brown to green. Since 2000 Spain has built 160 golf courses, mainly in the south.

RIGHT Almeria Province, Tabernas. Fake tents, part of a movie setting in an area called mini-Hollywood. Because the desert landscape is very similar to the American Far West, many cowboy movies were made here in the sixties and seventies.

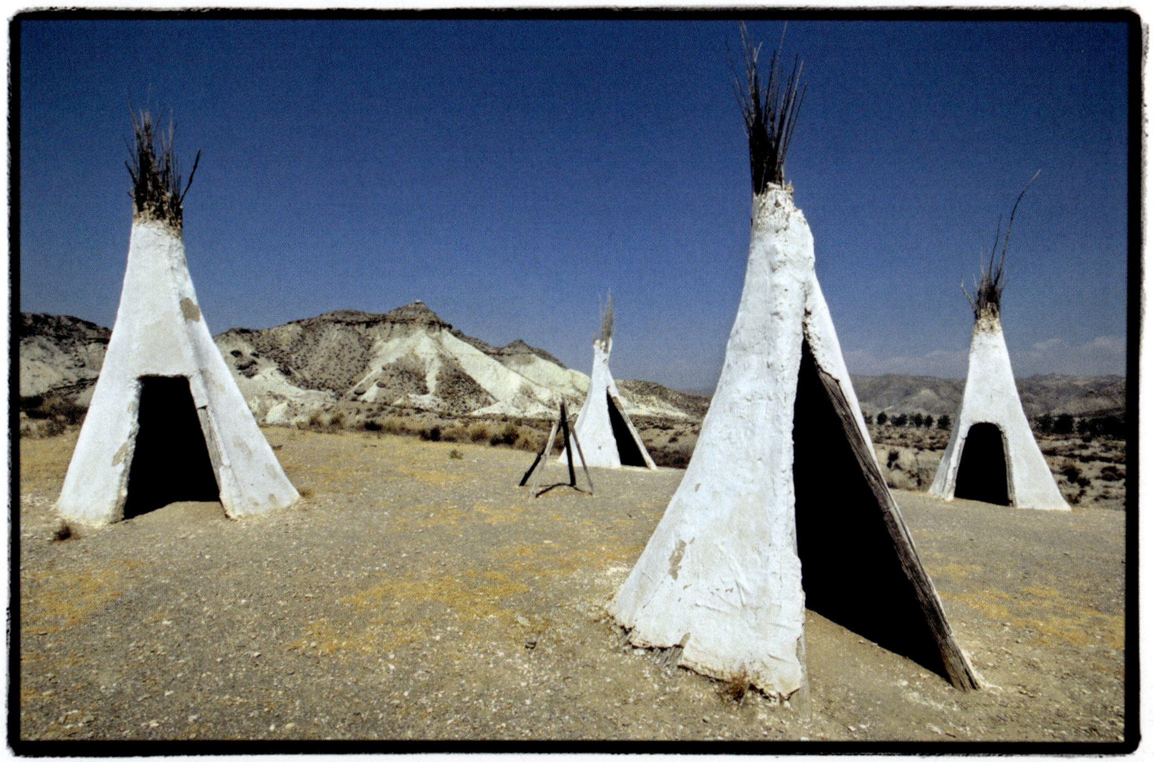

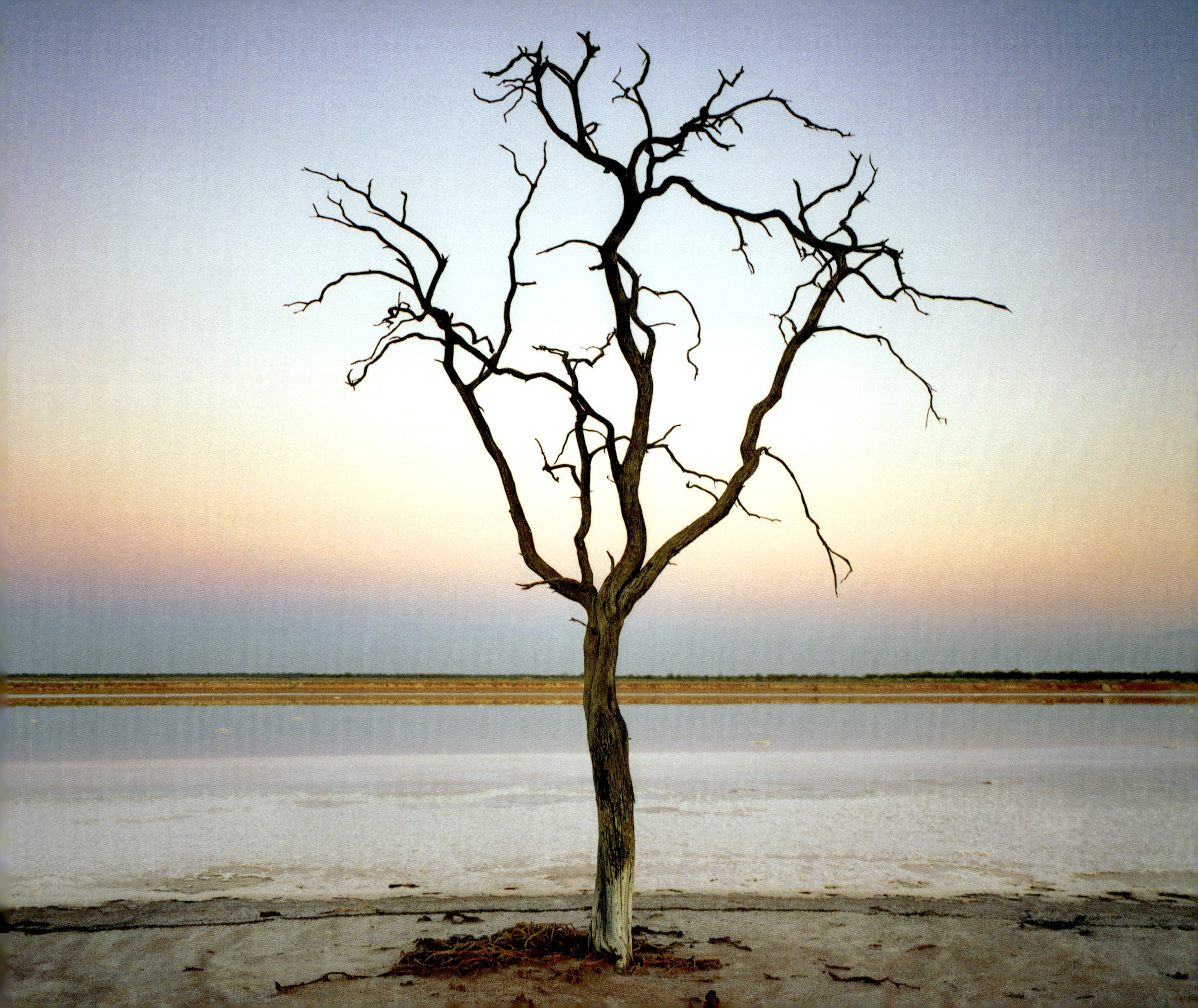

Australia
Fight against salinity

Since Europeans settled in Australia two centuries ago, an estimated 15 billion trees have been cut down. The trees used to suck up underground water and hold the water levels. Now the water table rises and the top soil becomes salty. This is exacerbated by over-farming and too much sheep trampling.

Water resources in Australia have been managed badly. The little water there is has often been wasted on water-hungry crops, like cotton and rice, which Australia should better leave to wetter countries. Furthermore, irrigation is mainly done by means of open channels, not pipes; so much of the water evaporates or is lost.

On top of this, there is a long-running drought that has resulted in massive agricultural losses. The Murray River and Darling River, which supply water for more than two-thirds of Australia's irrigation system, are flowing at their lowest levels in a century. In some states, the drought has been the worst in decades. In others, such as Western Australia, it is the worst on record. Scientists say that unless something is done a third of the continent could become a sterile white desert within 100 years.

Still, some Australians believe the climate change is temporary and part of a natural seasonal cycle. They point out that the continent has experienced very dry periods before. John Howard, the Prime Minister, refused to sign the Kyoto Protocol saying it would damage the economy. The opposite might be true: the Reserve Bank of Australia has warned that the drought is likely to slow economic growth, as a result of lower farming output.

The Australians have succeeded in transforming the driest inhabited continent into one of the world's most dynamic economies. But Jared Diamond, the American anthropologist and writer of the book "Collapse", believes Australia can only sustain a population of 8m, well under the present figure of 20m. With the population expected to at least double by the middle of the century, Australians are facing a new enormous challenge.

A December 2006 poll conducted for the newspaper The Australian shows that Australians might be up to the challenge: an overwhelming majority said they would be prepared to drink recycled sewage to help ease the national water crisis.

The government is working on the establishment of a single national water market. They want water to be priced on the whole continent according to national rather than local needs. This will make water users in the tropical north of Australia, where they still get plenty of rain, pay a higher price for it and force them to use water more carefully.

RIGHT South Australia Province. Footprints on Lake Torrens, a 5,700 km^2 salt lake. Australia's soil has been accumulating salt for up to a hundred thousand years, most probably blown in from the seas.

LEFT South Australia Province. Lake Hart, a dry salt lake. From about 6 million hectares in 2000, plant-killing salt is to cover 17 million hectares of Australia by 2050, according to a government forecast.

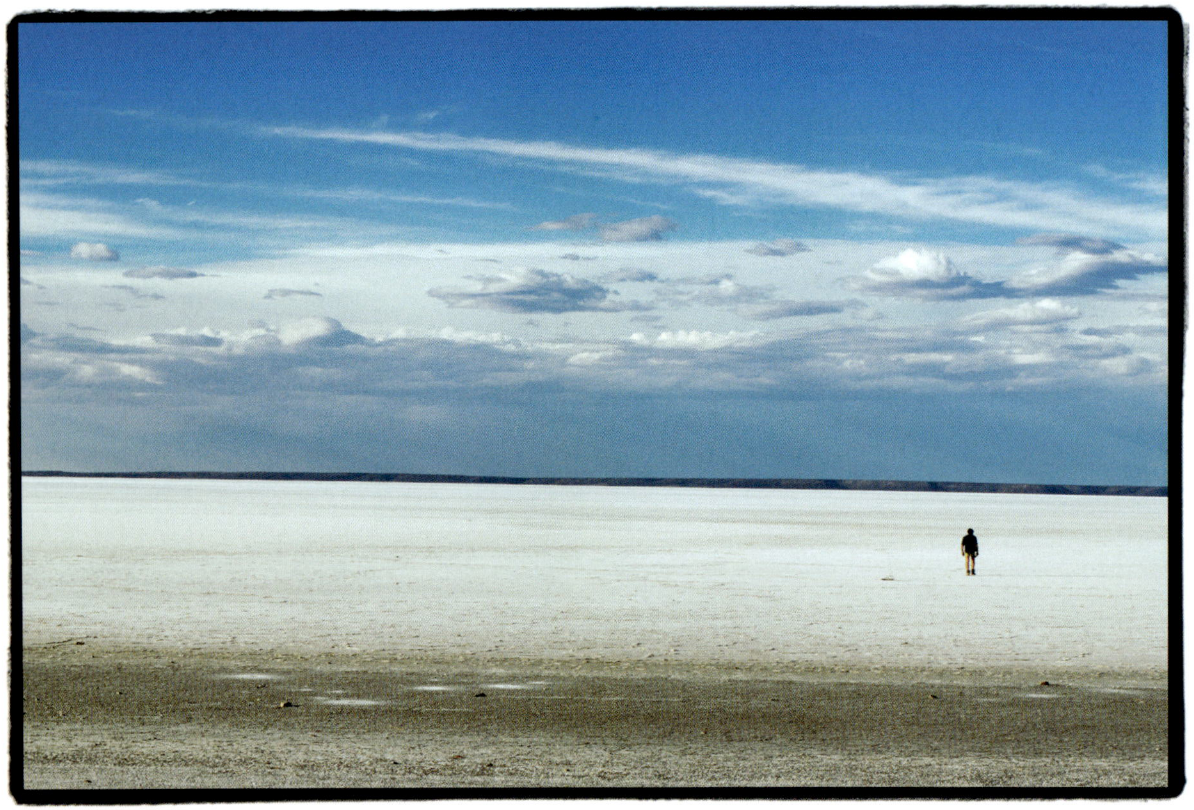

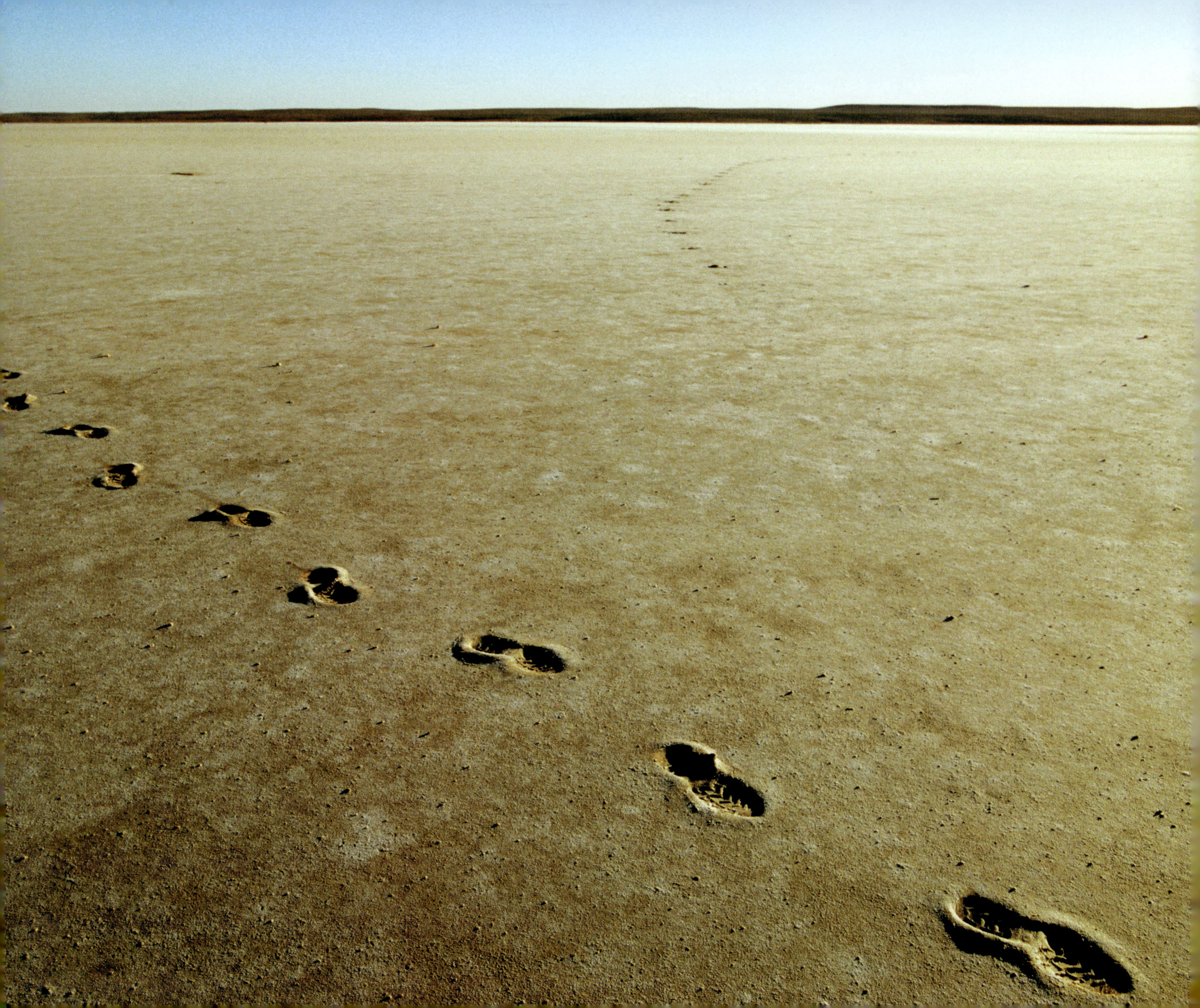

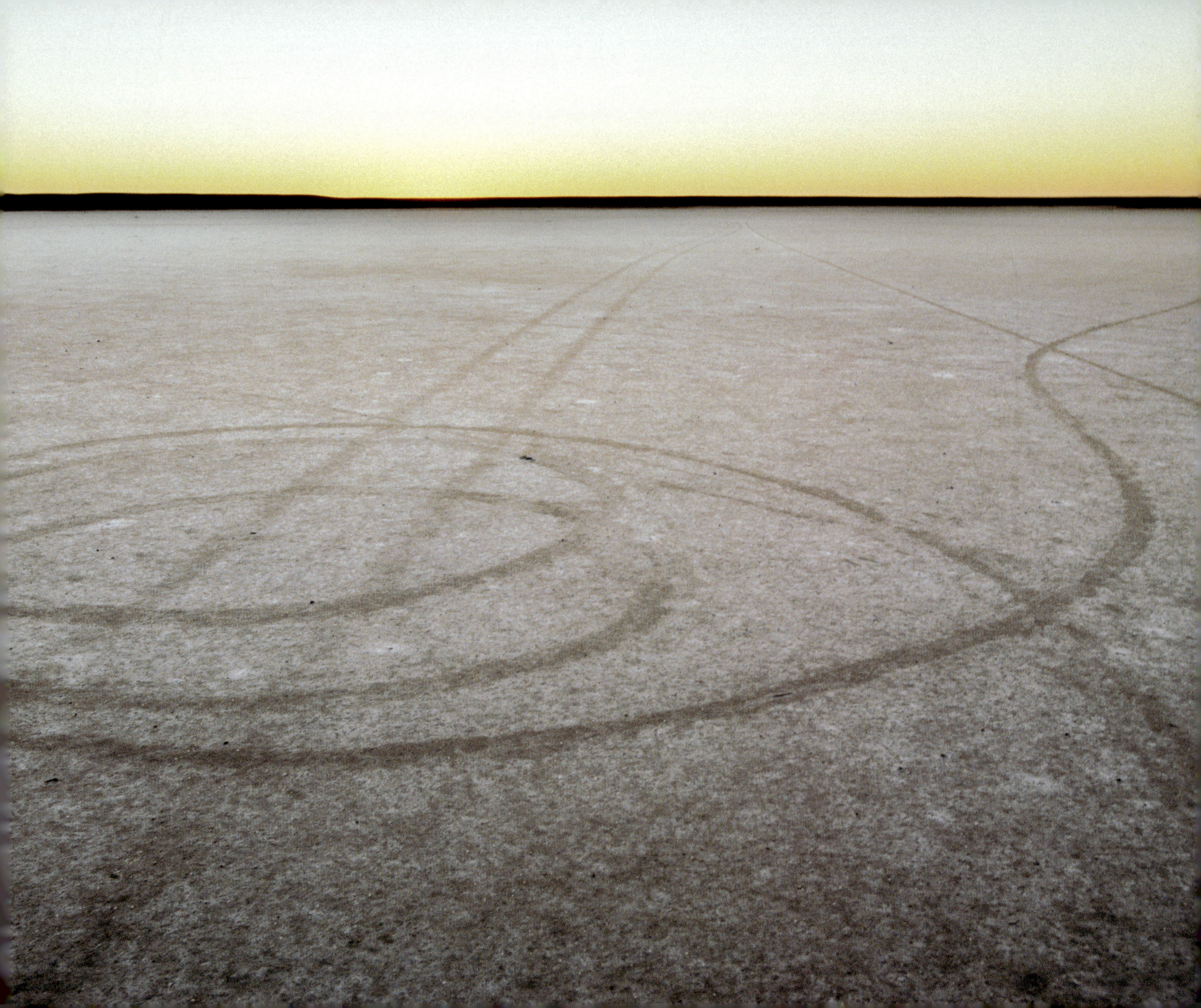

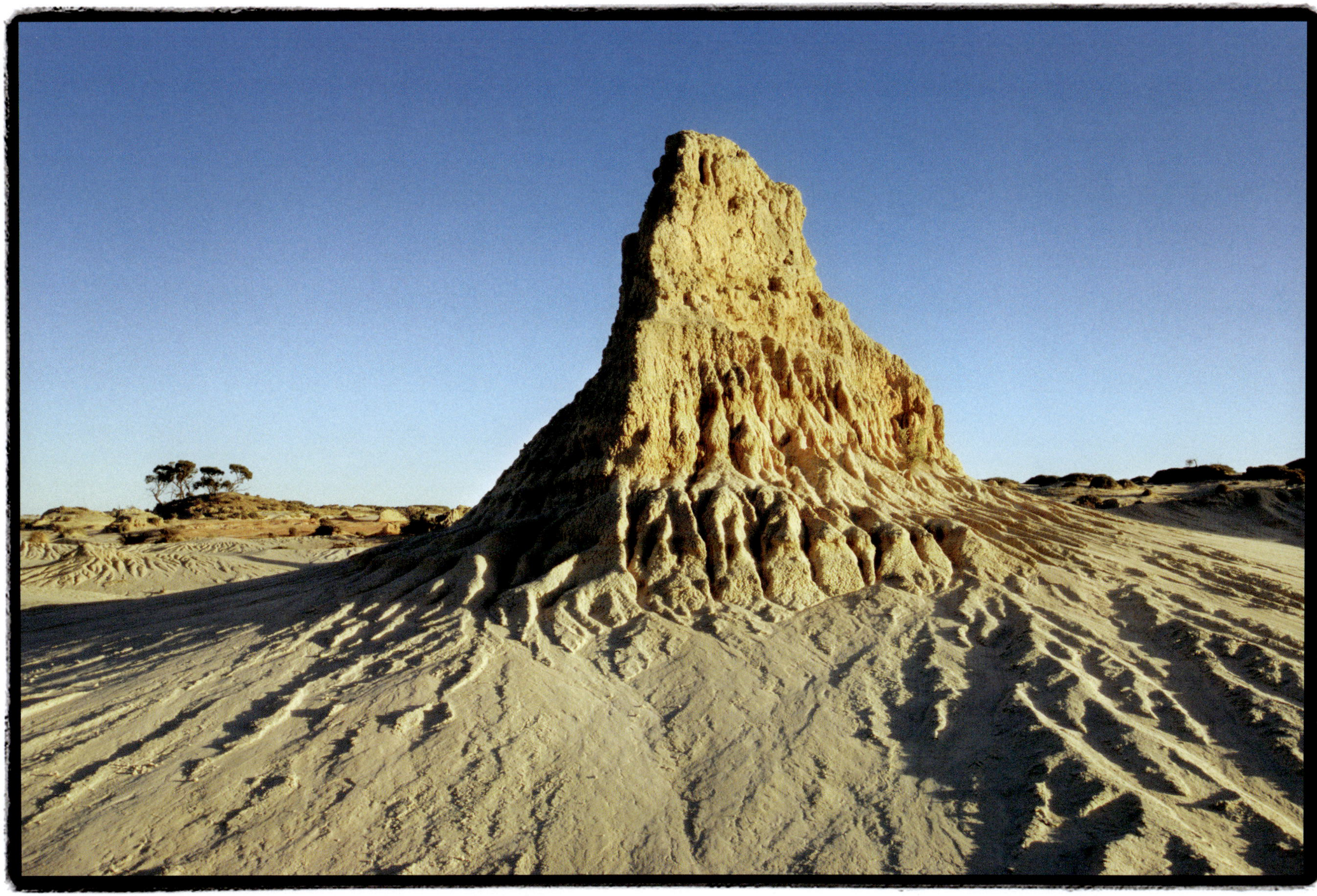

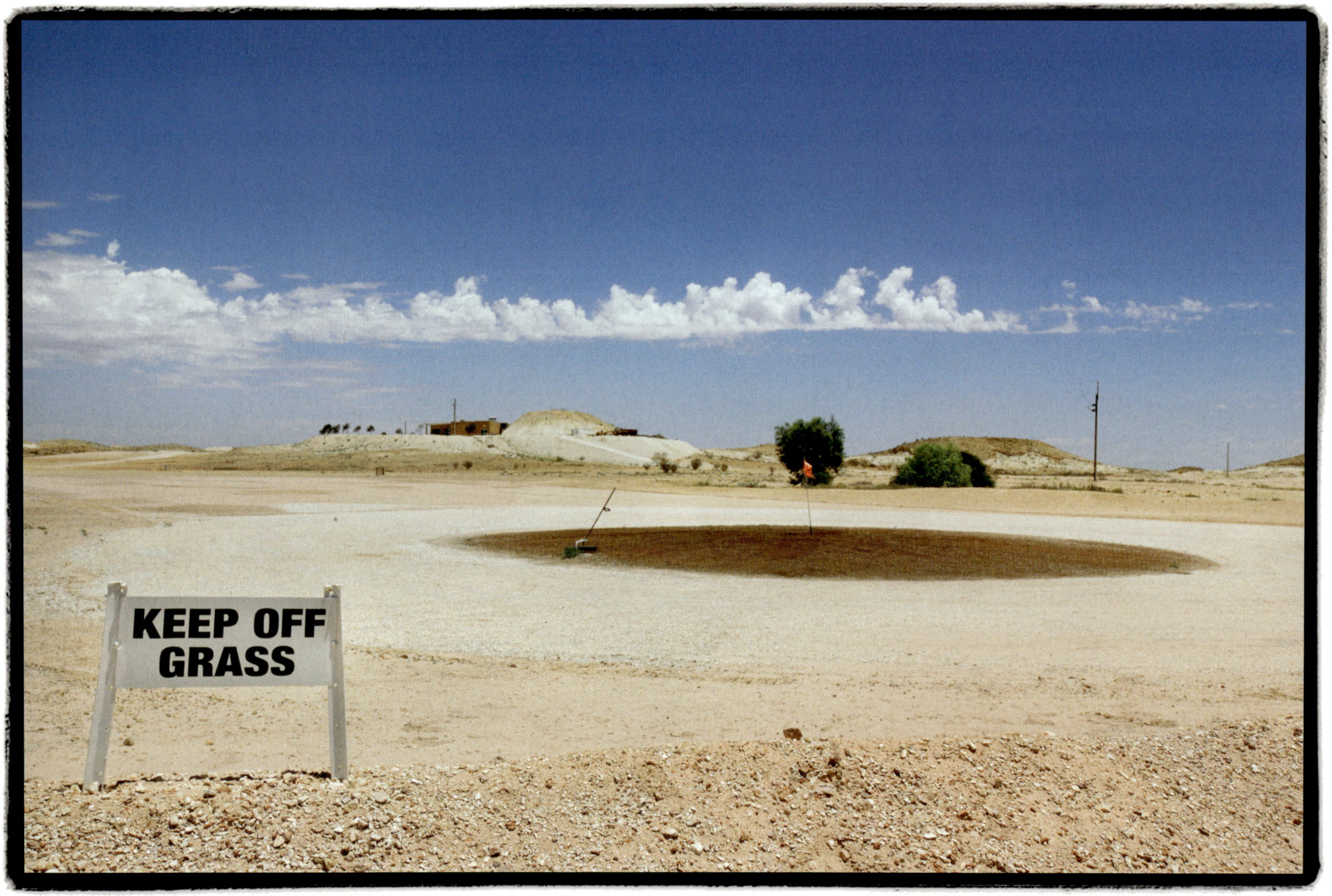

South Australia Province, Coober Pedy. A sign reading "Keep off the grass" stands at the edge of a desert golf course. Some players bring their own piece of artificial grass to use as a teeing ground.

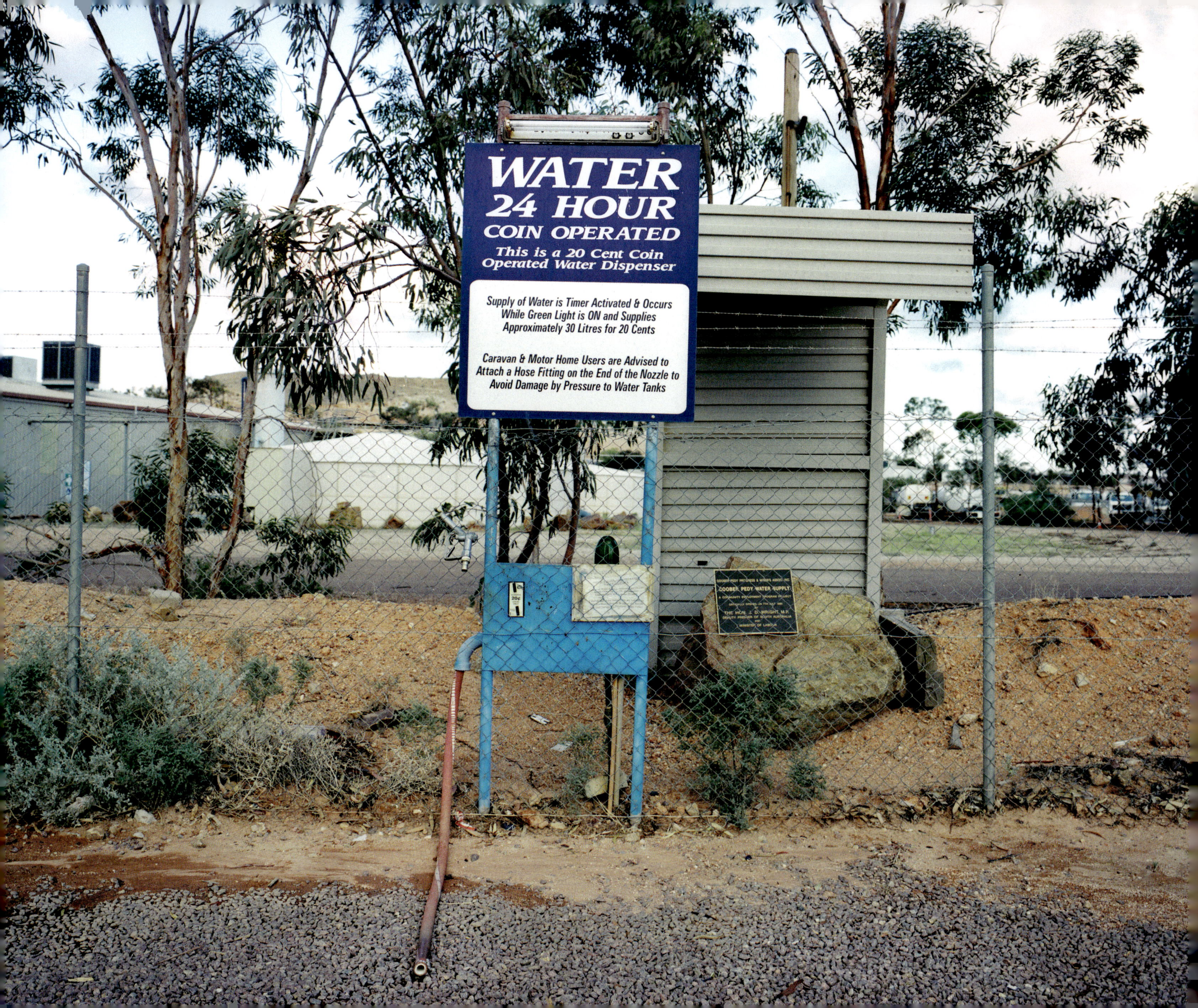
WATER
24 HOUR
COIN OPERATED
This is a 20 Cent Coin
Operated Water Dispenser
Supply of Water is Timer Activated & Occurs
While Green Light is ON and Supplies
Approximately 30 Litres for 20 Cents
Caravan & Motor Home Users are Advised to
Attach a Hose Fitting on the End of the Nozzle to
Avoid Damage by Pressure to Water Tanks
COOBER PEDY WATER SUPPLY
THE HON. J. D. WRIGHT M.P.

Conclusion
by PROTOS and GREEN Belgium

Dieter Telemans' photographs are disquieting. Another "Inconvenient Truth"? Al Gore's film is essentially about water as well. Climate change directly influences water conditions on Earth. Melting ice caps and glaciers make the sea levels rise. Low-lying areas are threatened by flood waters, whereas water scarcity becomes even more poignant in dry areas. No wonder water is increasingly on the political agenda. With the Millennium Goals of the United Nations, the International Community intends to halve the number of people who have no access to drinking water or proper sanitation by 2015. This initiative –if realised– still leaves 800 million people without clean drinking water and 1.8 billion without sanitation.

2015 is an important target date, symbolically as well as practically. It is the deadline for necessary investments and government measures without which 203,000 children's lives will be at stake immediately, and another one million children's lives by 2025. The EU is to increase the budget for development aid to 0.7% of the gross national product by 2015, whereas the G8 (the seven richest countries on Earth plus Russia) will double the aid package for Africa by 2010. All this, however, is insufficient to realise the Millennium Goals. Without a strong determination on the part of the global community to tackle the problems effectively and definitively, nothing much will happen. The 80s were declared the decade of water and sanitation for everyone. Unfortunately, in 1990 the situation was worse than ever before.

From now on, 10 billion US dollars should be invested every single year. That sounds like a huge amount, but actually, it is not. 10 billion US dollars is less than what the world spends on military purchases every five days. It is less than half of what Westerners spend on mineral water per year. Furthermore, the water lost per day through all the leaking taps in the rich countries represents a volume of water that is larger than what is currently available to more than 1 billion people daily. Investing in water as a 'human right' all over the world makes economic sense as well. The World Health Organization (WHO) has figured out that, because of reduced health costs and increased productivity, every dollar invested in 'water projects' will be worth 8 dollars in the long run. The Millennium Goals concerning water are crucial to all the other goals. Indeed, clean drinking water and appropriate sanitation infrastructure not only save many lives, especially children's, they are basic to schooling, healthcare and the fight against poverty as well.

What can we do on a personal level? We can begin by using drinking water sparingly. Rainwater is perfect for the washing machine and for flushing the toilet. The latter should be equipped with a dual flush system featuring two separate flush volumes. And by the way, fresh rainwater does not belong in the sewers. It should be collected, and after usage, be allowed to infiltrate the earth again in a natural water cycle. Another major point is to limit our use of detergents and chemical products. Finally, control one's tendency to consume, since for each item produced and consumed water has been used and polluted. You have made a first step in the right direction by purchasing this book. 10% of the money you spent to buy the book goes straight to water projects in Mali and Morocco. Of course, as an individual you can do much more: at home, at work, in the pub, at school. It demands some courage and creativity, but if everybody contributes his/her little bit, maybe a time will come when the pictures of "Troubled Waters" belong to the past for good.

Numerous are the organisations that are engaged in the pursuit of sustainable water management, provide help to developing countries to acquire clean drinking water and appropriate sanitation, or try to put water issues higher on the political agenda. These organisations need your moral and financial support. Actions or campaigns are very welcome as well, such as letting school kids run a car wash with rainwater, to raise car users' awareness of the water problematic and to support a water project in Haiti with the proceeds. Or a club of youngsters, rock bands and an open stage, organising a benefit day for the water carriers in Morocco. Or a business offering tap water on the occasion of its New Year's reception and donating the money saved to charities. These are not wild ideas, but examples of recent successful actions in Belgium. Above all, these actions are clear signals to our leaders and politicians to do more for the 'water in the world' issue.

Unrelated at first sight, but symbolic for the future is the 'Jupiter Ice Moons Project' that NASA intends to launch by 2015. A space ship is to explore the composition of the ice found on three moons of the planet Jupiter and the concomitant question on whether life is possible on Jupiter. What are the priorities of the international community? Will projects like the one just mentioned help to realise the Millennium Goals? All the photographs in this book were made on Earth. Let's stay on Earth and proceed to action here and now!

Your support

The proceeds from this book go to water projects in Morocco and Mali:

GREEN WANTS TO FREE THE WATER-CARRIER CHILDREN

Millions of children in Africa and elsewhere do not go to school, or do so irregularly. They are water carriers. Their task is to fetch the water for the family from springs, wells or rivers, often kilometres away, on foot, pushing carts or pulling donkeys. The quality of the water is not guaranteed, the cleanliness of the plastic recipients is questionable, the transport of the water and its preservation pose problems of hygiene, but it is the only available water around. Furthermore, the water is free for all. This water-carrying chore takes hours, leaving the children no time for school or play. The energy wasted on the task is a drain on development of villages and is one of the reasons for the rural exodus.

In 2005, GREEN, with the help of Rotary, Vivaqua, the 'Intercommunale Bruxelloise de Distribution d'Eau' (Greater Brussels Water Distribution Company) and IRGT, launched a campaign: "Let's free them from the water-carrying chore". Three villages in different regions of Morocco, namely Zaouit, Karia and Tighfert, received financing for the purchase and installation of water fountain units with the possibility of individual connections. Many youngsters were given scholarships in order to enable them to go back to school. Finally, with the help of 'Vlaams Partnerschap - Water voor Ontwikkeling' (Flemish Partnership - Water for Development) and with the direct collaboration of the local population, the most appropriate solutions are being sought for the treatment of used waters.

GREEN Belgium intends to continue this task of providing a maximum of children with drinking water so that they can enjoy their basic human right of going (back) to school and getting an education.

Info: www.porteursdeau.be

INTEGRAL WATER MANAGEMENT IN THE INNER DELTA OF THE NIGER IN MALI

The water programme PROTOS in Mali aims at improving the living standards of 40,000 people in a sustainable way by the year 2009. In consultation with the local authorities, the water committees and the actual users, PROTOS sets up projects concerning drinking water, sanitation, irrigation and water management. The chances of success and long-term effects are enhanced when local authorities and users participate in determining the contents and modalities of a project. The local communities are ultimately the ones that have to continue the projects on their own. The water programme focuses on the inner delta of the River Niger in the Mopti region, more particularly, Mopti City, the suburb Socoura and the rural communities Kewa and Soye. The programme includes the construction of 22 new water pits, the drilling of 20 new water wells and the installation of the necessary pumps, the repair of 20 existing water pits and pumps and the construction of 200 sanitation facilities. Finally, infrastructure works will be executed for the irrigation of 1,000 hectares of farmland.

The inner delta of the Niger covers a plain the size of Belgium. The population consists of fishermen, farmers and cattle raisers. The villages are built on natural dunes, which, during floods, remain above water level as little islands. From June to September, the whole region is flooded. From December to March the water recedes into some tributaries and lakes. The region then becomes one vast dry plain. Only one village out of three has access to a drinking water point. Cholera-epidemics are reported in several villages each year and diarrhoea caused by unclean water or insufficient sanitation facilities remains the number one killer of small children.

Info: www.protosh2o.org

Rotary

The printing of this book was made possible thanks to the financial support of Rotary Belgium-Luxemburg.

Rotary is an independent organisation founded on February 23, 1905 by Paul Harris, an American attorney. Rotary International is present in 169 countries with more than 32,000 clubs and 1,200,000 members. The Belgian-Luxemburg Rotary is subdivided into three districts with 254 clubs and more than 10,000 members.

"Service Above Self," is the basic principle of the Rotary philosophy, respected by every Rotarian. Among the wide variety of activities run by Rotary commissions, the following can be mentioned: Youth Actions (e.g. scholarships), the World Peace Forum, and the 3 H Programme (Health, Hunger and Humanity) which aims to combat hunger, improve health and contribute to better living conditions in developing countries.

The most important Rotary project has been the "Polio Plus" project, launched in 1985 in cooperation with UNICEF, the WHO (World Health Organization) and sponsors from the business and industrial world. The goal was to eradicate polio worldwide by 2005. Unfortunately, the objective has not yet been realised in seven countries, mostly because of local conflicts and traditions.

In 2004, Rotary International adopted one of the Millennium Goals as its focus of urgent action, namely the one that intends, by 2015, to reduce by half the world population that does not have access to drinking water. The Belgian-Luxemburg districts have been very active towards this goal. Districts 1620 and 2170 will provide 19 villages in Niger and 3 villages in Morocco with drinking water. Under the name " Water Bridge ," district 1630 will do the same for 25 villages in East Africa and Cambodia. Besides these, there are numerous initiatives taken by individual clubs, such as financing the installation of water pumps in India.

As long as part of the world population has no access to clean drinking water, as long as there are hungry children, as long as illiteracy and misery persist, as long as conflicts and violence continue, and as long as there are benevolent people, Rotary will have its reason to exist.

For District Governors 2006-2007, Paul Coppens (D-1620), Guido Vangansewinkel (D-1630) and Alain Van De Poel (D-2170) and for the Interdistrict Commission "Troubled Waters", Robert Gekiere, Past Governor and Ludo Feyen, Past Governor, Jan-Lodewijk Dujardin, Past Governor - Coordinator for Europe Rotary International "Water Resource Group," 2004-2007.

Translations

Vertaling van de teksten / Traduction des textes.

WOORD VOORAF / Dieter Telemans p.7

"Aral kaputt, Aral kaputt", zuchtte Sergei toen ik hem fotografeerde op zijn Zwaan, de boot waar hij acht jaar lang kapitein op was. Vandaag kwijnt de Zwaan weg tussen de wrakken van een vissersvloot die wel driehonderd schepen telde. De lokale bevolking verloor met de verdwijning van de Aral Zee, ooit de vierde grootste binnenzee ter wereld, meteen ook haar levensbron.

De ecologische ramp van de Aral Zee bewijst dat water essentieel is voor het menselijke bestaan en dat we er zorgzaam mee moeten omspringen. Het verhaal van Sergei inspireerde mij om dieper in te gaan op de waterproblematiek wereldwijd. Sindsdien heb ik landbouwers gefotografeerd in India, die hulpeloos toekijken hoe al hun waterbronnen stilaan uitdrogen. Duizenden boeren hebben al zelfmoord gepleegd door de pesticides, bestemd voor hun akkers, in te slikken. In de Hoorn van Afrika verloren de meeste herders in 2006 hun veestapel door de droogte. Volgens de oudere generatie staat hun eeuwenoude nomadische traditie op een keerpunt. Ik bezocht ook gebieden waar een toevloed van water mensen dakloos en ziek maakte. Zoals in Bangladesh, waar ik een zee van zoet water vond, maar geen druppel drinkbaar water. Sommige overheden doen nog altijd verwoede pogingen om de natuur te bedwingen. Gigantische waterprojecten worden vandaag uitgevoerd in onder meer Egypte en Turkije met als doel in de woestijn dorstige gewassen te telen. De plaatselijke bevolking kan daar misschien wel bij varen, maar stroomafwaarts zal men er de pijnlijke gevolgen van ondervinden.

De reportages in dit boek heb ik ingedeeld volgens de belangrijkste waterthema's. Maar de realiteit laat zich niet zo makkelijk in vakjes onderbrengen. Waar vandaag droogte heerst, kan er morgen watersnood zijn. In Ethiopië en Kenia zag ik mensen om water smeken, terwijl ze enkele maanden later te kampen kregen met overstromingen. En overal waar drinkwater schaars is, verandert leven in overleven en wordt de gezondheid van de mensen be-dreigd.

Klimaatverandering, bevolkingstoename en vooral de mentaliteit van 'het kan niet op' vormen de grootste bedreigingen voor een eerlijke en duurzame toegang tot water. De landbouw gebruikt meer dan 70 procent van het beschikbare water. Israël is een meester in woestijnlandbouw en zuinige irrigatietechnieken. Deze technologieën kunnen het waterverbruik wereldwijd drastisch verminderen. Maar in Israël zelf wordt de hebzucht om water er niet minder om. De Palestijnen op de Westelijke Jordaanoever moeten machteloos toezien hoe Israël 80 procent van hun water aftapt.

Deskundigen waarschuwen er al jaren voor: de grote conflicten van de 21ste eeuw zullen niet om olie, maar om water gaan. Een duurzaam en rechtvaardig beheer van de watervoorraden kan dit doemscenario voorkomen. Of om het met de woorden van Gandhi te zeggen: "Er is genoeg water voor ieders behoefte, maar niet voor ieders hebzucht."

HET RECHT OP LEVEN / Riccardo Petrella p.9

Het feit dat meer dan 1 miljard mensen het zonder drinkwater moet stellen, hoewel dat een levensnoodzakelijk goed is, tart alle verbeelding. Toch is het risico groot dat dat aantal zal aangroeien tot meer dan 3 miljard wanneer tegen 2020 de wereldbevolking ongeveer 8 miljard mensen zal tellen. Dat is onaanvaardbaar. Wij kunnen én moeten voorkomen dat dit rampscenario ook realiteit wordt. Maar hoe?

Water is de essentiële en onvervangbare 'bron van leven' van het ecosysteem. Daarom is het een levensnoodzakelijk goed, waar alle bewoners van de Aarde samen recht op hebben. Niemand, noch individueel noch in groep, heeft het recht om het zich toe te eigenen. Water is het erfgoed van de mensheid. Zowel de persoonlijke als maatschappelijke gezondheid is ervan afhankelijk. In tegenstelling tot elke andere natuurlijke grondstof, is water geen verruilbaar of verkoopbaar product.

Hoewel de waterverdeling vroeger vaak sociale ongelijkheid in de hand werkte, erkent men in de huidige samenleving dat toegang tot drinkwater een fundamenteel, onvervreemdbaar, individueel en collectief recht is. Het recht op water behoort tot de basisethiek van een 'goede' samenleving en een 'goede' economie. Zowel de maatschappij als geheel als de verschillende maatschappelijke beleidsniveaus moeten toegang tot drinkwater voor iedereen garanderen, volgens het dubbele principe van medeverantwoordelijkheid en subsidiariteit, en zonder onderscheid van ras, sekse, godsdienst, inkomen of sociale klasse.

De watervoorraden op Aarde zijn ongelijk verdeeld, evenals de inkomens. Maar dat impliceert niet dat dit ook moet gelden voor de toegang tot water. Volkeren die over veel water en een volwaardig inkomen beschikken, mogen niet onbeperkt water verbruiken, laat staan water verkopen of aankopen voor hun eigen maximale profijt of plezier. In de gebieden waar de vrouwen moeten instaan voor de watervoorziening van hun familie, blijft water een discriminatiefactor tussen mannen en vrouwen. Die ongelijkheden moeten verdwijnen. Iedereen zou ernaar moeten streven om de noodzakelijke voorwaarden te scheppen voor een efficiënte en duurzame toegang tot water. Dat is ook een intergenerationele kwestie. De huidige generaties moeten de watervoorraden op zo'n manier gebruiken, valoriseren, beschermen en in stand houden, dat de generaties na ons dezelfde handelings- en keuzevrijheid hebben die wij vandaag zo koesteren.

AVANT-PROPOS / Dieter Telemans p.7

"Aral kaputt, Aral parti", soupire Sergei tandis que je le photographie sur le Swan, le navire dont il a été le capitaine durant huit longues années. À présent, le Swan n'est plus qu'une de ces nombreuses carcasses de navires rongés par la rouille, qui formaient autrefois une superbe flottille de pêche, forte de 300 embarcations. La mer d'Aral était auparavant la quatrième plus grande mer intérieure au monde. Avec sa disparition, la population de la région a perdu le moteur de son économie.

La catastrophe écologique de la mer d'Aral illustre à quel point l'eau est essentielle à la survie de l'homme et dans quelle mesure elle devrait être protégée. L'histoire de Sergei m'a incité à me pencher davantage encore sur le problème de l'eau à l'échelle de la planète. C'est ainsi que j'ai photographié des agriculteurs indiens, résignés, observer leurs ressources en eau se tarir jour après jour. Des milliers d'entre eux se sont déjà donné la mort en absorbant certains des pesticides qu'ils destinaient à leurs cultures. En 2006, la plupart des bergers de la Corne de l'Afrique ont perdu leur bétail à cause de la sécheresse. Les anciens pensent que le mode de vie nomade ancestral est appelé à disparaître. Dans certaines régions, les inondations privent à jamais les habitants de leurs logements et propagent les maladies. Tel est le cas au Bangladesh, où j'ai trouvé de l'eau douce à volonté, mais pas une goutte d'eau potable. Or, certains dirigeants en poste croient pouvoir maîtriser Mère Nature. De ce fait, de gigantesques projets hydro-agricoles ont été mis en chantier, notamment en Égypte et en Turquie, dans le but de produire des cultures d'exportation dans les zones désertiques. Les populations locales en tirent peut-être des bénéfices, mais, en aval, les conséquences se font lourdement sentir.

J'ai voulu organiser ces reportages photo autour des grandes thématiques de l'eau. En réalité, les zones aujourd'hui touchées par la sécheresse peuvent se transformer demain en zones inondées. En Éthiopie et au Kenya, des zones où j'avais vu des hommes quémander de l'eau ont été inondées quelques mois plus tard. Parallèlement, partout où l'accès à l'eau potable est difficile, l'existence se mue en lutte pour la survie et la santé publique est menacée. Les changements climatiques, la croissance démographique et, surtout, cette mentalité du "on n'en manquera jamais" sont les principales menaces planant sur un accès à l'eau équitable et durable pour tous.

L'agriculture absorbe plus de 70% de l'eau disponible sur la Terre. Israël est un acteur de premier plan dans le domaine des techniques d'irrigation goutte à goutte et de l'agriculture en milieu désertique. Partout, ces technologies pourraient permettre de réduire de manière très significative la consommation d'eau. Pourtant, en Israël, la demande en eau ne fléchit pas: 80% de l'eau disponible dans les territoires palestiniens de Cisjordanie prennent ainsi la direction de l'État d'Israël.

Depuis des années, les experts et les scientifiques tirent la sonnette d'alarme: l'eau sera l'un des enjeux des principaux conflits du 21ème siècle.. Seule une gestion durable et équitable des ressources mondiales d'eau permettra de conjurer ce scénario catastrophe. Pour citer Gandhi: "Il y a assez d'eau dans le monde pour satisfaire aux besoins de l'homme, mais pas assez d'eau pour assouvir son avidité."

LE DROIT À LA VIE / Riccardo Petrella p.9

Plus d'un milliard d'êtres humains n'ont pas accès à l'eau potable, cet élément fondamental à la vie. Cette situation est intolérable. Or, le risque est grand de voir ce chiffre grimper à plus de trois milliards en 2020, lorsque la population mondiale aura franchi la barre des huit milliards d'hommes. C'est tout simplement inacceptable. Nous pouvons et nous devons empêcher que l'inacceptable devienne réalité.

En tant que source de vie fondamentale et irremplaçable pour l'écosystème, l'eau est un bien vital qui appartient à l'ensemble des habitants de la planète. Aucun individu ou groupe d'individus n'a le droit d'en revendiquer la propriété exclusive. L'eau appartient au patrimoine de l'humanité. La santé publique et le bien-être individuel en dépendent. L'eau est différente des autres ressources, car elle n'est pas une marchandise échangeable ou monnayable.

Alors que l'approvisionnement en eau pour tous a souvent été, par le passé, à l'origine de grandes inégalités sociales, les civilisations actuelles admettent désormais que l'accès à l'eau est un droit fondamental, inaliénable, individuel et collectif. Le droit à l'eau fait partie de l'éthique de base d'une société et d'une économie 'bienveillantes.' Il revient à la société dans son ensemble de garantir ce droit d'accès à l'eau sans discrimination de race, de sexe, de religion, de revenus ou de statut social.

Les ressources d'eau douce sur Terre sont réparties de façon inégale. Il en va de même des revenus. Cela ne veut pas dire pour autant qu'il doive également y avoir inégalité en matière d'accès à l'eau. Les populations riches en eau et en devises ne peuvent pas l'utiliser à leur guise, a fortiori la vendre ou l'acheter afin d'en tirer un maximum de bénéfices ou d'agrément. Or, dans de nombreuses régions du monde, l'eau demeure une source d'inégalité entre hommes et femmes, ces dernières étant contraintes de devoir subvenir aux besoins en eau de leurs familles. Toutes ces inégalités doivent disparaître. Nous devrions tous nous atteler à réunir les conditions nécessaires afin de pouvoir garantir un accès à l'eau efficace et durable. En outre, il s'agit d'une problématique intergénérationnelle, puisqu'il

Een sleutelrol is weggelegd voor de watergebruikers, in hun keuzes en gedrag. Zij moeten voldoende draagkracht garanderen op het vlak van milieu, economie en maatschappij. De 'partnerschappen voor water' vormen het basisprincipe van alle projecten. Zo voorzagen de 'rivierverdragen' in een doeltreffende oplossing van de conflicten tussen die bevolkingsgroepen die hetzelfde rivierbekken delen. De maatschappij moet alle kosten in verband met water collectief op zich nemen. Alleen op die manier kan de toegang tot water voorzien én gegarandeerd worden in die hoeveelheden en kwaliteit die beschouwd worden als het levensnoodzakelijke minimum. Eenmaal boven dat minimumniveau moet men de kostprijs progressief aanpassen: hoe meer water men verbruikt, hoe meer men betaalt per watereenheid. Ten slotte zou men elk misbruik en buitenmatig gebruik als illegaal moeten beschouwen.

Riccardo Petrella is voorzitter van de Groep van Lissabon, secretaris van het Internationale Comité voor het Wereldwatercontract en professor Menselijke Ecologie aan de Academie voor Architectuur van de USI (Università della Svizzera Italiana), Lugano, Zwitserland.

KENIA - ETHIOPIË / Droogte in de Hoorn van Afrika p.13

In het begin van 2006 kende Oost-Afrika de zwaarste droogte in meer dan tien jaar tijd. Voor ongeveer elf miljoen mensen in Djibouti, Ethiopië, Kenia, Somalië en Tanzania dreigde de hongerdood. Na jaren van weinig regen werd 2005 een rampjaar, doordat het regenseizoen nagenoeg volledig uitbleef. Miljoenen nomadische veehoeders verloren zo'n 70% van hun veestapel. Volgens de Keniaanse regering was dit de ergste natuurramp die het land geteisterd heeft sinds de droogte van 1971.

De droogteperiode vindt nu niet meer om de acht jaar plaats, maar al om de drie jaar. Door deze klimaatverandering ondergaat het land een ware transformatie. Daarin speelt ook de overbevolking een grote rol, met als gevolg overbegrazing en boomkap voor het winnen van brandstof en houtskool. Graslanden zijn woestijnen geworden, en de nomaden vinden geen weilanden meer om naar terug te keren. Plaatselijke leiders vragen zich af of hun veeteelttraditie nog wel stand kan houden. Maar het lijkt onmogelijk om volken zoals de Masaï en de Somali's ertoe te overhalen minder vee te hoeden. Hun levensonderhoud hangt immers voor 90% af van hun veestapel. Dat is al vele generaties een traditie. Om met deze traditie te kunnen breken, moeten eerst andere activiteiten ontwikkeld worden waaruit men een inkomen kan halen.

Vele jongeren snakken naar verandering en zijn ervan overtuigd dat een opleiding een betere investering is dan een veestapel. Maar de basisinfrastructuur ontbreekt in de door de droogte geteisterde regio's. Een degelijke opleiding en een job blijven zo een utopie. In de hoofdsteden debatteren de politici wel over deze problemen in de verafgelegen gebieden, maar ze komen niet tot het opzetten van ontwikkelingsprojecten op langere termijn. Zolang deze regio's geïsoleerd blijven, zijn de jongeren gedoemd tot de harde levenswijze van hun voorvaderen. Seizoensdroogtes zullen hun dodentol blijven eisen.

BANGLADESH / Slachtoffers van overstromingen p.33

Overstromingen zijn een terugkerend natuurverschijnsel in Bangladesh. Het land is eigenlijk één grote delta van drie rivieren: de Ganges (Padma), de Jamuna (Brahmaputra) en de Meghna. Het overtollige water van de moessonregens en het smeltwater van de Himalaya vloeien via deze rivieren naar de Golf van Bengalen. Door de overstromingen blijft vruchtbaar slib achter dat de bodem hernieuwt. Deze rivieren vormen de belangrijkste watervoorraad van Bangladesh, maar ze betekenen ook een groot gevaar.

De Bengalezen hebben leren leven met de overstromingen. Toch zijn er overstromingen die te veel vernieling zaaien. Tot nu toe gebeurden de ergste overstromingen om de tien jaar, bijvoorbeeld in 1998 en 1988. Maar in 2004 kwam ze drie jaar eerder. Tegen het einde van het moessonseizoen was bijna de helft van het land overstroomd. Zo'n 30 miljoen mensen moesten hun huizen verlaten en hun toevlucht zoeken in hogergelegen gebieden. 10 miljoen Bengalezen waren dakloos en zo'n 700 mensen stierven. Het Wereldvoedselprogramma van de Verenigde Naties schatte de schade aan boerderijen, fabrieken en infrastructuur op 7 miljard US dollar. Niet alleen het platteland overstroomde. Ook 40% van de hoofdstad Dhaka, een grootstad met een bevolking van 13 miljoen, kwam onder water te staan. Het vloedwater kwam terecht in het rioolsysteem van de stad en liet een zwarte stinkende modder achter in de straten. Door water overgebrachte ziekten verspreidden zich snel en een cholera-epidemie brak uit. Vele kinderen stierven door diarree en ademhalingsinfecties. Maar de crisis duurde voort, zelfs na het terugtrekken van het vloedwater. Mensen keerden terug naar hun vernielde huizen, maar hadden geen inkomen of bezittingen meer. Een enorme voedselcrisis was het gevolg, die pas overwonnen werd door de verdeling van voedselhulp. Volgens sommige experts is de oorzaak van de overstromingen te wijten aan de irrigatie voor de landbouw. Daardoor slibben de rivieren dicht en kunnen ze minder vloedwater verwerken of doen afvloeien naar de zee. Anderen beweren dat de opwarming van de aarde leidt tot een toename van smeltwater uit de Himalaya en een grotere regenval tijdens de moessons. Volgens hun voorspellingen zouden ook de kustgebieden van Bangladesh in de toekomst af te rekenen krijgen met zware overstromingen.

Als de CO_2-uitstoot niet daalt, zal de globale temperatuur de komende vijftig jaar stijgen met twee tot drie graden Celsius. De kust van Bangladesh, die zich nu grotendeels op zeeniveau bevindt, zal dan geconfronteerd worden met een stijging van de zeespiegel, met enorme overstromingen tot gevolg.

incombe aux générations actuelles d'utiliser, de valoriser, de protéger et de conserver les ressources en eau de telle façon que les générations à venir puissent bénéficier d'une liberté d'action et de choix identique à celle dont nous souhaitons profiter aujourd'hui.

À travers leurs choix et leurs habitudes, les consommateurs ont un rôle déterminant à jouer afin d'assurer la viabilité environnementale, économique et sociétale. L'ensemble des projets en la matière doit être guidé par le principe des 'partenariats pour l'eau', tels les 'contrats de rivière'. Ceux-ci ont permis de résoudre efficacement des conflits entre les populations riveraines d'un même bassin hydrographique. Il revient à la société d'assumer collégialement l'ensemble des coûts liés à l'eau afin de fournir et de garantir l'accès à l'eau, dans des quantités et selon un niveau de qualité considérés comme le minimum indispensable. Au-delà de ce minimum vital, la tarification appliquée doit être progressive. Plus grande est la consommation, plus élevé en sera le prix. Enfin, tous les abus et les excès de consommation doivent être qualifiés d'actes illégaux.

Riccardo Petrella est président du Groupe de Lisbonne, secrétaire du Comité international pour le Contrat mondial de l'eau et professeur d'écologie humaine auprès de l'Académie d'architecture de l'USI (Università della Svizzera Italiana), Lugano.

KENYA - ÉTHIOPIE / Sécheresse dans la Corne de l'Afrique *p.13*

Au début de l'année 2006, l'Afrique orientale a subi sa pire sécheresse depuis plus de dix ans. À Djibouti, au Kenya, en Somalie et en Tanzanie, la famine a menacé de mort près d'onze millions de personnes. Des années de faibles précipitations ont été aggravées par l'absence quasi-totale de pluie lors de la saison pluvieuse 2005. Selon les estimations, des millions de bergers nomades ont perdu 70% de leurs troupeaux. Le gouvernement kenyan a déclaré qu'il s'agissait là de la pire catastrophe naturelle qui ait touché le Kenya depuis la sécheresse de 1971.

La fréquence des sécheresses est passée d'une tous les huit ans à une tous les trois ans. Ce changement climatique, associé à la surpopulation qui entraîne le surpâturage et la déforestation (les arbres étant transformés en combustible et en charbon de bois), a transformé le pays. D'anciens pâturages sont maintenant devenus désertiques. Les nomades ne savent plus où mener leurs troupeaux. Les chefs de la communauté se demandent même si la tradition du nomadisme est encore viable. Cependant, l'on peut difficilement demander à des communautés telles que les Massaïs et les Somalis de réduire la taille de leurs troupeaux. En effet, leurs moyens de subsistance en sont à 90% tributaires. Depuis nombre de générations, cette pratique n'a guère évolué. Si l'on veut rompre avec cette tradition, il convient de proposer en échange des activités rémunératrices.

Beaucoup d'adolescents aspirent au changement et sont convaincus que l'éducation constitue un meilleur investissement qu'un troupeau. Cependant, ces régions frappées par la sécheresse manquent cruellement d'infrastructures. Avoir un bon bagage scolaire et un emploi semblent y relever de l'utopie. Reclus dans les capitales, les hommes politiques débattent de ces problèmes bien éloignés de leur quotidien sans apporter de projets de développement à long terme. Aussi longtemps que ces régions resteront isolées, les adolescents n'auront d'autre choix que la rude existence de leurs parents. Et les sécheresses saisonnières continueront à semer la désolation parmi les communautés.

BANGLADESH / Victimes des inondations *p.33*

Les inondations sont un phénomène naturel récurrent au Bangladesh. Le pays est situé dans les deltas du Gange (Padma), du Jamuna (Brahmapoutre) et du Meghana. Ces fleuves acheminent l'eau excédentaire provenant des pluies de mousson et de la fonte des neiges himalayennes vers le golfe du Bengale. Les inondations déposent du limon qui renouvelle la fertilité du sol. Cette situation fait de ce réseau hydrographique la principale ressource du pays aussi bien que son plus grand danger.

Les habitants du Bangladesh ont appris à vivre au rythme des inondations, sauf quand celles-ci occasionnent des dégâts majeurs. Récemment, les inondations sévères survenaient selon un cycle de dix ans. Les dernières grandes inondations ont eu lieu par exemple en 1998 et 1988. En 2004, le cycle a été rompu: elles sont arrivées avec trois ans d'avance. Vers la fin de la mousson, près de la moitié du pays était sous eau. Environ trente millions de personnes ont été obligées d'abandonner leurs habitations pour trouver refuge dans des zones situées en altitude. Dix millions de Bangladeshis se sont retrouvés sans abri et près de 700 ont trouvé la mort. Le Programme alimentaire mondial de l'ONU a évalué à sept milliards de dollars les dégâts occasionnés aux exploitations agricoles, aux industries et aux infrastructures. Les zones rurales n'ont pas été les seules à être inondées. 40% de la capitale, Dacca, une mégapole de treize millions d'âmes, a été également sous eau. Les eaux des fleuves, mélangées aux eaux usées de la ville, se sont retirées en laissant une boue noire et nauséabonde dans les rues. Des maladies hydriques n'ont pas tardé à se propager et une épidémie de choléra s'est déclarée en ville. Nombre d'enfants sont décédés de diarrhées et d'infections respiratoires. Le retrait des eaux n'a pas marqué pour autant la fin de la crise. Les habitants ont dû en effet réintégrer leurs maisons ravagées, privés de revenus ou de biens. Les autorités ont dû faire face à une crise alimentaire sans précédent. Le peuple n'a dû son salut qu'à l'aide alimentaire.

Les experts pointent un doigt accusateur sur l'irrigation des terres cultivées, à l'origine de l'ensablement des rivières. Ce phénomène réduit leur capacité à retenir les crues ou à les acheminer vers la mer. D'autres experts

HAÏTI / Overstromingsschade na tropische storm p.43

Haïti is het armste land van het westelijke halfrond. Zijn economische en sociale indicatoren zijn vergelijkbaar met die van vele landen in Sub-Sahara Afrika: 80% van de plattelandsbevolking leeft onder de armoedegrens, meer dan de helft van de bevolking is analfabeet en 50% is ondervoed. De gezondheidszorg is slecht georganiseerd en ontoereikend. Een groot deel van de bevolking heeft geen toegang tot de basisgezondheidszorg. Eén op twintig Haïtianen heeft hiv/aids en hun levensverwachting bij geboorte is minder dan vijftig jaar. Amper de helft van de bevolking beschikt over zuiver water en nog minder over sanitaire voorzieningen.

Vroeger was Haïti volledig bedekt door regenwoud, maar nu is het een van de meest ontboste landen ter wereld. De Franse kolonisators begonnen al met het kappen van duizenden hectare bos. In de plaats kwamen suikerrietplantages, en het betere timmerhout werd naar Europa verscheept. Wat overblijft van dat regenwoud, valt vandaag ten prooi aan de arme bevolking. Van het hout maken ze houtskool voor in de keuken. Zo gaat het hele woud letterlijk en figuurlijk in rook op. Door die ontbossing is Haïti heel kwetsbaar bij stormweer. Het water dat nu in grote hoeveelheden van de bergen stroomt, verandert in heuse modderstromen. Er zijn immers geen bossen meer die de overvloedige regen kunnen opnemen, en bijgevolg ook geen boomwortels die de grond vasthouden. Ook rivieren treden buiten hun oevers en veroorzaken overstromingen. Met dramatische gevolgen voor de dorpen die in dat overstromingsgebied liggen of bedolven worden door de modderstromen.

In september 2004 overstroomde de kuststad Gonaives in Noordwest- Haïti nadat het gebied dertig uur lang te lijden had gehad onder zware regenval tijdens de tropische storm Jeanne. De stad werd bedolven onder een laag modder, die tot aan het middel reikte. Duizenden mensen zaten vast op de daken van hun huizen, zevenhonderd mensen stierven en honderden waren vermist. De stad zat weken zonder elektriciteit of medische hulp. Het gevaar voor het uitbreken van epidemieën was groot. Terwijl de vrachtwagens met noodhulp met veel moeite bij de geteisterde gebieden probeerden te geraken, zaten zo'n 175.000 mensen zonder eten of zuiver water. Deze ramp gebeurde vier maanden nadat andere hevige regens in het zuiden van Haïti daar het leven kostten aan vijfhonderd mensen. Die overstromingen beschadigen niet alleen de al gebrekkige infrastructuur, maar ook de weinige landbouwgronden die er nog waren, wat resulteert in nog meer armoede. Een vicieuze cirkel, die echter nog tot ergere gevolgen zal leiden omdat door de opwarming van de aarde meer tropische orkanen het gebied zullen teisteren.

INDIA / Tsunami p.53

Op 26 december 2004 was de wereld getuige van de vreselijke kracht van de tsunami in de Indische Oceaan. Een zeebeving van 9.0 op de schaal van Richter deed de zeebodem verschuiven tussen India en Zuidoost-Azië. Miljarden tonnen zeewater werden erdoor opgestuwd en vormden reusachtige vloedgolven, die de kustlijn overspoelden en een spoor van vernieling achterlieten. Over een afstand van Sri Lanka, Thailand tot Somalië werden plaatsen getroffen door de ramp. Meer dan 220.000 mensen stierven en miljoenen waren dakloos. De Indonesische provincie Atjeh werd het zwaarst getroffen. Twee derde van de doden vielen hier. De vloedgolven bereikten Atjeh met een hoogte van twintig meter en vernielden de stad Banda Atjeh in amper vijftien minuten. Meer naar het westen doodden de vloedgolven meer dan 31.000 mensen in Sri Lanka en 10.000 mensen in India. Deze muur van water vernielde ook vissersboten, oogsten, bomen en huizen. Duizenden mensen in de visserij en het toerisme waren op slag werkloos. Vandaag leven nog altijd tienduizenden mensen in tenten of opvangplaatsen.

Volgens de IUCN (Wereldunie voor Natuurbehoud) legde de tsunami pijnlijk de gevolgen bloot van de verregaande verloedering van de ecosystemen aan de kust over de afgelopen decennia: op plaatsen waar de mangrovebossen verdwenen zijn, was de schade het grootst. In het dorp Kapuhenwala, in het zuiden van Sri Lanka, dat omgeven is door 200 hectare bos, dichtbegroeid met mangroven en kreupelhout, vielen maar twee slachtoffers. Maar bijna 6.000 mensen lieten het leven in het buurdorp Wanduruppa, waar het overgrote deel van het mangrovebos gekapt werd.

Bijna 40% van de mangrovebossen wereldwijd is gesitueerd in Azië. Maar sinds de jaren 1980 werd een vierde van deze natuurlijke beschermingsgordel gekapt voor de bouw en snelle uitbreiding van garnaal- en viskwekerijen. Vooral die grotendeels onbeschermde gebieden werden zwaar geteisterd door de tsunami.

De eerste inspanningen van hulporganisaties om de mangrovebossen weer aan te planten, mislukten omwille van een gebrek aan technische expertise. Maar in 2005 lanceerden milieugroeperingen in samenwerking met de regeringen van de landen aan de Indische Oceaan een aantal projecten die zich richten op de heraanleg van de ecosystemen langs de kustlijn en de plaatselijke economie weer zullen doen heropleven. De IUCN lanceerde bijvoorbeeld het project 'Mangroves voor de toekomst', waarmee men een natuurlijke barrière van mangrovebossen wil aanleggen in twaalf landen in Azië en Afrika. Kostenplaatje: 45 miljoen US dollar.

MALI / Watertransport p.63

Een groot deel van Mali - ruim tweemaal de oppervlakte van Frankrijk - is gelegen in de Sahara. Deze woestijn breidt zich langzaam maar zeker uit, met alle rampzalige gevolgen van dien voor een land waar 80% van de beroepsbevolking actief is in de landbouw en visserij. De rivier Niger is de levensader van Mali, maar die stroom is al helemaal aan het uitdrogen in de oude stad Timboektoe.

pensent que le réchauffement planétaire accélère la fonte des neiges himalayennes et augmente les précipitations de la mousson. Toujours selon les experts, il se pourrait, qu'à l'avenir, le Bangladesh doive faire face à des inondations à grande échelle de ses zones côtières.

Si aucune mesure n'est prise pour réduire les émissions de carbone, la température mondiale devrait augmenter de deux à trois degrés Celsius dans les cinquante prochaines années. La côte du Bangladesh, qui se situe pour l'essentiel au niveau de la mer, serait alors confrontée à une élévation des eaux et à des inondations de grande ampleur.

HAÏTI / Un pays ravagé après le passage d'une tempête tropicale p.43

Haïti est le pays le plus pauvre du continent américain. Sa situation économique et sociale est comparable à celle de nombreux pays subsahariens. 80% de la population rurale vit sous le seuil de pauvreté, plus de la moitié est illettrée et 50% souffre de malnutrition. Le système de santé est mal organisé et insuffisant. Une grande partie de la population n'a pas accès aux soins de santé de base. Un Haïtien sur 20 est séropositif ou atteint du SIDA. L'espérance de vie stagne en dessous de cinquante ans. Seule une moitié de la population a accès à l'eau potable et une proportion encore moindre, aux installations sanitaires.

Haïti était jadis recouvert d'une dense forêt tropicale. Il est à présent devenu l'un des pays du monde les plus en proie à la déforestation. Les colons français furent les premiers à élaguer des milliers d'hectares de forêt pour produire du sucre et fournir l'Europe en bois d'ébénisterie de qualité. De nos jours, la population appauvrie doit se contenter de ce qui subsiste de la forêt originelle. Le bois est transformé en charbon de bois pour la cuisine. La forêt disparaît en fumée. La déforestation rend Haïti particulièrement vulnérable aux tempêtes. En l'absence de forêts pour absorber les précipitations importantes et de racines pour accrocher le sol, l'eau dévale en cascade de la montagne et peut provoquer des glissements de terrain. Les rivières sortent de leurs lits et donnent naissance à des plaines inondables. Les conséquences sont terribles pour les villages situés dans ces zones ou sur le passage des coulées de boue.

En septembre 2004, la ville côtière de Gonaïves, dans le nord-ouest d' Haïti, a été inondée dans le sillage du cyclone Jeanne et de ses pluies diluviennes, qui se sont déchaînées 30 heures durant. Les habitants se sont retrouvés dans la boue jusqu'à la taille. Des milliers de personnes se sont réfugiées sur les toits. On a recensé 700 décès et des centaines de disparus. Pendant des semaines, la ville fut privée d'électricité et d'aide médicale. Le risque d'épidémie était grand. Alors que des camions d'aide humanitaire cheminaient péniblement vers les régions sinistrées, près de 175.000 personnes se retrouvaient sans nourriture ni eau potable. Ce désastre est survenu quatre mois après que d'autres pluies abondantes eurent coûté la vie à 500 personnes dans le sud d' Haïti. Chaque inondation détruit les infrastructures de fortune encore disponibles. Les terres cultivées se réduisent comme peau de chagrin, ce qui plonge encore davantage les habitants dans la pauvreté. Ce cercle vicieux pourrait encore s'amplifier. En effet, l'on s'attend à une recrudescence des tempêtes tropicales sous l'effet du réchauffement climatique.

INDE / Tsunami p.53

Le 26 décembre 2004, le monde entier assiste impuissant au déchaînement des forces de la nature dans l'océan Indien. Un tremblement de terre sous-marin, évalué à 9 sur l'échelle de Richter, fait trembler le plancher océanique entre l'Inde et l'Asie du Sud-Est et soulève des milliards de tonnes d'eau de mer. Plusieurs vagues gigantesques déferlent sur les côtes, semant la désolation sur leur passage. Des lieux aussi éloignés que le Sri Lanka, la Thaïlande ou la Somalie sont déclarés zones sinistrées. La catastrophe fait 220.000 morts et des millions de sans-abri. La province indonésienne d'Aceh est la région la plus touchée. Plus de deux tiers des décès sont concentrés dans cette zone. Des vagues hautes d'une vingtaine de mètres frappent les côtes d'Aceh et détruisent la ville de Banda Aceh en seulement quinze minutes. À l'ouest, les vagues tuent plus de 31.000 personnes au Sri Lanka et 10.000 personnes en Inde. L'eau emporte des bateaux de pêche, des récoltes, des arbres et des maisons dans son sillage. Des milliers de personnes perdent leur emploi dans les secteurs de la pêche et du tourisme. À présent, des dizaines de milliers de réfugiés vivent toujours dans des tentes ou des abris de fortune.

Selon l'Union mondiale pour la nature (UICN), le tsunami a mis en lumière les conséquences des graves déprédations subies par les écosystèmes côtiers durant les dernières décennies. Ainsi, les vagues ont été particulièrement meurtrières dans les zones jadis occupées par la mangrove, aujourd'hui disparue. Dans le village sri lankais de Kapuhenwala, situé au sud de l'île et entouré de 200 hectares de mangrove luxuriante et de strate arbustive, le tsunami n'a tué que deux personnes. En comparaison, près de 6.000 personnes ont trouvé la mort dans le village voisin de Wanduruppa, où la mangrove avait largement été détruite.

Près de 40% des mangroves de la planète sont situées en Asie. Cependant, depuis les années 1980, un quart de cette barrière naturelle a été sacrifié sur l'autel du développement et de la croissance des fermes d'aquaculture. C'est donc dans ces zones côtières dépourvues de toute protection que le tsunami a été le plus dévastateur.

Les efforts initiaux entrepris par les organisations humanitaires pour replanter la mangrove ont été voués à l'échec, beaucoup ne possédant pas un savoir-faire technique en la matière. En 2005, une coalition d'associations de défense de l'environnement a collaboré avec les différents gouvernements du bassin de l'océan Indien afin de mettre en oeuvre des projets visant à réhabiliter les écosystèmes littoraux et à redynamiser les économies locales.

Vooral vrouwen en meisjes zijn het slachtoffer van dit watertekort. Volgens de traditie staan zij immers in voor de waterbevoorrading van hun familie. Elke dag opnieuw moeten zij bij zonsopgang op stap naar de dichtstbijgelegen waterbron, vaak meer dan 16 kilometer ver. In het droogteseizoen kan die afstand zelfs verdubbelen. In de Dogonvallei kan het waterpeil wel 80 meter zakken, wat catastrofaal is. De meeste dorpen hebben maar één bron die diep genoeg is, en sommige dorpen zijn zelfs afhankelijk van een en dezelfde bron. Voor het watertransport alleen al zijn de vrouwen vaak zo'n zes uur onderweg, maar daar komt gemakkelijk nog eens vijf uur bij als ze ook moeten aanschuiven aan de bron zelf. Om dat te vermijden, starten vele vrouwen hun tocht al in het midden van de nacht. Zo kunnen ze tijdig bij de bron geraken. Die watertaak vergt een enorme inspanning van de vrouwen. Hun watercontainers bevatten meestal 20 liter water (20 kilo). Elke dag zo'n zwaar gewicht torsen, eist op de duur zijn tol voor hun gezondheid. Het water putten zelf is bovendien ook niet zo eenvoudig. Kijk maar naar de diepe groeven die de touwen achterlaten in de houten balken.

Door deze vermoeiende taak en tocht mislopen de vrouwen en meisjes in ontwikkelingslanden een basisopleiding en krijgen ze geen kansen om mee beslissingen te nemen in hun gemeenschap. Voorzien in zuiver water op een goed bereikbare plaats betekent voor die vrouwen niet alleen het einde van dit zinloze en vernederende slavenwerk. Het komt ook hun gezondheid en die van hun familie ten goede. Pas dan kunnen de vrouwen immers meer tijd vrijmaken voor opleiding, landbouw of ander werk dat een inkomen verschaft. Of ze kunnen beter voor de kinderen zorgen, of gewoon wat ontspannen.

INDIA / Van plattelandsarmoede tot stedelijke ellende p.83

De belangrijkste waterbronnen in India worden bedreigd. Vele boeren kunnen vrij over grondwater beschikken en telen liever waterintensieve gewassen bestemd voor export, zoals katoen, in plaats van gewassen voor de lokale markt die minder water behoeven. Gevolg: een buitensporig verbruik van grondwater en een waterpeil dat snel zakt. Zo zien al heel wat dorpen zich geconfronteerd met uitgedroogde waterbronnen of wordt het water te zout zodat het enkel nog gebruikt kan worden om te wassen. Door droogte en verwoestijning is ook regenwater geen betrouwbaar alternatief meer. Punjab bijvoorbeeld was vroeger de vruchtbaarste landbouwstreek van India, maar nu zijn hele gebieden veranderd in woestijn.

Voor de meerderheid van de 700 miljoen boeren in India is landbouw geen rendabele activiteit meer. In Gujarat en Rajasthan worden 50 miljoen mensen bedreigd door een gebrek aan water. Duizenden boeren in de staat Maharashtra hebben de laatste jaren zelfmoord gepleegd. Ze zagen geen uitweg meer uit de vicieuze cirkel van misoogsten, te wijten aan de droogte, en hun stijgende schulden doordat ze nieuwe zaden en pesticiden moesten aankopen. Ze pleegden uiteindelijk zelfmoord door die pesticiden in te slikken, waardoor ze hun weduwen en kinderen op een schuldenberg achterlieten.

Miljoenen wanhopige boeren hebben het platteland al verlaten op zoek naar een beter leven in de stad. Voor hen de enige manier om hun gezinnen te kunnen onderhouden. Het bevolkingsaantal in de Indiase steden is in de loop van de vorige eeuw vertienvoudigd. Daardoor komt het land op de tweede plaats in de wereldrangorde te staan, na China. Volgens de regering zal de helft van de Indiase bevolking in 2025 in de steden leven. De oorzaak van die verschuiving is dat de economische activiteiten zich concentreren in de stedelijke gebieden. Maar de meerderheid van de migranten is analfabeet en behoort tot de laagste kaste. Daardoor kunnen ze ook enkel aan de slag in jobs met een lage verloning. Omdat ze automatisch buitengesloten worden uit de veiligere en betere stadsbuurten, hebben ze geen andere keuze dan te gaan wonen in de overbevolkte krottenwijken.

Ook milieu- en sociale problemen doen de armoede in de steden toenemen. De meeste krottenwijken beschikken niet over basisvoorzieningen zoals zuiver water, afvoersystemen en elektriciteit. Deze diensten worden in India geleverd door zwakke overheidsinstellingen, waar corruptie en vriendjespolitiek welig tieren. In de 1.100 krottenwijken van New Delhi is er ofwel helemaal geen kraantjeswater, ofwel maar een paar uur per dag. Als er al toiletten zijn, dan bevinden die zich in zo'n slechte toestand dat de mensen nog liever in openlucht hun behoefte doen en zo een roepie per bezoek aan het toilet kunnen uitsparen. Door het gebrek aan dergelijke basisvoorzieningen lijden vele bewoners van de krottenwijken aan huidziekten, ziekten die via water overgedragen kunnen worden en tuberculose. Dat is helemaal niet wat ze zich hadden voorgesteld toen ze het leven op het platteland inruilden tegen dat in de stad.

CHINA / Vervuiling vergiftigt de vooruitgang p.97

Met een gemiddelde groei van meer dan 9% per jaar in de afgelopen twintig jaar, behoort China tot de snelst groeiende economieën ter wereld. Het energieverbruik stijgt navenant en zal tegen 2025 bijna 15% van het totale energieverbruik op wereldschaal bedragen. Het gemiddelde inkomen stijgt en de armoede wordt ingedijkt.

Maar ook de uitstoot van fossiele brandstoffen is negen keer groter geworden in de afgelopen vijftig jaar. Volgens de Wereldbank zijn zestien van de twintig meest vervuilde steden ter wereld gelegen in China. Steenkool is de belangrijkste bron van vervuiling. Drie vierde van de primaire energie wordt gewonnen uit steenkool. Daarmee stookt men doorgaans ongefilterd, en op het vlak van vervuiling ontbreekt elke reglementering. In China tekenen we de hoogste emissies van zwaveldioxide ter wereld op, en ook de toenemende uitstoot van uitlaatgassen ten gevolge van het explosief stijgende autobezit begint problematische proporties aan te nemen.

L'UICN a notamment lancé le projet 'Mangroves for the Future', un programme de 45 millions de dollars visant à planter des mangroves, véritables barrières naturelles, dans douze pays d'Asie et d'Afrique.

MALI / Transport de l'eau *p.63*

La majeure partie du Mali, dont la superficie équivaut à plus du double de celle de la France, se situe dans le désert du Sahara. Celui-ci gagne peu à peu du terrain sur les zones habitables. Cette expansion a des conséquences graves pour un pays dont 80% de la population active est employée dans l'agriculture ou la pêche. Le fleuve Niger, qui constitue le principal cours d'eau du Mali, s'assèche rapidement dans l'antique cité de Tombouctou.

Le manque d'eau touche particulièrement les femmes et les adolescentes. Comme le veut la tradition, c'est à elles qu'échoit d'assurer l'approvisionnement d'eau de leur famille. Chaque matin, elles se lèvent donc dès l'aube pour se rendre au puits le plus proche. Il n'est pas rare qu'elles parcourent seize kilomètres, voire davantage, pour y parvenir. Lors de la saison sèche, les femmes doivent parfois effectuer le double de cette distance. Dans le pays Dogon, le niveau de l'eau peut baisser de quatre-vingts mètres. La situation devient alors catastrophique. En effet, la plupart des villages ne disposent que d'un puits suffisamment profond, qu'ils sont parfois amenés à partager avec d'autres villages. Outre les six heures nécessaires au transport de l'eau, les femmes doivent également attendre leur tour au puits, ce qui peut prendre cinq heures de plus. Afin d'éviter cette attente, nombre de femmes se lèvent en pleine nuit pour parvenir au puits à temps.

La corvée de l'eau réclame beaucoup d'énergie de la part des femmes. Leur réservoir peut habituellement contenir vingt litres d'eau, ce qui représente un poids de vingt kilos. Le transport de telles charges, jour après jour, est lourd de conséquences pour la santé. En outre, prendre de l'eau au puits n'est pas une tâche aisée comme le prouvent les entailles profondes laissées par les cordes sur les poutres.

Cette tâche éprouvante empêche les femmes et les adolescentes des pays en développement de bénéficier de programmes d'éducation ou de peser sur les décisions prises au sein de leurs communautés. En facilitant l'accès à l'eau potable, on met non seulement un terme à une corvée inutile, mais on contribue aussi à améliorer leur état de santé et celui de leurs familles. Elles pourront alors se consacrer à leur éducation, à l'agriculture ou à un autre travail rémunéré. Elles auront du temps pour élever leurs enfants ou simplement se relaxer.

INDE / De la pauvreté rurale à la misère urbaine *p.83*

Les principales ressources d'eau de l'Inde sont menacées. En effet, de nombreux agriculteurs disposant du libre accès aux nappes aquifères privilégient les cultures marchandes gourmandes en eau, destinées à l'exportation (tel le coton), au détriment des cultures vivrières susceptibles de répondre aux besoins locaux. Cette situation a entraîné une surexploitation des eaux souterraines et la baisse rapide du niveau des nappes phréatiques. Les puits communautaires se sont asséchés ou l'eau est devenue trop saline pour couvrir les besoins, à l'exception de la lessive. L'approvisionnement en eau de pluie est aussi moins fiable en raison de la sécheresse et de la désertification. Au Pendjab, la région agricole jadis la plus prospère du sous-continent, de vastes étendues de terres ont ainsi cédé la place au désert.

Pour bon nombre des 700 millions de paysans indiens, l'agriculture est devenue une activité non viable. Au Gujarât et au Rajasthan, 50 millions d'êtres humains sont confrontés à une grave pénurie d'eau. Ces dernières années, des milliers de paysans de l'État du Maharastra se sont donné la mort, poussés au désespoir par les ravages répétés causés à leurs cultures par la sécheresse et étouffés par le poids des dettes contractées pour financer l'achat de nouvelles graines et de pesticides. Ils se sont suicidés en avalant des pesticides, léguant leurs dettes à leurs veuves et à leurs enfants.

Prêts à tout pour subvenir aux besoins de leurs familles, des millions d'agriculteurs ont d'ores et déjà abandonné leurs terres dans l'espoir d'une vie meilleure à la ville. En Inde, la population urbaine a été multipliée par dix au cours du siècle dernier. Elle est désormais la deuxième plus importante du monde, après la Chine. Le gouvernement estime qu'à l'horizon 2025, la moitié de la population vivra dans les zones urbaines. Le moteur de cette évolution réside dans les perspectives économiques qui y sont réunies. Cependant, la majorité des migrants sont analphabètes et appartiennent aux castes inférieures de la société indienne. Ils ne peuvent donc accéder qu'à des postes à bas salaire. Automatiquement exclus des quartiers sûrs et huppés des villes, ils n'ont d'autre choix que de vivre dans les bidonvilles surpeuplés.

La pauvreté urbaine galopante a aussi un coût environnemental et social. La plupart des bidonvilles sont dépourvus de services publics de base (approvisionnement d'eau potable, égouttage et électricité). En Inde, ces services sont fournis par des systèmes rongés par la mal gouvernance, la corruption et le copinage. Dans les 1.100 bidonvilles de New Delhi, l'eau courante est soit inexistante, soit uniquement disponible quelques heures par jour. Si, d'aventure, des toilettes sont accessibles, elles sont si mal entretenues que les habitants préfèrent économiser leurs roupies en allant faire leurs besoins en plein air. Privés de ces services de base, les habitants de bidonvilles sont victimes d'affections cutanées, de maladies hydriques et de tuberculose. Ce n'est pas franchement ce qu'ils s'attendaient à trouver en quittant la campagne pour la ville.

Ademhalingsinfecties zijn momenteel de meest voorkomende kinderziekte en een van de belangrijkste doodsoorzaken in heel China.

De watervoorraden van China per inwoner behoren tot de laagste ter wereld en zijn geconcentreerd in het zuiden. Het noorden en het westen worden regelmatig geteisterd door droogtes. En zelfs op plaatsen waar en in periodes wanneer er geen waterschaarste is, is het water zelden zuiver. Het Ministerie van Milieu kwalificeerde meer dan 70% van het water in vijf van de zeven belangrijkste rivieren van China als "ongeschikt voor menselijk contact". In 2005 kondigden de autoriteiten aan dat de Gele Rivier zo erg vervuild was dat het niet meer veilig was om dat water te gebruiken. Men schat dat zo'n 300 miljoen Chinezen geen zuiver drinkbaar water hebben.

Ondertussen kondigde de regering een wet af die moet aanzetten tot een milieuvriendelijkere productie, de 'Cleaner Production Promotion Law'. Via pilootprojecten wil men zo de vervuiling tegengaan in tien grote Chinese steden, en een aantal riviervalleien worden als voorrangsgebied aangewezen. Om de luchtvervuiling te verminderen werd een belasting ingevoerd op steenkool met een hoog zwavelgehalte. Men heeft steenkoolvrije zones ingepland en ook de bouw van gaspijpleidingen staat op het programma.

De vervuiling is niet alleen schadelijk voor de gezondheid van de Chinese bevolking en de lucht- en waterkwaliteit, maar heeft ook een nefaste invloed op de economie. Volgens de Wereldbank kost de directe schade van de vervuiling 8 tot 12% van het BNP van China. Daaronder vallen bijvoorbeeld de invloed van zure regen op de oogst, medische kosten, werkverlet wegens ziekte, kosten voor rampenhulp na overstromingen en de impliciete kost door de uitputting van natuurlijke bronnen.

ARAL ZEE / Op zoek naar de verdwenen zee p.111

Op de waterexperts na kunnen waarschijnlijk weinig mensen Karakalpakstan aanduiden op de wereldkaart. In deze vergeten regio in Oezbekistan voltrok zich een van de ergste ecologische rampen van de vorige eeuw: de verdwijning van de Aral Zee. Een droge bedding vol roestende scheepswrakken en zeeschelpen zijn de enige overblijfselen van wat ooit een bedrijvige haven aan de vierde grootste binnenzee ter wereld was. In 1918 besloot de Sovjet-Unie om twee rivieren die uitmondden in de Aral Zee, namelijk de Amu Darya en de Syr Darya, om te leiden om de omliggende woestijn te irrigeren. Daar wou men immers op grote schaal katoen verbouwen, het zogenaamde 'witte goud' van Centraal-Azië en het landbouwproduct van de toekomst.

Tegen 1960 begon de zee kleiner te worden. Op een periode van één generatie was het water al 100 kilometer teruggetrokken. In 2004 bleef er amper een vierde van de oorspronkelijke zeeoppervlakte over, en was de zee opgesplitst in twee kleinere zeeën: de Kleine Aral Zee en de Grote Aral Zee. Over een gebied dat even groot was als België en Nederland samen veranderde het reusachtige waterreservoir in een vervuilde woestijn vol slib. Duizenden tonnen pesticiden en meststoffen, die gebruikt werden op de katoenvelden, werden door de wind verspreid, wat bij de plaatselijke bevolking een gevoeligheid voor oogirritaties en ernstige gezondheidsproblemen tot gevolg had. De combinatie van armoede, slechte voeding en besmet drinkwater door het vervuilde zeezout zorgde in de afgelopen twintig jaar voor een toename van 30 à 40% van bronchitis, kanker, lever- en nierziekten.

De vissers van vroeger dromen nog altijd van de getijden en zachte golven die nu voorgoed verdwenen zijn, en sturen hun kinderen naar vruchtbaardere regio's of het buitenland. Maar toch is er nog hoop voor de mensen wier leven vervlochten blijft met het lot van de zee. Door de bouw van een nieuwe dam en een aantal dijken, die in 2006 afgerond werd, stijgt het waterpeil opnieuw in de Kleine Aral Zee. Bovendien werd er op de bodem van de uitgedroogde Aral Zee ook geboord naar olie. Zal het 'zwarte goud' de redding betekenen voor Karakalpakstan?

EGYPTE - ETHIOPIË / Het Nijlverdrag p.125

De Nijl is van levensbelang voor de overleving van 160 miljoen mensen in tien landen die het Nijlbekken delen. Vooral voor Egypte is de rivier een kwestie van leven en dood, want dat land heeft nauwelijks andere natuurlijke watervoorraden. Een verdrag tussen Egypte en koloniaal Engeland (dat Oeganda, Tanzania, Kenia en Soedan vertegenwoordigde) uit 1929 stelde dat geen enkele ingreep op de rivier mocht worden uitgevoerd die het waterdebiet in Egypte zou beperken. Het land van de farao's ziet elke verandering aan dit verdrag als een strategisch dreigement.

Door de bevolkingsexplosie en klimaatverandering stijgen de spanningen in de regio en stellen de buurlanden dit koloniale verdrag ter discussie. Vooral Ethiopië, waar 86% van het Nijlwater uit het Tanameer komt, eist zijn deel van het water op. Omdat het Tanameer zo uitgestrekt is, is het moeilijk te begrijpen waarom de 2,5 miljoen boeren in dit deel van Ethiopië voor hun overleving afhankelijk zijn van voedselhulp. Volgens de Ethiopische regering leeft de overgrote meerderheid van de bevolking er van de landbouw, maar die lijdt onder de grillige seizoenregens en het verdrag van 1929.

Het water dat onaangeroerd door Ethiopië stroomt, wordt in Egypte gebruikt in grootschalige irrigatieprojecten. Met de bouw van de Hoge Aswandam controleert Egypte het stromen van de Nijl, waardoor de boeren grotere oogsten hebben kunnen binnenhalen. Maar dat volstond blijkbaar niet. Daarom lanceerde president Hosni Mubarak het ambitieuze Toshka Project, dat een tweede Nijlvallei zou creëren in het midden van de westelijke woestijn. Het Sheikh Zayed-kanaal zal een tiende van het Nijlwater overbrengen van het kunstmatige Nassermeer naar een woestijngebied van zo'n 230.000 hectare. Het project, dat afgerond zou moeten worden in 2017,

CHINE / La pollution: un obstacle au développement *p.97*

Forte d'une croissance annuelle moyenne de plus de 9% au cours des deux dernières décennies, la Chine compte parmi les économies les plus dynamiques de la planète. Ses besoins énergétiques ne cessent d'augmenter et devraient représenter près de 15% du total de la consommation énergétique mondiale à l'horizon 2025. Le revenu moyen est à la hausse et la pauvreté est en régression. Cependant, les émissions de combustibles fossiles ont été multipliées par neuf au cours des cinquante dernières années. Selon la Banque mondiale, 16 des 20 villes les plus polluées au monde se trouvent en Chine. La principale source de pollution est le charbon, qui fournit trois quarts de l'énergie primaire et qui est brûlé brut et en l'absence de toute réglementation de lutte contre la pollution.

La Chine détient également le triste record mondial du taux d'émission de dioxyde de soufre le plus élevé, alors que les émissions dégagées par les voitures de tourisme, toujours plus nombreuses, commencent seulement à poser problème. Les affections respiratoires constituent à présent la première maladie infantile et une des principales causes de décès dans l'ensemble du pays.

Les ressources en eau par habitant de la Chine comptent parmi les plus basses au monde et sont concentrées dans le sud du pays. Le nord et l'ouest subissent régulièrement des sécheresses. Même dans les endroits ou durant les périodes où elle n'est pas rare, l'eau est rarement potable. Le ministère de l'Environnement a ainsi déclaré "impropre au contact humain" plus de 70% de l'eau de cinq des sept principaux bassins hydrographiques de Chine. En 2005, les autorités ont annoncé que le fleuve Jaune était si pollué que son eau était "impropre à la consommation". Selon les estimations, 300 millions de Chinois n'ont pas accès à l'eau potable.

Le gouvernement a donc institué la 'Cleaner Production Promotion Law' (loi pour la promotion d'une production propre). Il a organisé des programmes pilotes pour remédier à la pollution dans dix grandes villes chinoises et classé plusieurs vallées en zone prioritaire. Afin de réduire la pollution atmosphérique, le gouvernement a décidé d'instaurer une taxe sur le charbon à haute teneur en soufre et a circonscrit des zones où le charbon est interdit. Il projette également la construction de gazoducs.

La pollution ne provoque pas seulement de graves atteintes à la santé humaine et à la qualité de l'air et de l'eau; elle a également des effets néfastes sur l'économie. La Banque mondiale estime que chaque année, de 8 à 12% du produit intérieur brut chinois sert à financer les dommages directs de la pollution: impact des pluies acides sur les récoltes, frais médicaux, absentéisme pour cause de maladie, opérations humanitaires lors les inondations et coût engendré par l'épuisement des ressources.

MER D'ARAL / À la recherche de la mer disparue *p.111*

Rares sont ceux qui peuvent situer la région du Karakalpakstan sur une carte, à moins d'être experts en hydrologie. Cette région oubliée d'Ouzbékistan porte les stigmates d'une des pires catastrophes naturelles du siècle dernier: la disparition de la mer d'Aral. Du port autrefois animé sur la quatrième mer intérieure du monde ne subsiste qu'un lit asséché, parsemé de carcasses de bateaux rouillées et de coquillages. En 1918, l'Union soviétique décide de détourner les cours de l'Amou-Daria et du Sir-Daria, les deux fleuves alimentant la mer d'Aral, afin d'irriguer le désert environnant et de permettre, à cet endroit, la culture à grande échelle du coton, 'l'or blanc' d'Asie centrale et produit agricole d'avenir.

En 1960 déjà, le niveau de la mer d'Aral avait commencé à baisser. En l'espace d'une génération, l'eau se retire sur une centaine de kilomètres. En 2004, la surface de la mer est réduite à un quart de sa taille initiale et se divise en deux mers plus petites, appelées petite et grande mer d'Aral. Couvrant une région aussi grande que la Belgique et les Pays-Bas réunis, l'immense réservoir d'eau douce s'est transformé en un désert vaseux et pollué. Des milliers de tonnes de pesticides et d'engrais, utilisés dans la culture du coton, sont dispersés par le vent et provoquent chez la population locale des irritations oculaires ainsi que de graves problèmes de santé. Ces 20 dernières années, la pauvreté, la malnutrition et la contamination de l'eau potable par les sels marins toxiques ont augmenté de 30 à 40 fois la prévalence des bronchites, des cancers et des pathologies hépatiques et rénales.

Les anciens pêcheurs rêvent de marées et de vagues à jamais disparues. Ils envoient leurs enfants dans des régions plus prospères ou à l'étranger. Néanmoins, il reste de l'espoir pour ceux dont la vie est intimement liée au destin de la mer d'Aral. La construction d'un nouveau barrage et d'une série de digues, terminée en 2006, a entraîné une hausse du niveau des eaux de la petite mer d'Aral. En outre, le lit asséché de la mer d'Aral fait l'objet de prospections pétrolières. 'L'or noir' sauvera-t-il la région du Karakalpakstan?

ÉGYPTE - ÉTHIOPIE / Le traité du Nil *p.125*

Ce fleuve est d'une importance vitale pour la survie de 160 millions de personnes, réparties sur les dix pays que baigne le Nil. L'Egypte est particulièrement vulnérable, car le fleuve constitue presque la seule ressource en eau du pays. Conclu entre l'Égypte et les colonies britanniques, à savoir l'Ouganda, la Tanzanie, le Kenya et le Soudan, le traité de 1929 stipule qu'aucun chantier susceptible de réduire le volume d'eau arrivant en Égypte ne peut être réalisé sur le fleuve. Le pays des Pharaons considère d'ailleurs tout amendement à ce traité comme une menace stratégique.

Face à l'explosion démographique et aux changements climatiques, les tensions entre l'Égypte et ses voisins

zou 2,8 miljoen nieuwe jobs creëren en meer dan 16 miljoen mensen aantrekken naar de nieuwe steden die er gebouwd zullen worden.

Rekening houdend met het acute watertekort in Afrika, vragen andere landen in de regio zich af of het wel verantwoord is dat Egypte tot elke prijs gewassen wil telen in de woestijn. Slechts 3% van Ethiopië wordt geïrrigeerd, en de Ethiopische regering wil dan ook het water van de Blauwe Nijl zelf gebruiken om uit de armoede te geraken. In 1999 hebben tien Nijlstaten het Initiatief voor het Nijlbekken ondertekend, waarbij ze zich ertoe engageerden om samen te werken op het vlak van watermanagement. Dialoog zou 'het nieuwe geschenk van de Nijl' zijn, zo beloofden ze elkaar. Er werden al vele gesprekken gevoerd, maar een overeenkomst werd nog niet bereikt. Voor de negen tot vijftien miljoen Ethiopiërs die afhankelijk zijn van voedselhulp, mogen die gesprekken niet eindeloos aanslepen, aldus het Wereldvoedselprogramma van de Verenigde Naties. De enige oplossing is een vermindering van het waterverbruik en wederzijdse samenwerking.

TURKIJE / Het grootste irrigatieproject ter wereld p.137

In vergelijking met de rest van het Midden-Oosten beschikt Turkije over een rijke watervoorraad. Toch bedraagt die maar een zesde van wat een West-Europeaan ter beschikking heeft. Door de snelle bevolkingsgroei, de industrialisatie en de hogere levensstandaard worden de watervoorraden in Turkije steeds minder toereikend. Turkije streeft naar economische groei en sociale stabiliteit door een optimaal gebruik van het land en de watervoorraden. In 1977 startte men een van de grootste bouwprojecten ter wereld in Zuidoost-Anatolië, het meest achtergestelde gebied van Turkije. Dat project omhelsde de bouw van 22 dammen, 19 elektriciteitscentrales en 2 irrigatietunnels.

Het GAP-project, dat staat voor Groot-Anatolië-Project en 32 miljard dollar kostte, zou een regio zo groot als Oostenrijk moeten voorzien van zuiver water en ook een vierde van de elektriciteit moeten leveren die het land nodig heeft. Het totale gebied van het project beslaat ongeveer 10% van de oppervlakte van Turkije en is van belang voor 9,5% van de Turkse bevolking. Met deze dammen hoopt Ankara 3,3 miljoen jobs te creëren en zelfvoorzienend te worden op het vlak van voedselproductie. Turkije schat dat het nationale inkomen zal stijgen met 12% en dat het BNP van Zuidoost-Anatolië minstens zal verviervoudigen.

Maar niet iedereen is opgezet met het project dat ontwikkeld wordt in het gebied van de Eufraat en de Tigris. Deze waterlopen stromen door Irak en Syrië vooraleer ze uitmonden in de Perzische Golf. Syrië en Irak vrezen dat het GAP-project, waarvan de Atatürkdam en het tunnelsysteem van Sanliurfa al operationeel zijn, uiteindelijk zal leiden tot een inperking van hun eigen watervoorziening. Turkije stemde al in met een waterdebiet van 500 m^3 per seconde voor Syrië. Maar Syrië wil meer, want het land kampt met een watertekort en heeft een grotere watertoevoer nodig om zijn economie te kunnen handhaven. In het verleden heeft Ankara er al mee gedreigd om de watertoevoer naar Syrië te verminderen als Damascus de Koerdische PKK bleef steunen. En Irak dreigt ermee Turkije voor een internationaal tribunaal te brengen wegens schendingen van de internationale overeenkomsten over damprojecten met de landen stroomafwaarts.

Het GAP-project zou voor Turkije een enorm strategisch voordeel betekenen in een regio die constant bedreigd wordt door waterschaarste. Vele experts voorspellen dat de landen die de watervoorziening controleren in het Nabije Oosten, in de 21ste eeuw grotendeels de economische en politieke agenda zullen bepalen. Voormalig VN-secretaris-generaal Boutros Boutros-Ghali waarschuwde ooit een Amerikaanse denktank ervoor dat "de volgende oorlog in het Nabije Oosten niet over politieke kwesties gevoerd zal worden, maar over water".

ISRAËL / Het recept voor een conflict p.145

In het Midden-Oosten probeert 5% van de wereldbevolking te overleven met 1% van de wereldwatervoorraad. Sinds de oprichting van de staat Israël is water altijd een prioriteit geweest voor de Israëlische regering. De watervoorraden voldoen amper aan de behoeften van de 6 miljoen Israëli's. En dat is een probleem voor de overige 13 miljoen inwoners van die regio. Droogte en woestijnvorming zijn dan ook heel reële dreigingen.

Dit is het ideale recept voor een conflict. Koning Hoessein van Jordanië heeft nog gezegd dat hij nooit een oorlog zou beginnen tegen Israël, tenzij water de inzet is. Yitzhak Rabin stelde dat water een kwestie van leven en dood is voor Israël. En in 2002 dreigde Ariel Sharon zelfs met een oorlog tegen Libanon als de Libanese regering water zou aftappen van een van de zijrivieren van de Jordaan om zo de dorpen in Zuid-Libanon van water te voorzien.

In 1967 viel Israël Jeruzalem binnen en bezette de Westelijke Jordaanoever. Alle watervoorraden werden geklasseerd als militair gebied. Sinds de bouw van de muur in 2002 werden duizenden Palestijnen afgesloten van hun landbouwgronden en watervoorraden. Onder de Westelijke Jordaanoever ligt een belangrijke waterhoudende grondlaag. Hoewel het voor de Palestijnen verboden is er water op te pompen, wordt de kolonisten integendeel wel toegelaten de meest geavanceerde pompsystemen te gebruiken. Volgens de Palestijnen zouden de Joodse kolonies 80% van dat water gebruiken, hoewel ze maar 10% van de bevolking uitmaken.

Israël haalt ook meer dan een derde van zijn watervoorraad uit de Jordaanrivier. Terwijl elke Israëli beschikt over 350 liter water per dag, is dat voor elke Jordaniër maar 80 liter water per dag. Volgens de FoEME (Friends of the Earth Middle East) wordt meer dan 90% van het water van de Jordaan nu afgeleid, waardoor de rivier volledig

se sont aggravées. Ces derniers remettent en effet en question le traité colonial. C'est particulièrement le cas de l'Éthiopie, dont le lac Tana alimente 86% du Nil, et qui exige sa part. En contemplant l'étendue du lac Tana, l'on a du mal à croire que, dans cette seule région d'Éthiopie, 2,5 millions d'agriculteurs ne doivent leur survie qu'à l'aide alimentaire. À en croire le gouvernement éthiopien, l'agriculture, dont dépend la majorité de la population, pâtirait de l'irrégularité des pluies saisonnières et du traité de 1929.

L'eau qui coule paresseusement, libre de toute entrave en Éthiopie, est utilisée en Égypte pour des projets d'irrigation importants. Grâce à la construction du barrage d'Assouan, l'Égypte maîtrise désormais les crues du Nil et les agriculteurs voient leurs rendements se multiplier. Comme si cela ne suffisait pas encore, le président égyptien Hosni Moubarak a donné son accord à la réalisation de l'ambitieux projet Tochka, visant la création d'une seconde vallée du Nil dans le désert occidental. Grâce au canal Cheikh Zayed, un dixième de l'eau du Nil sera transféré du lac artificiel Nasser à 230.000 hectares de désert. Ce projet, dont la fin est prévue en 2017, prévoit de créer 2,8 millions d'emplois et d'attirer plus de seize millions de personnes dans les nouvelles villes qui y seront construites.

Face à l'important manque d'eau dont souffre le continent africain, certains pays voisins se demandent s'il est raisonnable que l'Égypte s'attaque au désert pour y faire pousser des récoltes à un tel prix. En Éthiopie, où seulement 3% du pays bénéficie de systèmes d'irrigation, le gouvernement est résolu à utiliser le Nil bleu afin de se sortir de la pauvreté. En 1999, les dix États du bassin du Nil ont signé l'initiative éponyme et se sont engagés en faveur d'une gestion concertée de l'eau. Chacun a promis à son voisin que le dialogue serait 'le nouveau don du Nil'. Les pourparlers sont en cours, mais un accord se fait toujours attendre. Le Programme alimentaire mondial de l'ONU a lancé une mise en garde. Face à la situation catastrophique des neuf à quinze millions d'Éthiopiens tributaires de l'aide alimentaire, les débats ne peuvent se prolonger éternellement. Seules une diminution de la consommation d'eau et la coopération mutuelle pourront venir à bout de ce problème.

TURQUIE / Le projet d'irrigation le plus ambitieux du monde p.137

Selon les critères du Moyen-Orient, la Turquie dispose de réserves suffisantes d'eau. Cependant, celles-ci ne s'élèvent qu'à un sixième de celles disponibles par habitant en Europe occidentale. Compte tenu de la rapide croissance démographique, de l'industrialisation et de la hausse du niveau de vie, les ressources en eau de la Turquie ne pourront pas longtemps faire face au besoin croissant. Le pays entend d'ailleurs stimuler la croissance économique ainsi que la stabilité sociale par une exploitation optimale des terres et des ressources en eau. En 1977, le gouvernement a ainsi entrepris un des plus ambitieux projets de construction au monde en Anatolie du Sud-Est, une des régions les plus reculées du pays. Il s'agit d'ériger 22 barrages, 19 centrales hydroélectriques et deux tunnels d'irrigation.

Le projet de développement de l'Anatolie du Sud-Est, baptisé GAP, d'un montant de 32 milliards de dollars, vise à couvrir près d'un quart de la production électrique du pays et à alimenter en eau une région qui en manque cruellement. L'Anatolie du Sud-Est possède une superficie équivalente à celle de l'Autriche. La région concernée par ce projet occupe environ 10% de la superficie totale de la Turquie et abrite 9,5% de la population totale turque. Grâce à ces barrages, Ankara espère créer 3,3 millions d'emplois et gagner son autosuffisance alimentaire. Le revenu national turc devrait augmenter de 12% et le produit régional brut de l'Anatolie du Sud-Est devrait au moins quadrupler.

Néanmoins, ce projet n'a pas l'heur de plaire à tout le monde. En effet, il aura des implications sur le bassin du Tigre et de l'Euphrate. Ces deux fleuves traversent l'Irak et la Syrie avant de se jeter dans le golfe Persique. Or, ces deux pays ont exprimé leurs craintes quant à une éventuelle baisse du débit des deux fleuves induite par le projet GAP, dont le barrage Atatürk et le réseau de tunnels Sanliurfa ont déjà été mis en service. La Turquie a donné son accord afin de fournir 500 mètres cubes d'eau par seconde à la Syrie. Cette dernière juge néanmoins cette concession insuffisante. La Syrie subit déjà un manque d'eau et ne peut maintenir son économie à son niveau actuel sans ressources plus importantes. Jadis, Ankara avait menacé de réduire l'approvisionnement en eau de la Syrie si Damas continuait à apporter son soutien au parti kurde du PKK. Le gouvernement irakien avait répliqué en menaçant de traduire la Turquie en justice devant un tribunal international pour non-respect des accords internationaux conclus avec les pays en aval des barrages. Le projet devrait conférer à la Turquie un énorme avantage stratégique dans une partie du monde qui souffre chroniquement du manque d'eau. Nombre d'experts prédisent qu'au 21e siècle, le pays qui contrôlera l'approvisionnement en eau au Proche-Orient se rendra maître de la situation économique et politique de la région. L'ancien secrétaire général de l'ONU, Boutros Boutros-Ghali avait averti un comité d'experts américains que "la prochaine guerre au Proche-Orient se mènerait sur le front de l'eau et non de la politique".

ISRAËL / Les ingrédients du conflit p.145

Au Moyen-Orient, 5% de la population de la planète tente de survivre avec 1% des réserves mondiales d'eau. Depuis la création de l'État d'Israël, l'eau a toujours constitué une des priorités du gouvernement. En effet, les réserves d'eau se révèlent à peine suffisantes pour faire face aux besoins de six millions d'Israéliens, sans tenir compte des treize millions d'habitants disséminés dans l'ensemble de la région. La menace de désertification et de sécheresse est plus que jamais réelle.

On dispose là de tous les ingrédients d'un conflit potentiel. Feu le roi Hussein de Jordanie avait dit qu'il ne déclarerait jamais la guerre à Israël, sauf si l'eau en était l'enjeu. Yitzhak Rabin avait également insisté sur l'importance de

dreigt te verdwijnen. Ook de Dode Zee is 30% gekrompen over de laatste vijftig jaar, en het waterpeil van het Meer van Galilei in de Golanhoogten is de laatste jaren gezakt. Er wordt dan ook gevreesd voor een verzilting van het grootste waterreservoir van Israël.

Door de aanhoudende conflicten in Gaza kunnen de Palestijnen geen rioolstelsel of waterzuiveringsinstallaties bouwen. Daardoor is het grondwater vervuild door landbouwchemicaliën en het storten en verbranden van afval. Voor Israël is een herverdeling van de watervoorraden niet aan de orde, de Palestijnen moeten hun watertekort maar zelf oplossen door een efficiënter waterverbruik. De Palestijnen daarentegen willen hun rechten op water doen gelden.

TSJAAD / Oorlogsvluchtelingen p.165

Soedan beschikt over een van de grootste veestapels in Afrika. 50% van de plattelandsbevolking zijn herders. De kortere droogtecycli en de expansie van de Sahara in zuidelijke richting hebben de laatste jaren geleid tot een afbraak van de bodemkwaliteit, met een daling van de veeteelt tot gevolg. Bovendien kapt de plattelandsbevolking te veel bomen voor brandhout, wat nog meer schade toebrengt aan het ecosysteem.

Door het gebrek aan water en land ontstaan er conflicten tussen de nomadische en sedentaire stammen. Wanhopige, hongerige boeren uit Darfoer hebben in 2003 zelfs een aantal regeringsposten aangevallen. Volgens hen zou de regering in Khartoem de Darfoerregio sterk verwaarlozen, een trend die al bestaat van in de koloniale periode. De Janjaweed, een militiegroep van Arabische kameeldrijvers, beantwoordde deze aanvallen. De Soedanese regering ontkent elke betrokkenheid met de Janjaweed, maar ze leverde wel wapens en ondersteuning, en heeft deelgenomen aan gezamenlijke aanvallen.

De CIJ (Coalition for International Justice) schat het aantal slachtoffers in dat conflict op ongeveer 400.000. Zo'n 2,5 miljoen mensen zouden op de vlucht zijn. Het oosten van Tsjaad biedt onderdak aan meer dan 234.000 Soedanese vluchtelingen. De internationale pers gewaagt zowel van een etnische zuivering als van genocide. Ondanks een staakt-het-vuren in april 2004 heeft een groot deel van de vluchtelingen nog altijd geen toegang tot humanitaire hulp. Er zijn ook grote problemen op het vlak van veiligheid en mensenrechten.

Het conflict in Darfoer zou wel eens de eerste oorlog kunnen worden, veroorzaakt door de globale klimaatverandering. Volgens het Raamverdrag van de Verenigde Naties inzake klimaatveranderingen zou tegen 2085 tussen 25 en 40% van Afrika's natuurlijke habitat verdwenen zijn. Hoewel een conflict nooit het gevolg is van één specifieke factor, menen steeds meer hulpverleners, gezagdragers en experts dat de klimaatverandering de conflicten tussen de nomaden en sedentaire boeren in veel regio's zal doen escaleren. Als de beschikbare weilanden afnemen en water schaarser wordt om het land te bevloeien of de dorstige kuddes te drenken, kan de strijd voor dat land of water uitdraaien op een geweldadig conflict.

SPANJE / De kostprijs van golf p.177

De Spaanse provincie Murcia wordt overspoeld door prestigieuze bouwprojecten. Enkel al in 2005 werden 350.000 vakantieverblijven gebouwd, en er zijn plannen voor nog zo'n 1,65 miljoen extra. In de buurt van deze nieuwe gebouwen werden meer dan honderd nieuwe golfterreinen aangelegd, die elk evenveel water verbruiken als een stad van 16.000 inwoners. Al twee miljoen buitenlanders hebben een vakantiewoning aan de kust gekocht, en men schat dat tegen 2020 zo'n zes miljoen gepensioneerden - vooral uit Groot-Brittannië en Duitsland - in de regio zullen komen wonen.

Tezelfdertijd heeft Spanje de zwaarste droogtes gekend sinds 1947. Het waterpeil van de rivieren en stuwmeren staat historisch laag en in een aantal plaatsen heeft het al vijftien jaar niet meer geregend. De watercommissies van twee kustregio's hebben al laten weten dat zij onmogelijk de watervoorraad kunnen garanderen voor de 450.000 nieuwe woningen die gepland zijn voor 2006. In 2005 waren enkele Spaanse boeren blijkbaar zo wanhopig dat ze ongezuiverd afvalwater hebben gebruikt voor het besproeien van hun exportgewassen. Salmonellabesmettingen in het Verenigd Koninkrijk en Finland waren het gevolg.

Ondanks de waarschuwingen tonen politici weinig bereidheid om de projectontwikkelaars een halt toe te roepen. Het toerisme is immers goed voor 10% van het Spaanse BNP. Integendeel, bouwfirma's genieten aanzienlijke belastingverlagingen en de wetgeving op de ruimtelijke ordening wordt er heel flexibel geïnterpreteerd. Alleen al de golfsector was in 1997 goed voor een omzet van 2 miljard euro, een bedrag dat sindsdien verdrievoudigd is. Om meer water naar het gebied rond Valencia, Almeria en Murcia te krijgen, werd een project voorgesteld om water van de Ebro in het noorden te laten afvloeien. Kostenplaatje: 4,2 miljard euro. Na tienduizenden klachten werd dit plan echter afgevoerd. Men vreesde vooral dat deze ingreep verkeerd watergebruik in de hand zou werken, en dat de kwetsbare Ebrodelta veel schade zou ondervinden. Daarom heeft de regering de opdracht gegeven voor de bouw van 17 ontziltingsinstallaties. Eens in gebruik, zouden zij water moeten leveren voor 800.000 mensen.

Maar misschien is het al te laat. Het economische succes van Spanje legt nu al een enorme hypotheek op het milieu. Volgens de Verenigde Naties zal een derde van Spanje over 50 jaar uit woestijn bestaan als de bouwwoede en het overmatige waterverbruik in de landbouw geen halt toegeroepen wordt. Dat is slecht nieuws voor de plaatselijke bevolking en de fauna. Maar dat kan ook het einde betekenen van de toeristische sector en de distributiesector lamleggen die goedkoop fruit en groenten verdeelt aan de Europese supermarkten.

l'eau pour Israël: "une question de vie ou de mort". En 2002, Ariel Sharon avait menacé de déclarer la guerre au gouvernement libanais, celui-ci projetant d'utiliser l'eau d'un des affluents du Jourdain afin de la fournir aux villages du Sud Liban.

En 1967, Israël a pris le contrôle de Jérusalem et de la Cisjordanie. L'État a fait alors passer toutes les sources d'eau en 'zones militaires'. Depuis la construction du mur en 2002, des milliers de Palestiniens sont séparés de leurs terres cultivées et leurs sources d'eau. La Cisjordanie abrite une importante nappe aquifère souterraine, mais les Palestiniens ne peuvent y pomper de l'eau. En revanche, les colons peuvent mettre en oeuvre des équipements de pompage ultramodernes. Les Palestiniens arguent que les colonies juives s'attribuent 80% de la nappe aquifère, alors qu'elles ne représentent que 10% de la population.

Israël puise jusqu'à un tiers de son approvisionnement en eau dans le fleuve Jourdain. Alors que chaque citoyen israélien dispose de 350 litres d'eau par jour, les Jordaniens n'en ont que 80 litres. Selon les Amis de la Terre au Moyen-Orient, plus de 90% du cours du Jourdain a été détourné et la rivière est à présent menacée de disparition. Durant les cinquante dernières années, la mer Morte a aussi perdu 30% de sa surface. Le niveau de la mer de Galilée, sur le plateau du Golan, a récemment baissé, attisant les craintes d'une salinisation du principal réservoir d'eau d'Israël.

À Gaza, les rivalités perpétuelles empêchent les Palestiniens de construire des systèmes d'égouttage et de traitement des eaux. Par conséquent, la nappe phréatique est polluée par les engrais et pesticides chimiques, les déchets des décharges publiques et l'incinération des ordures ménagères. Selon Israël, le manque d'eau des Palestiniens devrait être résolu par une utilisation plus rationnelle de l'eau, sans que la distribution soit mise en cause. Néanmoins, le peuple palestinien entend faire valoir son droit d'accès à l'eau.

TCHAD / Réfugiés de guerre p.165

Le Soudan possède la plus importante concentration de bétail d'Afrique. L'élevage en pâturages occupe pas moins de 50% de la population rurale du pays. Récemment, la fréquence accrue des cycles de sécheresse, ainsi que l'expansion méridionale du Sahara ont induit une dégradation de la qualité des sols, les rendant moins adaptés à l'élevage des troupeaux. En outre, les habitants des zones rurales pratiquent une déforestation intensive afin de se procurer du bois de chauffage, ce qui constitue une atteinte supplémentaire des écosystèmes.

Le manque d'eau et de terres a provoqué des conflits entre tribus nomades et sédentaires. En 2003, des groupes de paysans affamés, laissés pour compte et vivant une situation désespérée au Darfour, ont attaqué des avant-postes du gouvernement. Selon eux, le gouvernement de Khartoum ne se soucierait nullement de la région du Darfour. Une absence d'intérêt déjà observée à l'époque coloniale. La contre-attaque fut menée par les Janjaweed, une milice composée d'Arabes, éleveurs de chameaux. Le gouvernement soudanais dément tout lien avec les Janjaweed, mais il leur a fourni des armes ainsi qu'un soutien logistique. Il les a même fait participer à des offensives de l'armée soudanaise.

Selon la 'Coalition pour la justice internationale' (CIJ), ce conflit aurait fait quelque 400.000 morts. Plus de 2,5 millions de personnes auraient été déplacées. L'est du Tchad a accueilli plus de 234.000 réfugiés soudanais. Les médias du monde entier ont qualifié ce conflit "d'épuration ethnique" et de "génocide". En dépit d'un cessez-le-feu signé en avril 2004, nombre de réfugiés ne bénéficient toujours pas de l'aide humanitaire. Des problématiques essentielles en matière de sécurité et de droits de l'homme restent très préoccupantes.

Le Darfour pourrait être le premier conflit provoqué par le changement climatique mondial. Selon un rapport établi par la Convention-cadre de l'ONU sur les changements climatiques, de 25 à 40% des habitats naturels africains pourraient avoir disparu à l'horizon 2085. Il est risqué d'imputer la cause d'un conflit à un facteur isolé. Néanmoins, de plus en plus de travailleurs humanitaires, d'autorités et d'experts s'accordent pour dire que les changements climatiques peuvent accentuer, dans nombre de régions, les tensions préexistantes entre nomades et paysans, jusqu'au point de non retour. Si les terres sont de moins en moins disponibles et si l'eau vient à manquer pour irriguer les cultures ou abreuver les animaux, on risque d'assister à un embrasement des conflits ayant l'eau ou la terre pour enjeu.

ESPAGNE / Le coût du golf p.177

La province espagnole de Murcie est devenue la proie des promoteurs immobiliers. L'an dernier, 350.000 résidences de vacances y sont sorties de terre. 1,65 million de plus devraient suivre. À côté de ces nouveaux quartiers, plus d'une centaine de terrains de golf ont vu le jour, chacun consommant autant d'eau qu'une ville de 16.000 habitants. Deux millions de touristes ont déjà acquis une résidence de vacances sur le littoral. On estime que d'ici à 2020, près de six millions d'entre eux, principalement des Britanniques et des Allemands, s'établiront dans la région.

Entre-temps, l'Espagne a vécu sa pire sécheresse depuis 1947. Les fleuves et les réservoirs ont atteint des niveaux d'étiage records. Certains endroits n'ont pas vu une goutte de pluie depuis quinze ans. Dans ce contexte, les offices des eaux de deux régions côtières se sont déclarés incapables de garantir l'approvisionnement en eau des 450.000 maisons neuves dont la construction est programmée en 2006. En 2005, par dépit, quelques agricul-teurs espagnols sont même allés jusqu'à irriguer leurs cultures avec des eaux usées non traitées, provoquant des épidémies de salmonellose en Grande-Bretagne et en Finlande.

AUSTRALIË / Strijd tegen verzilting p.187

Sinds de Europeanen twee eeuwen geleden voet aan wal zetten in Australië, werden zo'n 15 miljard bomen geveld. Deze bomen zogen grondwater op, waardoor het waterpeil constant bleef. Nu stijgt de grondwaterspiegel en is de bovenste grondlaag verzout. Deze negatieve evolutie wordt nog verergerd door te intensieve landbouw en te veel schapenkuddes.

De watervoorraden worden in Australië slecht beheerd. Het weinige beschikbare water wordt vaak verkwist aan waterintensieve teelten, zoals katoen en rijst, die Australië beter zou overlaten aan nattere landen. De irrigatie gebeurt ook meestal via open kanalen in plaats van leidingbuizen, waardoor heel wat water verdampt of in de aarde wegsijpelt. Daarbovenop wordt het land ook geteisterd door een langdurige droogte, wat resulteert in een aanzienlijke daling van de oogst. De Murray en de Darling, twee rivieren die meer dan twee derde van het irrigatiesysteem in Australië van water voorzien, noteren het laagste waterniveau van de eeuw. Een aantal regio's kent momenteel de ergste droogte in tientallen jaren. In West-Australië gewaagt men zelfs van een record. Wetenschappers voorspellen dat over honderd jaar een derde van het land herschapen zal zijn in een onvruchtbare witte woestijn, als niet snel ingegrepen wordt.

Toch denken heel wat Australiërs dat de klimaatverandering tijdelijk is en deel uitmaakt van een natuurlijke seizoenscyclus. Ze verwijzen daarvoor naar de talrijke droogteperiodes in het verleden. Eerste minister John Howard weigerde het Kyotoprotocol te ondertekenen omdat het de economie zou schaden. Het tegenovergestelde zou wel eens waar kunnen zijn: de Reserve Bank van Australië heeft ervoor gewaarschuwd dat de droogte de economische groei zou kunnen vertragen, als gevolg van lagere opbrengsten in de landbouwsector.

De Australiërs hebben van het droogste bewoonde continent een van de meest dynamische wereldeconomieën gemaakt. Maar volgens Jared Diamond, een Amerikaanse antropoloog en auteur van het boek 'Ondergang', kan Australië maar een bevolking van 8 miljoen mensen dragen, en dat is een pak minder dan de huidige 20 miljoen. In het vooruitzicht dat de bevolking minstens zal verdubbelen tegen halfweg de 21ste eeuw, staat Australië voor een enorme uitdaging. Uit een opiniepeiling die in december 2006 gehouden werd voor de krant The Australian, blijkt dat de Australiërs die uitdaging met beide handen willen aannemen. Een overweldigende meerderheid is immers bereid om gezuiverd afvalwater te drinken als dat een oplossing zou betekenen voor de nationale watercrisis. De regering werkt momenteel aan de invoering van één nationale watermarkt. De prijszetting van water zou dan in het hele land gebeuren op basis van de nationale noden in plaats van lokale behoeften. Zo zouden waterverbruikers in het tropische noorden van Australië, waar nog altijd genoeg regen valt, meer moeten betalen, waardoor ze zorgvuldiger met water zullen omspringen.

CONCLUSIE / PROTOS en GREEN Belgium p.195

De foto's in dit boek laten geen mens onbewogen. Ze houden onze maatschappij een spiegel voor. Another inconvenient truth? Ook de film van Al Gore gaat in essentie over problemen rond water. Klimaatveranderingen hebben rechtstreekse gevolgen voor de waterhuishouding. Smeltende ijskappen en gletsjers doen de zeespiegel stijgen. Voor wie in laaggelegen gebieden woont, geen prettig vooruitzicht. In droge gebieden daarentegen wordt de waterschaarste nog schrijnender.

Gelukkig komt water de laatste tijd al iets vaker op de politieke agenda. Met de Millenniumdoelstellingen van de Verenigde Naties engageert de internationale gemeenschap zich om tegen 2015 het aantal mensen dat geen toegang heeft tot drinkbaar water en sanitatie (sanitaire infrastructuur) met de helft te reduceren. Maar dan nog zullen 800 miljoen mensen het dan moeten doen zonder drinkwater en 1,8 miljard zonder sanitatie!
2015 is zowel symbolisch als praktisch van groot belang. Het is de deadline voor de nodige investeringen en beleidsmaatregelen. Tegen 2015 staat het leven van 203.000 kinderen op het spel en de daaropvolgende tien jaar zijn dat er nog eens een miljoen.
De Europese Unie zou tegen 2015 het budget voor ontwikkelingssamenwerking verhogen tot 0,7% van het bruto nationaal product en de G8 (de zeven rijkste landen plus Rusland) wil tegen 2010 de hulp aan Afrika verdubbelen. Toch is dat niet voldoende om de doelstellingen te halen! Zonder een breed maatschappelijk draagvlak om de problemen daadwerkelijk aan te pakken, zal er niet veel veranderen. De jaren '80 werden door de internationale gemeenschap al uitgeroepen tot decennium van water en sanitatie voor iedereen. Na tien jaar zou alle ellende van de baan zijn. Helaas, na 1990 was de wereld er nog erger aan toe.

Vanaf nu moet elk jaar 10 miljard dollar extra worden geïnvesteerd. Dat lijkt veel, maar het is minder dan wat wereldwijd in vijf dagen gespendeerd wordt voor militaire doeleinden. En minder dan de helft van wat westerlingen jaarlijks uitgeven aan mineraalwater. Overigens, alle druppende kranen in rijke landen verspillen meer water dan wat per dag beschikbaar is voor meer dan 1 miljard mensen... Investeren in water als 'mensenrecht' voor iedereen op Aarde is ook economisch voordelig. De Wereldgezondheidsorganisatie heeft becijferd dat elke dollar die wordt geïnvesteerd er op termijn 8 waard zal zijn, omdat de kosten voor gezondheidszorg dalen en de productiviteit stijgt. De Millenniumdoelstelling met betrekking tot water is van cruciaal belang voor de andere doelstellingen. Zuiver water en sanitaire voorzieningen redden niet alleen ontelbaar veel kinderlevens, maar zijn ook een basisvoorwaarde voor onderwijs, gezondheidszorg en armoedebestrijding.

Wat kunnen we zelf doen? Niets zo simpel als zuinig omspringen met ons eigen drinkwater. Regenwater is net zo geschikt voor het toilet en de wasmachine, en een spaartoets zou tegenwoordig een vanzelfsprekendheid

Les signes avant-coureurs sont bel et bien présents. Cependant, les hommes politiques se montrent peu enclins à tempérer l'ardeur des promoteurs immobiliers, le secteur touristique représentant un dixième du produit intérieur brut espagnol. Au contraire, les entrepreneurs se sont vu accorder d'importants allègements fiscaux et bénéficient d'une interprétation laxiste de la législation sur les lotissements. En 1997, le secteur du golf générait à lui seul deux milliards d'euros de bénéfices. Cette somme a triplé depuis. Afin d'accroître l'approvisionnement en eau de la région de Valence, Almeria et Murcie, un projet de 4,2 milliards d'euros visant à détourner le cours de l'Ebre, fleuve situé au nord de la péninsule, a été mis sur pied. Il a dû être abandonné sous la pression de dizaines de milliers de citoyens hostiles. Le projet risquait en effet d'encourager une consommation d'eau abusive et d'avoir de graves répercussions sur le fragile delta de l'Ebre. Le gouvernement a donc décidé de créer 17 sites de dessalement d'eau de mer, susceptibles de répondre aux besoins de 800.000 personnes.

Cette décision pourrait bien intervenir trop tard. En effet, le triomphe économique de l'Espagne a déjà occasionné de graves dégâts écologiques. Selon l'ONU, dans cinquante ans, un tiers de la superficie de l'Espagne sera transformée en désert si l'explosion immobilière et l'utilisation abusive d'eau par des agriculteurs continuent à ce rythme. La population et la faune locales ne seront pas les seules victimes. La désertification risque aussi de sonner le glas du secteur touristique et de causer la faillite des entreprises qui approvisionnent les supermarchés européens en fruits et légumes bon marché.

AUSTRALIE / Combat contre la salinisation p.187

On estime à quinze milliards le nombre d'arbres abattus depuis l'installation des Européens en Australie il y a deux siècles. Or, les racines des arbres absorbaient l'eau des nappes phréatiques et contribuaient à les maintenir à un niveau constant. À présent, le niveau de la nappe phréatique monte et l'on assiste à une salinisation des terres arables. Cette situation est aggravée par la surexploitation agricole et le tassement des sols induit par l'élevage ovin.

L'Australie a mené une mauvaise politique de gestion de l'eau. Le peu d'eau disponible a souvent été gaspillé pour des cultures gourmandes telles que le coton ou le riz, que l'Australie ferait mieux de laisser aux régions plus humides. L'irrigation est principalement réalisée par des chenaux au lieu de tuyaux, avec pour corollaire une évaporation importante et de grosses pertes d'eau. Pour couronner le tout, une sécheresse de longue durée s'est installée dans le pays et a provoqué d'importantes pertes agricoles. Les fleuves Murray et Darling, dont est tributaire l'approvisionnement de plus de deux tiers du système d'irrigation australien, ont atteint un niveau d'étiage record depuis un siècle. Dans certains états, la sécheresse actuelle est la pire que la population ait jamais connue depuis de décennies. Dans d'autres, tels que l'Australie occidentale, cette sécheresse bat tous les records. Selon des scientifiques, un tiers du continent pourrait se transformer dans une centaine d'années en un désert blanc et stérile si aucune mesure n'est prise d'ici là.

Pourtant, certains Australiens pensent que ces changements climatiques sont temporaires et font partie d'un cycle saisonnier naturel. Ils avancent que le continent a déjà vécu des périodes de sécheresse auparavant. Le Premier ministre australien, John Howard, a même refusé de ratifier le protocole de Kyoto, arguant qu'il hypothéquerait l'économie du pays. C'est pourtant bien le contraire qui pourrait se produire. En effet, la Reserve Bank (banque centrale australienne) a prévenu que la sécheresse risquait de ralentir la croissance économique, conséquence de la baisse de la production agricole.

Les Australiens ont réussi à transformer un continent aride et inhospitalier en une économie parmi les plus dynamiques du monde. Jared Diamond, anthropologiste américain et auteur du livre 'Effondrement. Comment les sociétés décident de leur disparition ou de leur survie', pense que l'Australie ne peut subvenir qu'aux besoins que huit millions de personnes, alors que la population actuelle s'élève à 20 millions. Dès lors qu'on annonce un doublement de la population au tournant du demi-siècle, les Australiens doivent faire face à un autre défi majeur. Un sondage réalisé en décembre 2006 par le quotidien 'The Australian' a révélé que les Australiens pourraient être à la hauteur de ce défi. En effet, une écrasante majorité d'entre eux se déclare prête à boire des eaux usées recyclées afin d'aider à résoudre la crise nationale de l'eau. Le gouvernement oeuvre à présent à l'instauration d'un marché national unique de l'eau. Il souhaite que le prix de l'eau sur l'ensemble du continent soit évalué en fonction des besoins nationaux plutôt que locaux. Cette mesure obligera les utilisateurs du nord de l'Australie, où règne un climat tropical et où les précipitations sont abondantes, à payer un prix plus élevé et ainsi à adopter une consommation d'eau plus responsable.

CONCLUSION / PROTOS et GREEN Belgium p.195

Les photos illustrant cet ouvrage ne laissent personne indifférent. Elles sont comme le miroir de notre société. Est-ce là une autre vérité qui dérange? Le film d'Al Gore touche lui aussi à la problématique de l'eau. Car les changements climatiques ont des répercussions directes sur le cycle de l'eau. La fonte de la banquise et des glaciers provoque une élévation du niveau des mers. La perspective est loin d'être réjouissante pour les populations installées dans les zones de basses terres. En revanche, dans les régions sèches, la pénurie d'eau s'aggrave.

Heureusement, cette problématique de l'eau est davantage prise en compte par le monde politique. Au travers des Objectifs du Millénaire des Nations unies, la communauté internationale s'engage à réduire de moitié le nombre

moeten zijn. Vers regenwater hoort sowieso niet thuis bij het rioolwater en moet weer kunnen infiltreren om de natuurlijke waterkringloop niet te verstoren. We kunnen ook rationeler omspringen met detergenten en chemische producten. En onze consumptiedrang beteugelen, want voor alles wat we produceren en consumeren wordt water gebruikt en vervuild.

U hebt alvast een steentje bijgedragen door u dit boek aan te schaffen. Tien procent van de verkoopprijs gaat rechtstreeks naar waterprojecten in Mali en Marokko. Uiteraard kunt u als individu veel meer doen: thuis, op de werkvloer, in het clublokaal, op school... Het vergt enige moed en creativiteit, maar als iedereen één enkel iets zou doen, worden de beelden van Troubled Waters misschien ooit voltooid verleden tijd. Er zijn tal van organisaties die zich inzetten voor duurzaam waterbeheer, die ontwikkelingslanden aan zuiver water en behoorlijk sanitair helpen of die de waterproblematiek hoger op de politieke agenda proberen te plaatsen. Die organisaties hebben uw morele, financiële én actieve steun hard nodig. Wat dacht u van een school waar de kinderen een carwash met regenwater opzetten, om autogebruikers waterbewust te maken en met de opbrengst een waterproject in Haiti steunen? Of van een groepje jongeren dat een benefietdag organiseert voor de waterdragers in Marokko, met de plaatselijke fanfare, rockbands en een vrij podium? Van een bedrijf dat zijn nieuwjaarsreceptie houdt met kraantjeswater en het uitgespaarde bedrag aan een goed doel schenkt? Dat zijn geen wilde ideeën, maar voorbeelden van succesvolle acties die nog niet zo lang geleden in België plaatsvonden. Bovenal zijn dit signalen naar onze beleidsverantwoordelijken om 'water in de wereld' meer ter harte te nemen.

Op het eerste gezicht onbelangrijk, maar erg symbolisch voor de toekomst is het 'Jupiter Ice Moons Project' dat de NASA in petto heeft voor 2015. Een ruimteschip zal drie manen van de planeet Jupiter exploreren om er de samenstelling van het ijs te onderzoeken en om na te gaan of er leven mogelijk is. Waar liggen de prioriteiten in onze internationale samenleving? Zullen dergelijke projecten de Millenniumdoelstellingen helpen waarmaken? Alle foto's in dit boek werden op aarde genomen. Laat ons met beide voeten op de grond blijven en hier en nu tot actie overgaan!

UW STEUN / De opbrengst van het boek gaat naar de waterprojecten van GREEN in Marokko en PROTOS in Mali p. 196

GREEN WIL KINDEREN VRIJSTELLEN VAN WATERWERK

Miljoenen kinderen gaan onvoldoende of helemaal niet naar school omdat ze, zeulend met een karretje of een ezel, meerdere kilometers moeten stappen naar de bronnen, putten en beekjes om er water te halen, vaak van een bedenkelijke kwaliteit. Helaas zijn dat de enige plaatsen die bereikbaar zijn en waar het water gratis is. Ook de waterbakken voor transport en opslag van water stellen problemen op het vlak van hygiëne. Dat waterwerk slorpt zoveel energie op, dat er geen tijd meer overblijft voor onderwijs of ontspanning. Het hypothekeert ook de ontwikkeling van de dorpen en is een van de redenen van de plattelandsvlucht.

Met de steun van Rotary, de Brusselse watermaatschappij Vivaqua, de Brusselse Intercommunale voor Waterdistributie en het KINT, lanceerde GREEN in 2005 de campagne 'Maak ons vrij van waterwerk': in de Marokkaanse dorpen Zaouit, Karia en Tighfert, elk uit een verschillende regio met een eigen socio-economische context, financiert men de plaatsing van waterfonteinen en individuele aansluitingen op het drinkwaterleidingnet, evenals de schoolhervatting van vele jongeren. Met de steun van het Vlaams Partnerschap Water voor Ontwikkeling en dankzij de medewerking van de lokale bevolking wil men er ook het afvalwater zuiveren.

GREEN Belgium wil zich inzetten om zoveel mogelijk kinderen toegang te geven tot drinkbaar water. Zo wordt hun recht op scholing en een menswaardige ontwikkeling gegarandeerd.

Info: www.waterdragers.be

INTEGRAAL WATERBEHEER IN DE BINNENDELTA VAN DE NIGER IN MALI

Met het waterprogramma in Mali wil PROTOS tegen 2009 de levensomstandigheden van 40.000 mensen duurzaam verbeteren. In overleg met de plaatselijke overheden, de watercomités en de gebruikers zet PROTOS projecten op rond drinkwater, sanitatie, irrigatie en waterbeheer. Doordat lokale bestuurders en gebruikers het project mee inhoud en vorm geven, verhogen de slaagkansen en het effect op langere termijn. Het waterprogramma concentreert zich op de binnendelta van de Niger in de regio van Mopti, meer bepaald Mopti-stad, Socoura (semistedelijk) en de landelijke gemeenten Kewa en Soye. Men voorziet in de aanleg van 22 waterputten, 20 nieuwe boringen en pompen, de herstelling van 20 putten en pompen, de installatie van 200 sanitaire voorzieningen en infrastructuurwerken voor de irrigatie van 1.000 hectare landbouwgrond.

Het gebied van de binnendelta van de Niger is even groot als België. De bevolking bestaat uit vissers, landbouwers en veetelers. Hun dorpen zijn gebouwd op natuurlijke duinen, die tijdens de overstromingsperiode (juni-september) als eilandjes boven het water uitsteken. Vanaf december tot maart trekt het water zich terug in de zijarmen en meren, waardoor de streek in één droge vlakte verandert. Slechts één op drie dorpen beschikt over een drinkwaterpunt. Cholera-epidemieën teisteren jaarlijks meerdere dorpen. Buikloop, door slecht water of onvoldoende sanitaire voorzieningen, blijft een van de belangrijkste doodsoorzaken bij kinderen.

Info: www.protosh2o.org, onder 'Onze programma's'

d'habitants ne disposant pas d'un accès à l'eau potable et à des infrastructures sanitaires à l'horizon 2015. Malgré cela, 800 millions d'êtres humains resteront encore privés d'eau potable et 1,8 milliard de dispositifs sanitaires.

2015: une date cruciale tant sur le plan pratique que symbolique. C'est en effet la date butoir pour les investissements et les mesures politiques. D'ici à 2015, la vie de 203.000 enfants est en jeu. Les dix années suivantes, c'est le sort d'un million supplémentaire qui sera scellé.

D'ici à 2015, l'Union européenne devrait porter le budget d'aide aux pays en développement à 0,7% du produit national brut. D'ici à 2010, le G8 (les sept pays les plus riches du monde plus la Russie) entend doubler le montant des aides accordées à l'Afrique. Néanmoins, ces montants ne suffiront pas pour atteindre les objectifs. Peu de choses changeront sans un engagement social élargi et une volonté de prendre les problèmes à bras le corps. Les années 1980 avaient déjà été proclamées décennie de l'eau et de l'assainissement par la communauté internationale. Dix ans plus tard, la misère ne devait plus avoir droit de cité. Après 1990, la situation mondiale avait cependant encore empiré.

Désormais, il convient, chaque année, d'investir dix milliards de dollars supplémentaires. Si cette somme peut sembler démesurée, elle est toutefois inférieure au budget consacré à l'échelle mondiale aux dépenses militaires en cinq jours. Elle représente moins de la moitié de ce que les Occidentaux dépensent chaque année en eau minérale. En outre, tous les robinets qui fuient dans les pays riches gaspillent plus d'eau que la quantité quotidienne accessible à un milliard d'êtres humains. Or, investir en faveur de l'eau en tant que 'droit de l'homme' pour tous les habitants de cette Terre présente un avantage économique. L'Organisation mondiale de la Santé a calculé que chaque dollar investi équivaudra à terme à huit dollars, en raison de la baisse des frais des soins de santé et de la hausse de la productivité. L'Objectif du Millénaire dans le domaine de l'eau revêt une importance cruciale pour les autres objectifs. Non contentes de sauver la vie d'un nombre incalculable d'enfants, la propreté de l'eau et la présence d'infrastructures sanitaires sont des prérequis pour l'éducation, les soins de santé et la lutte contre la pauvreté.

Comment pouvons-nous agir individuellement? La consommation rationnelle de l'eau potable est d'une rare simplicité. L'eau de pluie peut parfaitement alimenter les chasses et le lave-linge. L'installation d'une chasse économique devrait actuellement s'imposer comme une évidence. L'eau de pluie douce n'a rien à faire dans les égouts et doit pouvoir se réinfiltrer pour ne pas perturber le cycle naturel de l'eau. Nous pouvons aussi adopter une utilisation plus rationnelle des détergents et des produits chimiques. De même, il est urgent de mettre un frein à notre fièvre consommatrice, car pour tout ce que nous produisons et consommons, de l'eau est utilisée et polluée.

Sachez que vous avez d'ores et déjà apporté votre pierre à l'édifice en achetant cet ouvrage. En effet, dix pour cent du prix de vente servira à financer des projets dans le domaine de l'eau au Mali et au Maroc. En tant qu'individu, vous pouvez faire encore plus: à la maison, au travail, dans les associations de loisirs, à l'école, etc. Il suffit d'un peu d'efforts et de créativité. Si chacun fait ne fût-ce qu'un petit geste, les images de Troubled Waters seront peut-être à jamais reléguées au passé. Une foule d'organismes s'engagent déjà en faveur d'une gestion durable de l'eau. Ils aident les pays en développement à disposer d'un accès à l'eau potable et aux installations sanitaires ou s'emploient à sensibiliser les politiques à la problématique de l'eau. Ces organismes ont grandement besoin de votre soutien, à la fois moral, financier et actif. Que diriez-vous d'une école où les élèves mettent sur pied une station de lavage de voitures fonctionnant à l'eau de pluie afin de conscientiser les automobilistes à l'usage de l'eau et de soutenir, avec les bénéfices, un projet sur le thème de l'eau en Haïti? D'un groupement de jeunes qui organise une journée de solidarité au bénéfice des porteurs d'eau au Maroc, avec concert de la fanfare locale, groupes de rock et podium libre? D'une entreprise qui organise sa réception de Nouvel An à l'eau du robinet et verse le montant économisé à une association caritative? Il ne s'agit pas là d'idées en l'air, mais d'actions concrètes, menées avec succès il n'y a pas si longtemps en Belgique. Autant de signaux envoyés à nos responsables politiques afin de prendre davantage en compte 'l'eau dans le monde'.

Insignifiant de prime abord mais hautement symbolique pour l'avenir: le Jupiter Ice Moons Project que la NASA envisage pour 2015. Une navette spatiale explorera les trois lunes de Jupiter afin d'analyser la composition de la glace et d'observer si elles sont susceptibles d'abriter la vie. Où se situent les priorités de la communauté internationale? De tels projets pourront-ils permettre de concrétiser les Objectifs du Millénaire? Toutes les photos de cet ouvrage ont été prises sur notre planète. Gardons les pieds sur Terre et agissons, ici et maintenant!

VOTRE SUPPORT / Le produit de la vente de cet ouvrage servira à financer les projets 'eau' de GREEN au Maroc et de PROTOS au Mali *p.196*

GREEN VEUT LIBÉRER LES ENFANTS DE LA CORVÉE DE L'EAU

Des millions d'enfants d'Afrique et d'ailleurs ne fréquentent pas l'école ou de façon irrégulière, car ils doivent aller chercher de l'eau dans des sources, puits, cours d'eau, parfois distants de plusieurs kilomètres, à pied, en poussant une charrette ou en menant un âne. La qualité de cette eau n'est pas garantie, mais ce sont souvent les seuls points accessibles où l'eau est gratuite. Les conditions de transport et de conservation posent aussi des problèmes d'hygiène selon la propreté des récipients. Les heures consacrées à cette corvée empêchent donc de fréquenter l'école et de s'octroyer des loisirs. Ce temps et cette énergie hypothèquent aussi le développement des villages et expliquent en partie l'exode rural. Avec l'aide du Rotary, de Vivaqua, de l'Intercommunale bruxelloise de Distribution d'Eau et de l'IRGT, GREEN a lancé, en 2005, la campagne 'Libérons-les de la corvée eau'. Trois douars marocains situés dans des régions et des contextes socio-économiques très différents, Zaouit, Karia et Tighfert, bénéficient du financement de bornes fontaines et de raccordements individuels, tandis que de nombreux jeunes ont pu réintégrer la filière

ROTARY p..197

De druk van dit boek werd mogelijk gemaakt dankzij de financiële steun van Rotary België- Luxemburg.

Op 23 februari 1905 stichtte de Amerikaanse advocaat Paul Harris de niet-gebonden organisatie Rotary. Vandaag telt Rotary International ongeveer 1.200.000 leden en meer dan 32.000 clubs in 169 landen. De Belgisch-Luxemburgse Rotary is ingedeeld in drie districten en telt meer dan 10.000 leden en 254 clubs.

'Dienst aan de gemeenschap boven eigenbelang' vormt het basisprincipe in het gedachtegoed van Rotary, en is dan ook het motto van elk Rotarylid. Dat blijkt uit de brede waaier van activiteiten die de Rotarydistricten wereldwijd organiseren. Enkele spraakmakende voorbeelden: jeugdacties, waaronder studiebeurzen; het Forum voor Vrede in de Wereld; het 3H-programma, Health, Hunger & Humanity, waarmee men in de ontwikkelingslanden de honger wil bestrijden, de gezondheid wil bevorderen en een bijdrage wil leveren tot betere levensomstandigheden; Polio Plus, het grootste en meest uitdagende Rotaryproject.

Het project Polio Plus werd opgestart in 1985 in samenwerking met Unicef, de Wereldgezondheidsorganisatie (WHO) en sponsors uit de zakenwereld en de industrie. De doelstelling was om tegen 2005 polio de wereld uit te helpen. Hoewel het hele project een succes genoemd kan worden, telt men toch nog zeven landen waar die wens niet voor 100% waargemaakt kon worden, onder meer door conflicten, geweld en oude tradities.

In 2004 heeft Rotary International zich ook achter de millenniumdoelstelling geschaard dat het aantal mensen dat geen toegang heeft tot drinkwater, tegen 2015 wereldwijd gehalveerd zou moeten zijn. De Belgisch-Luxemburgse districten hebben daarvoor al heel wat projecten opgezet. Districten 1620 en 2170 zullen 19 dorpen in Niger en 3 dorpen in Marokko van water voorzien. Met het project 'Waterbrug' werkt ook district 1630 mee aan de waterbevoorrading van 25 dorpen in Oost-Afrika en Cambodja. Daarnaast zijn er de talrijke initiatieven van individuele clubs, zoals de financiering van waterpompen in India.

En toch is dat nog maar het begin. Dankzij de vele gemotiveerde mensen kan Rotary zich blijven inzetten voor het welzijn op aarde. En zolang niet iedereen toegang heeft tot drinkbaar water, zolang er kinderen zijn die honger lijden, zolang er analfabetisme, ellende en armoede heersen, zolang conflicten en geweld de wereld teisteren, zal Rotary een bestaansreden hebben.

Districtgouverneurs 2006-2007: Paul Coppens (D-1620), Guido Vangansewinkel (D-1630), Alain Van De Poel (D-2170). Interdistrictwerkgroep 'Troubled Waters': Robert Gekiere (past-gouverneur), Ludo Feyen (past-gouverneur), Jan-Lodewijk Dujardin (past-gouverneur, coördinator voor Europa 2004-2007, Rotary International 'Water Resource Group')

scolaire. De plus, avec le soutien du Vlaams Partnerschap Water voor Ontwikkeling et la participation directe de la population locale, des solutions appropriées sont étudiées pour assurer l'assainissement des eaux usées.
GREEN Belgium veut poursuivre la fourniture d'eau potable à un maximum d'enfants de manière à leur garantir le droit à la scolarisation et au développement humain.
Info: www.porteursdeau.be

GESTION INTÉGRÉE DES RESSOURCES EN EAU DANS LE DELTA DU NIGER AU MALI
Dans le cadre de son programme eau au Mali, PROTOS entend améliorer durablement les conditions d'existence de 40.000 personnes à l'horizon 2009. En concertation avec les pouvoirs locaux, les comités de l'eau et les utilisateurs, PROTOS déploie des projets axés sur l'eau potable, les installations sanitaires, l'irrigation et la gestion de l'eau. Le succès et la pérennité du projet sont accrus, car ce sont les gestionnaires et acteurs locaux qui lui confèrent son contenu et sa forme. Les collectivités locales doivent être, à terme, capables de s'autogérer. Ce programme concerne le delta intérieur du Niger, dans la région de Mopti, en particulier la ville de Mopti, Socoura (milieu semi-urbain) et les zones rurales de Kewa et Soye. Il prévoit l'aménagement de 22 puits, 20 nouveaux forages par pompes, la réhabilitation de 20 puits et pompes, l'installation de 200 sanitaires et des travaux d'irrigation de 1.000 hectares de terres agricoles.
Le delta intérieur du Niger occupe la superficie de la Belgique. Sa population se compose de pêcheurs, d'agriculteurs et d'éleveurs. Leurs villages sont installés sur des dunes naturelles, semi-immergées pendant la mousson (juin-septembre). De décembre à mars, l'eau se retire dans des bras secondaires et des lacs et provoque la sécheresse. Un village sur trois dispose d'un point d'eau potable. Chaque année, les épidémies de choléra touchent toujours plus de villages. La diarrhée due à la contamination de l'eau et au manque d'installations sanitaires, y reste l'une des principales causes de mortalité infantile.
Infos: www.protosh2o.org, sous la rubrique 'Nos programmes'

ROTARY p.197

Ce livre a pu être imprimé grâce au support financier de Rotary Belgique-Luxembourg.

Le Rotary est un organisme indépendant fondé le 23 février 1905 par Paul Harris, un avocat américain. Rotary International est présente dans 169 pays au travers de plus de 32.000 clubs et 1.200.000 membres. Subdivisé en trois districts, le Rotary belgo-luxembourgeois est fort de 254 clubs et plus de 10.000 membres.

Sa devise, 'Service Above Self' (Servir d'abord), est au coeur de sa philosophie, respectée par tous les rotariens. Parmi le large éventail d'activités gérées par les commissions du Rotary, citons des actions pour la jeunesse (bourses), le Forum mondial pour la Paix, ainsi que le programme 3 H (Health, Hunger and Humanity - santé, faim et humanité) qui vise à lutter contre la faim, à promouvoir la santé et à contribuer à de meilleures conditions d'existence dans les pays en développement.

Le projet le plus ambitieux du Rotary est le projet 'Polio Plus', lancé en 1985 en partenariat avec l'UNICEF, l'OMS (Organisation mondiale de la Santé) et des sponsors du monde des affaires et de l'industrie. L'objectif était d'éradiquer la polio partout dans le monde à l'horizon 2005. Malheureusement, celui-ci n'a pas encore pu être atteint dans sept pays, principalement en raison de conflits locaux et de traditions.

En 2004, Rotary International a fait sien l'un des Objectifs du Millénaire en l'intégrant dans ses actions prioritaires, à savoir réduire de moitié la population privée d'un accès à l'eau potable à l'horizon 2015. Les districts belgo-luxembourgeois n'ont pas ménagé leurs efforts dans ce sens. Les districts 1620 et 2170 ont ainsi fourni un accès à l'eau potable à 19 villages au Niger et trois villages au Maroc. Dans le cadre du projet baptisé 'Water Bridge' (un pont pour l'eau), le district 1630 fera de même pour 25 villages en Afrique de l'Est et au Cambodge. Parallèlement à ces actions, épinglons encore une foule d'initiatives déployées par les clubs, comme le financement de l'installation de pompes à eau en Inde.

Tant qu'une partie de la population mondiale sera privée d'un accès à l'eau potable, tant qu'il y aura des enfants affamés, tant que l'analphabétisme et la misère persisteront, tant que la violence et les conflits s'éterniseront, tant qu'il y aura des hommes charitables, le Rotary aura une raison d'exister.

Au nom des gouverneurs de district 2006-2007 Paul Coppens (D-1620), Guido Vangansewinkel (D-1630) et Alain Van De Poel (D-2170) et de la Commission interdistrict 'Troubled Waters' Robert Gekiere (ancien gouverneur), Ludo Feyen (ancien gouverneur), Jan-Lodewijk Dujardin (ancien gouverneur - coordinateur du 'Water Resource Group') 2004-2007 pour Rotary International Europe).

Acknowledgements

I dedicate this book to my parents who have worked for 40 years in development projects in Africa and Asia.

The photographs in this book were made possible thanks to the help of many organisations and people: Evelien Schotsman, Sofie Stevens, Ilse Casteels, Liesbeth Schockaert, Bruno De Cock and Yoerik Maes from Médecins Sans Frontières, Hilde Herssens at Caritas International and volunteers from Caritas India, Caritas Bangladesh and WaterAid in Mali, Evelien Lernout from Oxfam and Michelle from MA'AN in the West Bank, Brigitte Herremans from Broederlijk Delen, Ahikam Seri and Mary Schweizer who is a peace symbol on her own in Israel, Celia and Ana from Greenpeace Spain, Andrew Girvan in Australia, Catherine Vuylsteke and Ayfer Erkul from the daily De Morgen and Tom Wattez from the magazine Grande, Adrian Evans and Michael Regnier from Panos Pictures and Eike den Hertog from Hollandse Hoogte, and Spes Foundation and Fonds Pascal Decroos for their financial support. Many thanks also to my father and Hilde Devriese for the translations and to Eric de Mildt, my favourite graphic designer.

The global project Troubled Waters - www.troubledwaters.eu - would simply not exist without the exceptional help and support of Jo Van Cauwenberge at GREEN Belgium and Johan Verstraete at PROTOS. I am also most grateful for the trust that was given to me by Véronique Laureys and André Van Nieuwkerke from the Senate, by Kathleen Borms from BAI and by Paul Coppens and Jan-Lodewijk Dujardin from the Rotary.

Last but not least I wish to thank Mieke Deceuninck for her exceptional kindness, support and patience.

Dieter Telemans

About the author

Dieter Telemans was born and brought up in Burundi. He is now based as a freelance photographer in Brussels. He is a member of the photo agencies Panos Pictures in London, Hollandse Hoogte in Amsterdam, Cosmos in Paris and Grazia Neri in Milan. His pictures have been published worldwide in De Morgen, The New York Times, Newsweek, Libération, Le Monde and The Independent, amongst others. He is dedicated to continuing his work on Troubled Waters. More of his photography can be seen on www.dietertelemans.com and www.panos.co.uk.

Internet Sources

news.bbc.co.uk
environment.guardian.co.uk
www.economist.com
www.water-technology.net
www.unccd.int
www.abc.net.au
news.google.com
www.unesco.org/water
www.independent.co.uk
www.btselem.org/English
www.ecosan.nl
www.oursoil.org
climatecrisis.org
www.worldwaterday.org
www.greenpeace.org
www.stopthewall.org
www.china.org.cn
www.waterconserve.org
news.nationalgeographic.com
www.worldwaterforum4.org.mx
www.indiaenews.com
www.waterland.net
www.eia.doe.gov
www.argusmilieu.be
www.irc.nl
www.iht.com
www.bbc.co.uk/climate/impact/water.shtml
www.citymayors.com
www.water-ed.org
www.unep.org